Noteables™

Interactive Study Notebook with

Algebra 1

Contributing Author
Dinah Zike

FOLDABLES

Consultant
Douglas Fisher, PhD
Director of Professional Development
San Diego State University
San Diego, CA

New York, New York   Columbus, Ohio   Chicago, Illinois   Peoria, Illinois   Woodland Hills, California

The McGraw-Hill Companies

Send all inquiries to:
The McGraw-Hill Companies
8787 Orion Place
Columbus, OH 43240-4027

ISBN: 0-07-868210-X

Algebra 1 (Student Edition)
Noteables™: Interactive Study Notebook with Foldables™

5 6 7 8 9 10 047 09 08 07 06

Contents

Organizing Your Foldables

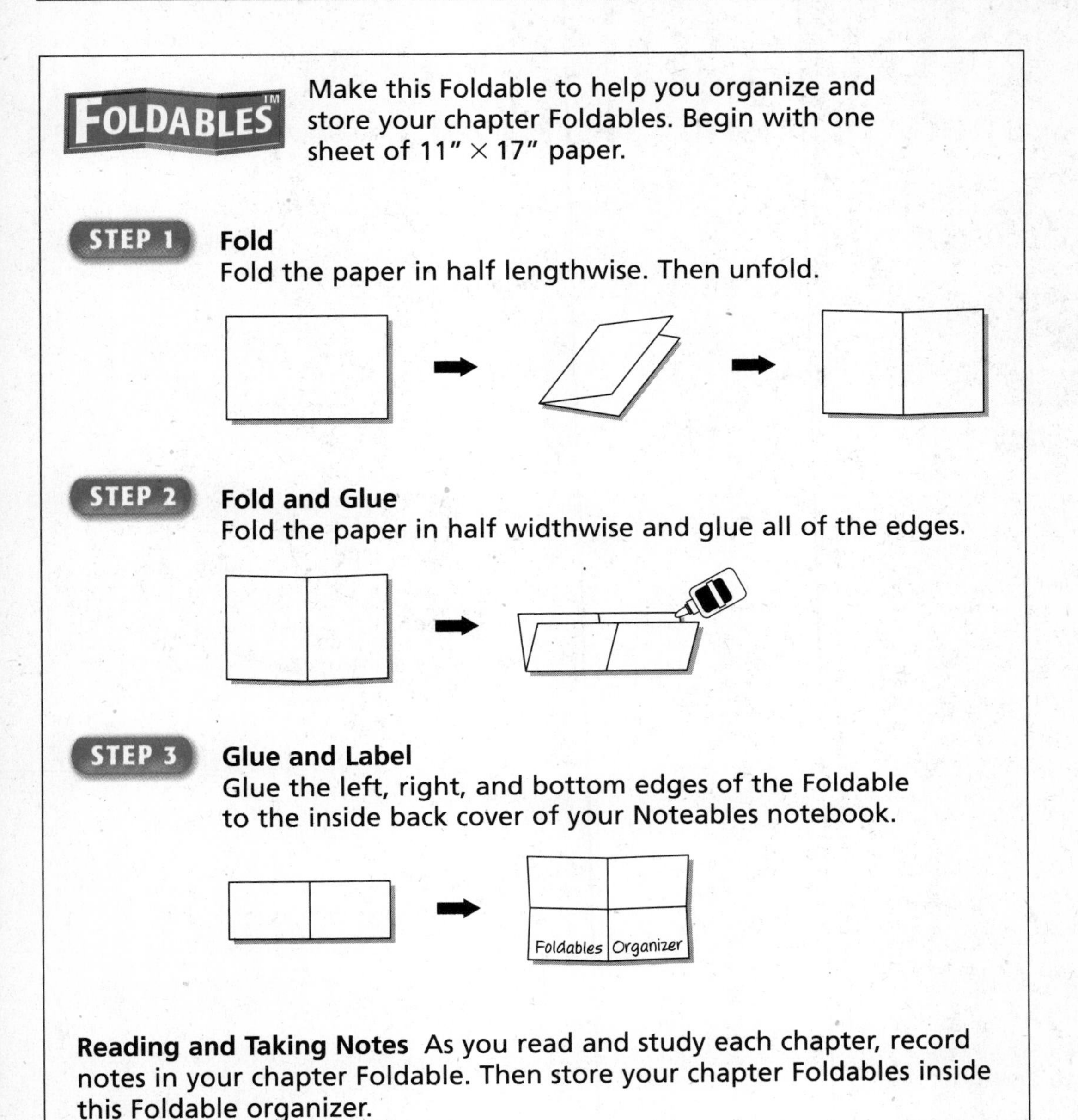

Make this Foldable to help you organize and store your chapter Foldables. Begin with one sheet of 11″ × 17″ paper.

STEP 1 **Fold**
Fold the paper in half lengthwise. Then unfold.

STEP 2 **Fold and Glue**
Fold the paper in half widthwise and glue all of the edges.

STEP 3 **Glue and Label**
Glue the left, right, and bottom edges of the Foldable to the inside back cover of your Noteables notebook.

Reading and Taking Notes As you read and study each chapter, record notes in your chapter Foldable. Then store your chapter Foldables inside this Foldable organizer.

Using Your Noteables™ with Foldables™ Interactive Study Notebook

This note-taking guide is designed to help you succeed in *Algebra 1*. Each chapter includes:

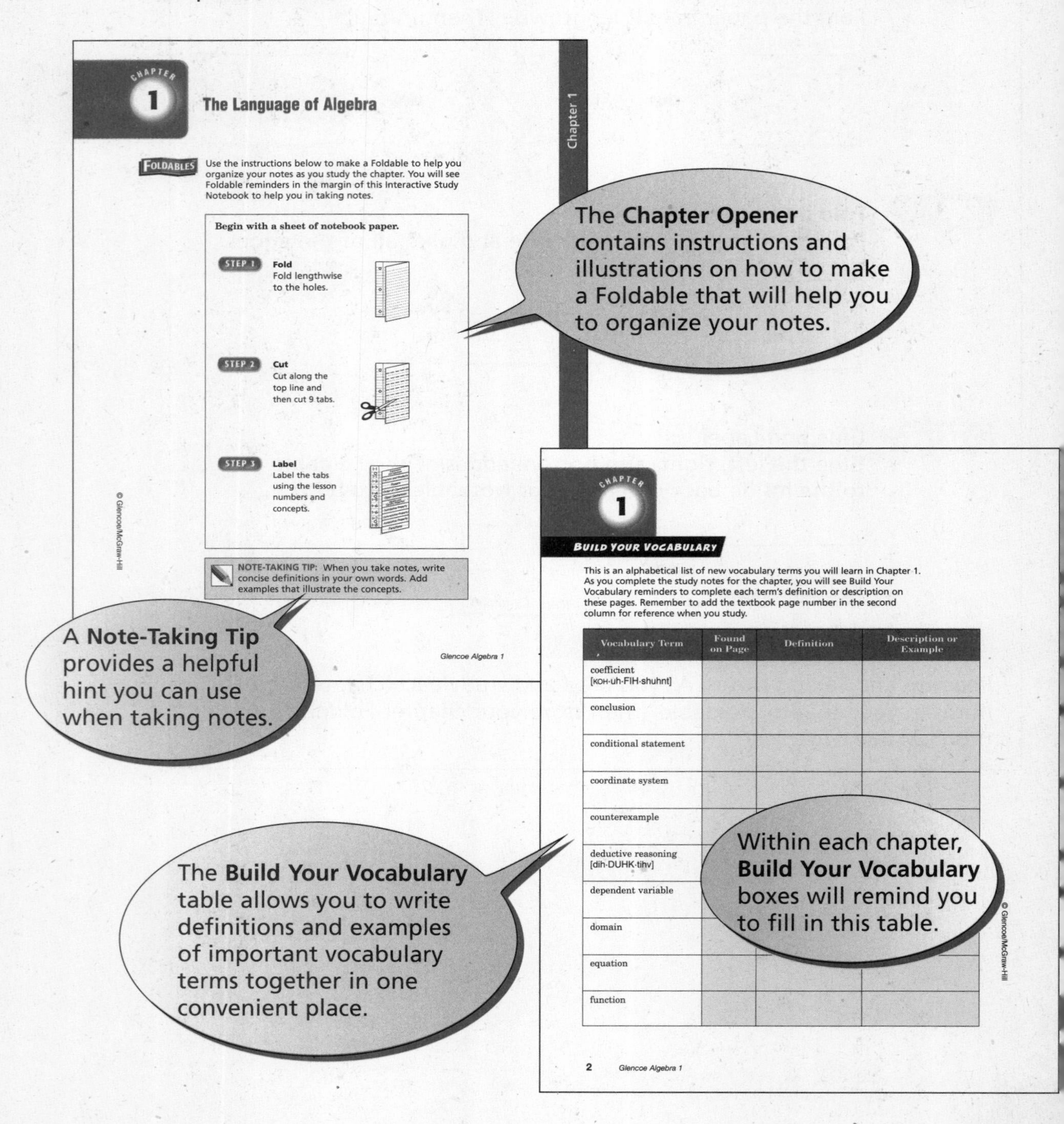

1–1 Variables and Expressions

BUILD YOUR VOCABULARY (page 3)

In algebra, **variables** are symbols used to represent unspecified ______ or ______.

An expression like ______ is called a **power** and is read ______ power.

WHAT YOU'LL LEARN

- Write mathematical expressions for verbal expressions.
- Write verbal expressions for mathematical expressions.

Lessons cover the content of the lessons in your textbook. As your teacher discusses each example, follow along and complete the **fill-in boxes.** Take notes as appropriate.

EXAMPLE Write Algebraic Expressions

1 **Write an algebraic expression for each verbal expression.**

a. five less than a number c

The words *less than* suggest ______.

a number c → c, less → $-$, five → 5

Thus, the algebraic expression is ______.

b. the sum of 9 and 2 times the number d

Sum implies ______ and *times* implies ______.

So, the expression can be written as ______.

FOLDABLES **ORGANIZE IT**

Under the tab for Lesson 1-1, take notes on writing expressions. Be sure to include examples.

Your Turn **Write an algebraic expression for each verbal expression.**

a. nine more than a number h

b. the difference of 6 and 4 times a number x

Foldables feature reminds you to take notes in your Foldable.

Examples parallel the examples in your textbook.

1–2

EXAMPLE Fraction Bar

3 **Evaluate** $\frac{2^5 - 6 \cdot 2}{3^3 - 5 \cdot 3 - 2}$.

$\frac{2^5 - 6 \cdot 2}{3^3 - 5 \cdot 3 - 2}$ means $(2^5 - 6 \cdot 2)$ ☐ $(3^3 - 5 \cdot 3 - 2)$.

$\frac{2^5 - 6 \cdot 2}{3^3 - 5 \cdot 3 - 2} = \frac{☐ - 6 \cdot 2}{3^3 - 5 \cdot 3 - 2}$ Evaluate the power in the numerator.

$= \frac{☐ - ☐}{3^3 - 5 \cdot 3 - 2}$ Multiply 6 and 2 in the numerator.

$= \frac{20}{3^3 - 5 \cdot 3 - 2}$ Subtract ☐ and ☐ in the numerator.

$= \frac{20}{☐ - 5 \cdot 3 - 2}$ Evaluate the power in the denominator.

$= \frac{20}{☐ - ☐ - 2}$ Multiply ☐ and ☐ in the denominator.

$= \frac{20}{☐}$ or ☐ Subtract from left to right in the denominator. Then simplify.

Your Turn Evaluate $\frac{3^3 - 4 \cdot 3}{2^5 - 5 \cdot 3 - 2}$.

Your Turn Exercises allow you to solve similar exercises on your own.

Glencoe Algebra 1 7

CHAPTER 1 **BRINGING IT ALL TOGETHER**

STUDY GUIDE

FOLDABLES	VOCABULARY PUZZLEMAKER	BUILD YOUR VOCABULARY
Use your **Chapter 1 Foldable** to help you study for your chapter test.	To make a crossword puzzle, word search, or jumble puzzle of the vocabulary words in Chapter 1, go to www.glencoe.com/sec/math/t_resources/free/index.php	You can use your **Vocabulary Build**[...] (pages 2–3) to he[...] solve the puzzle.

1-1 Variables and Expressions

Write the letter of the algebraic expression that best matches each phrase.

1. three more than a number n
2. five times the difference of x and 4
3. one half the number r
4. the product of x and y divided by 2
5. x to the fourth power

a. $5(x - 4)$
b. x^4
c. $\frac{1}{2}r$
d. $n + 3$
e. $\frac{xy}{2}$

1-2 Order of Operations

For each of the following expressions, write *addition, subtraction, multiplication, division,* or *evaluate powers* to tell what operation to use first when evaluating the expression.

6. $400 - 5[12 + 9]$
7. $26 - 8 + 14$
8. $17 + 3 \cdot 6$
9. $69 + 57 \div 3 + 16 \cdot 4$
10. $\frac{51 + 729}{9^2}$

Glencoe Algebra 1 25

Bringing It All Together Study Guide reviews the main ideas and key concepts from each lesson.

NOTE-TAKING TIPS

Your notes are a reminder of what you learned in class. Taking good notes can help you succeed in mathematics. The following tips will help you take better classroom notes.

- Before class, ask what your teacher will be discussing in class. Review mentally what you already know about the concept.
- Be an active listener. Focus on what your teacher is saying. Listen for important concepts. Pay attention to words, examples, and/or diagrams your teacher emphasizes.
- Write your notes as clear and concise as possible. The following symbols and abbreviations may be helpful in your note-taking.

Word or Phrase	Symbol or Abbreviation	Word or Phrase	Symbol or Abbreviation
for example	e.g.	not equal	$\neq$
such as	i.e.	approximately	$\approx$
with	w/	therefore	$\therefore$
without	w/o	versus	vs
and	+	angle	$\angle$

- Use a symbol such as a star (★) or an asterisk (*) to emphasis important concepts. Place a question mark (?) next to anything that you do not understand.
- Ask questions and participate in class discussion.
- Draw and label pictures or diagrams to help clarify a concept.
- When working out an example, write what you are doing to solve the problem next to each step. Be sure to use your own words.
- Review your notes as soon as possible after class. During this time, organize and summarize new concepts and clarify misunderstandings.

Note-Taking Don'ts

- **Don't** write every word. Concentrate on the main ideas and concepts.
- **Don't** use someone else's notes as they may not make sense.
- **Don't** doodle. It distracts you from listening actively.
- **Don't** lose focus or you will become lost in your note-taking.

The Language of Algebra

Use the instructions below to make a Foldable to help you organize your notes as you study the chapter. You will see Foldable reminders in the margin of this Interactive Study Notebook to help you in taking notes.

Begin with a sheet of notebook paper.

STEP 1 **Fold**
Fold lengthwise to the holes.

STEP 2 **Cut**
Cut along the top line and then cut 9 tabs.

STEP 3 **Label**
Label the tabs using the lesson numbers and concepts.

NOTE-TAKING TIP: When you take notes, write concise definitions in your own words. Add examples that illustrate the concepts.

CHAPTER 1

BUILD YOUR VOCABULARY

This is an alphabetical list of new vocabulary terms you will learn in Chapter 1. As you complete the study notes for the chapter, you will see Build Your Vocabulary reminders to complete each term's definition or description on these pages. Remember to add the textbook page number in the second column for reference when you study.

Vocabulary Term	Found on Page	Definition	Description or Example
coefficient [KOH·uh·FIH·shuhnt]			
conclusion			
conditional statement			
coordinate system			
counterexample			
deductive reasoning [dih·DUHK·tihv]			
dependent variable			
domain			
equation			
function			

Vocabulary Term	Found on Page	Definition	Description or Example
hypothesis [hy·PAH·thuh-suhs]			
independent variable			
inequality			
like terms			
order of operations			
power			
range			
replacement set			
solving an open sentence			
variables			

1–1 Variables and Expressions

WHAT YOU'LL LEARN

- Write mathematical expressions for verbal expressions.
- Write verbal expressions for mathematical expressions.

BUILD YOUR VOCABULARY (page 3)

In algebra, **variables** are symbols used to represent unspecified ______ or ______.

An expression like ______ is called a **power** and is read ______ power.

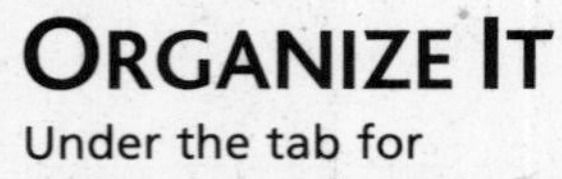

ORGANIZE IT

Under the tab for Lesson 1-1, take notes on writing expressions. Be sure to include examples.

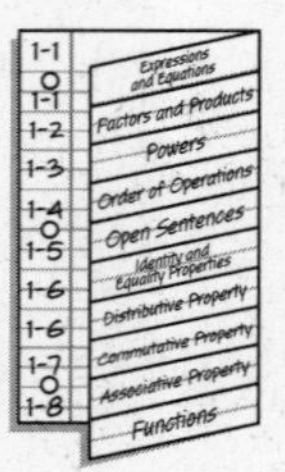

EXAMPLE Write Algebraic Expressions

1 **Write an algebraic expression for each verbal expression.**

a. five less than a number c

The words *less than* suggest ______.

a number c → c less → $-$ five → 5

Thus, the algebraic expression is ______.

b. the sum of 9 and 2 times the number d

Sum implies ______ and *times* implies ______.

So, the expression can be written as ______.

Your Turn **Write an algebraic expression for each verbal expression.**

a. nine more than a number h

______

b. the difference of 6 and 4 times a number x

______

EXAMPLE Write Algebraic Expressions with Powers

2 **Write each expression algebraically.**

a. the product of $\frac{3}{4}$ and a to the seventh power

The expression is ______.

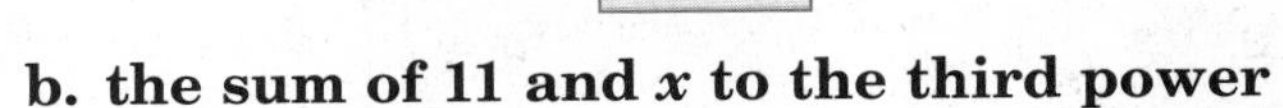

b. the sum of 11 and x to the third power

The expression is ______.

> **REMEMBER IT**
> When no exponent is shown, it is understood to be 1.

Your Turn **Write each expression algebraically.**

a. the difference of 12 and x squared

______

b. the quotient of 6 and x to the fifth power

______

EXAMPLE Evaluate Powers

3 **Evaluate 3^4.**

$3^4 = 3 \cdot 3 \cdot 3 \cdot 3$ Use ___ as a factor ___ times.

$=$ ___ Multiply.

Your Turn **Evaluate each expression.**

a. 5^4

b. 2^5

______

EXAMPLE Write Verbal Expressions

4 **Write a verbal expression for $\frac{8x^2}{5}$.**

the quotient of 8 times ______ and ___

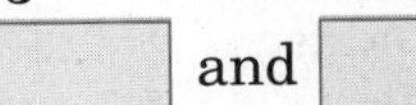

HOMEWORK ASSIGNMENT

Page(s):

Exercises:

Your Turn **Write a verbal expression for each algebraic expression.**

a. $7a^4$

______

b. $x^2 + 3$

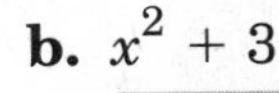

______

1–2 Order of Operations

WHAT YOU'LL LEARN

- Evaluate numerical expressions by using the order of operations.
- Evaluate algebraic expressions by using the order of operations.

KEY CONCEPT

Order of Operations

Step 1 Evaluate expressions inside grouping symbols.

Step 2 Evaluate all powers.

Step 3 Do all multiplications and/or divisions from left to right.

Step 4 Do all additions and/or subtractions from left to right.

FOLDABLES On the tab for Lesson 1–2, write the Order of Operations. Include examples.

EXAMPLE Evaluate Expressions

1 Evaluate $48 \div 2^3 \cdot 3 + 5$.

$48 \div 2^3 \cdot 3 + 5 = 48 \div \_\_\_ \cdot 3 + 5$	Evaluate powers.
$= \_\_\_ \cdot 3 + 5$	Divide ___ by ___.
$= \_\_\_ + 5$	Multiply ___ and ___.
$= \_\_\_$	Add ___ and ___.

Your Turn **Evaluate each expression.**

a. $18 + 2 \cdot 4 - 3$

b. $3 + 6^2 \div 4 - 5$

EXAMPLE Grouping Symbols

2 Evaluate $(8 - 3) \cdot 3(3 + 2)$.

$(8 - 3) \cdot 3(3 + 2) = 5 \cdot 3(5)$	Evaluate inside grouping symbols.
$= \_\_\_(5)$	Multiply ___ and ___.
$= \_\_\_$	Multiply ___ and ___.

Your Turn **Evaluate each expression.**

a. $2(4 + 7) \cdot (9 - 5)$

b. $3[5 - 2 \cdot 2]^2$

EXAMPLE Fraction Bar

3 Evaluate $\dfrac{2^5 - 6 \cdot 2}{3^3 - 5 \cdot 3 - 2}$.

$\dfrac{2^5 - 6 \cdot 2}{3^3 - 5 \cdot 3 - 2}$ means $(2^5 - 6 \cdot 2) \; \square \; (3^3 - 5 \cdot 3 - 2)$.

$\dfrac{2^5 - 6 \cdot 2}{3^3 - 5 \cdot 3 - 2} = \dfrac{\square - 6 \cdot 2}{3^3 - 5 \cdot 3 - 2}$ Evaluate the power in the numerator.

$= \dfrac{\square - \square}{3^3 - 5 \cdot 3 - 2}$ Multiply 6 and 2 in the numerator.

$= \dfrac{20}{3^3 - 5 \cdot 3 - 2}$ Subtract $\square$ and $\square$ in the numerator.

$= \dfrac{20}{\square - 5 \cdot 3 - 2}$ Evaluate the power in the denominator.

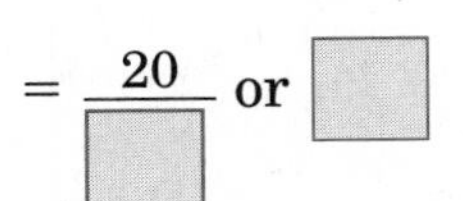

$= \dfrac{20}{\square - \square - 2}$ Multiply $\square$ and $\square$ in the denominator.

$= \dfrac{20}{\square}$ or $\square$ Subtract from left to right in the denominator. Then simplify.

Your Turn Evaluate $\dfrac{3^3 - 4 \cdot 3}{2^5 - 5 \cdot 3 - 2}$.

EXAMPLE Evaluate an Algebraic Expression

4 Evaluate $2(x^2 - y) + z^2$ if $x = 4$, $y = 3$, and $z = 2$.

$2(x^2 - y) + z^2 = 2(4^2 - 3) + 2^2$ Replace x with ___, y with ___, and z with ___.

$= 2(\_\_\_ - 3) + 2^2$ Evaluate ___.

$= 2(\_\_\_) + 2^2$ Subtract ___ and ___.

$= 2(13) + \_\_\_$ Evaluate ___.

$= \_\_\_ + 4$ Multiply ___ and ___.

$= \_\_\_$ Add.

Your Turn Evaluate $x^3 - y^3 + z$ if $x = 3$, $y = 2$, and $z = 5$.

HOMEWORK ASSIGNMENT

Page(s):

Exercises:

1–3 Open Sentences

WHAT YOU'LL LEARN

- Solve open sentence equations.
- Solve open sentence inequalities.

BUILD YOUR VOCABULARY (pages 2–3)

The process of finding a value for a variable that results in a ______ sentence is called **solving the open sentence.**

A sentence that contains an ______ sign is called an **equation**.

A set of numbers from which replacements for a ______ may be chosen is called a **replacement set**.

FOLDABLES

ORGANIZE IT

Under the tab for Lesson 1-3, explain how to solve open sentence equations and inequalities. Include examples.

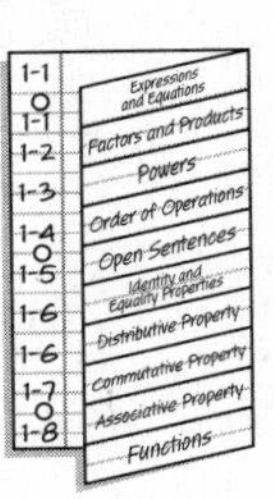

EXAMPLE Use a Replacement Set to Solve an Equation

1 Find the solution set for $4a + 7 = 23$ if the replacement set is {2, 3, 4, 5, 6}.

Replace a in $4a + 7 = 23$ with each value in the replacement set.

a	$4a + 7 = 23$	True or False?
2	$4(2) + 7 = 23 \rightarrow$ ______ $= 23$	
3	$4(3) + 7 = 23 \rightarrow$ ______ $= 23$	
4	$4(4) + 7 = 23 \rightarrow$ ______ $= 23$	
5	$4(5) + 7 = 23 \rightarrow$ ______ $= 23$	
6	$4(6) + 7 = 23 \rightarrow$ ______ $= 23$	

The solution set is ______.

Your Turn Find the solution set for the equation $6c - 5 = 7$ if the replacement set is {0, 1, 2, 3, 4}.

BUILD YOUR VOCABULARY (page 3)

An open sentence that contains the symbol ___, ___, ___, or ___ is called an **inequality**.

EXAMPLE Find the Solution Set of an Inequality

2 Find the solution set for $z + 11 \geq 32$ if the replacement set is {20, 21, 22, 23, 24}.

Replace z in $z + 11 \geq 32$ with each value in the replacement set.

a	$z + 11 \geq 32$	True or False?
20	$20 + 11 \geq 32 \rightarrow$ ___ ≥ 32	
21	$21 + 11 \geq 32 \rightarrow$ ___ ≥ 32	
22	$22 + 11 \geq 32 \rightarrow$ ___ ≥ 32	
23	$23 + 11 \geq 32 \rightarrow$ ___ ≥ 32	
24	$24 + 11 \geq 32 \rightarrow$ ___ ≥ 32	

The solution set for $z + 11 \geq 32$ is ___.

Your Turn Find the solution set for $2x - 3 > 6$ if the replacement set is {2, 3, 4, 5}.

REVIEW IT

Write in words how each of the following symbols is read: $>$, $<$, $\geq$, $\leq$.

HOMEWORK ASSIGNMENT

Page(s):

Exercises:

1–4 Identity and Equality Properties

BUILD YOUR VOCABULARY (page 3)

Two numbers whose ______ is 1 are called **multiplicative inverses** or **reciprocals**.

WHAT YOU'LL LEARN

- Recognize the properties of identity and equality.
- Use the properties of identity and equality.

KEY CONCEPTS

Additive Identity For any number a, the sum of a and 0 is a.

Multiplicative Identity For any number a, the product of a and 1 is a.

Multiplicative Property of Zero For any number a, the product of a and 0 is a.

Multiplicative Inverse For every number $\frac{a}{b}$, where $a, b \neq 0$, there is exactly one number $\frac{b}{a}$ such that the product of $\frac{a}{b}$ and $\frac{b}{a}$ is 1.

EXAMPLE Identify Properties

1 **Name the property used in each equation. Then find the value of n.**

a. $n \cdot 12 = 0$

Multiplicative Property of Zero

$n =$ ______, since ______ $\cdot\ 12 = 0$.

b. $n \cdot \frac{1}{5} = 1$

Multiplicative Inverse Property

$n =$ ______, since ______ $\cdot\ \frac{1}{5} = 1$.

c. $0 + n = 8$

Additive Identity Property

$n =$ ______, since $0 +$ ______ $= 8$.

Your Turn **Name the property used in each equation. Then find the value of n.**

a. $n \cdot \frac{1}{2} = 1$

______

b. $n + 0 = 11$

______

c. $n \cdot 4 = 0$

______

Key Concepts

Reflexive Any quantity is equal to itself.

Symmetric If one quantity equals a second quantity, then the second quantity equals the first.

Transitive If one quantity equals a second quantity and the second quantity equals a third quantity, then the first quantity equals the third quantity.

Substitution A quantity may be substituted for its equal in any expression.

FOLDABLES List the Identity and Equality Properties under the tab for Lesson 1–4. Include an example of each property.

EXAMPLE Evaluate Using Properties

1 Evaluate $\frac{1}{4}(12 - 8) + 3(15 \div 5 - 2)$. Name the property used in each step.

$\frac{1}{4}(12 - 8) + 3(15 \div 5 - 2)$

$= \frac{1}{4}(4) + 3(15 \div 5 - 2)$ ______ ; $12 - 8 = 4$

$= \frac{1}{4}(4) + 3(3 - 2)$ ______ ; $15 \div 5 = 3$

$= \frac{1}{4}(4) + 3(1)$ ______ ; $3 - 2 = 1$

$= 1 + 3(1)$ ______ ; $\frac{1}{4} \cdot 4 = 1$

$= 1 + 3$ ______ ; $3 \cdot 1 = 3$

$= 4$ ______ ; $1 + 3 = 4$

Your Turn Evaluate $\frac{1}{3}(10 - 7) + 4(18 \div 9 - 1)$. Name the property used in each step.

$\frac{1}{3}(10 - 7) + 4(18 \div 9 - 1)$

$= \frac{1}{3}(3) + 4(18 \div 9 - 1)$ ______

$= \frac{1}{3}(3) + 4(2 - 1)$ ______

$= \frac{1}{3}(3) + 4(1)$ ______

$= 1 + 4(1)$ ______

$= 1 + 4$ ______

$= 5$ ______

Homework Assignment

Page(s):

Exercises:

1–5 The Distributive Property

WHAT YOU'LL LEARN

- Use the Distributive Property to evaluate expressions.
- Use the Distributive Property to simplify expressions.

KEY CONCEPT

Distributive Property

For any numbers *a*, *b*, and *c*,

$a(b + c) = ab + ac$ and
$(b + c)a = ba + ca$ and
$a(b - c) = ab - ac$ and
$(b - c)a = ba - ca$

FOLDABLES Under the tab for Lesson 1–5, write the Distributive Property. Write a numeric and algebraic example of the property.

EXAMPLE Distribute Over Addition

1 **Rewrite 5(7 + 2) using the Distributive Property. Then evaluate.**

$5(7 + 2) = 5(7) + 5(2)$	Distributive Property.
$= \square + \square$	Multiply.
$= \square$	Add.

EXAMPLE Distributive Over Subtraction

2 **Rewrite (16 – 7)3 using the Distributive Property. Then evaluate.**

$(16 - 7)3 = 16 \cdot \square - 7 \cdot \square$	Distributive Property.
$= 48 - 21$	Multiply.
$= \square$	Subtract.

EXAMPLE Use the Distributive Property

3 **Use the Distributive Property to find $27\left(3\frac{2}{3}\right)$.**

$27\left(3\frac{2}{3}\right) = 27\left(3 + \frac{2}{3}\right)$	Think: $\left(3\frac{2}{3}\right) = 3 + \frac{2}{3}$
$= 27(3) + 27\left(\frac{2}{3}\right)$	Distributive Property.
$= \square + \square$	Multiply.
$= \square$	Add.

Your Turn

Rewrite each expression using the Distributive Property. Then evaluate.

a. $4(11 + 6)$

b. $(12 - 7)2$

c. Use the Distributive Property to find $15\left(1\frac{2}{5}\right)$.

BUILD YOUR VOCABULARY (pages 2–3)

Like terms are terms that contain the same variables, with corresponding variables having the same ______.

The **coefficient** of a term is the ______ factor.

EXAMPLE Algebraic Expressions

4 **Rewrite $4(y^2 + 8y + 2)$ using the Distributive Property. Then simplify.**

$4(y^2 + 8y + 2) = \square(y^2) + \square(8y) + \square(2)$ Distributive Property

$= \square + \square + \square$ Multiply.

REMEMBER IT

When you simplify expressions, first identify like terms.

EXAMPLE Combine Like Terms

5 **Simplify $17a + 21a$.**

$17a + 21a = (17 + 21)a$ Distributive Property

$= \square$ Substitution

WRITE IT

Give an example of two like terms. Then give an example of two terms that are not like terms.

______

______

______

______

Your Turn

a. Rewrite $3(x^3 + 2x^2 - 5x + 7)$ using the Distributive Property. Then simplify.

b. Simplify $14x - 9x$.

HOMEWORK ASSIGNMENT

Page(s):

Exercises:

1–6 Commutative and Associative Properties

What You'll Learn

- Recognize the Commutative and Associative Properties.
- Use the Commutative and Associative Properties to simplify expressions.

Key Concepts

Commutative Property The order in which you add or multiply numbers does not change their sum or product.

Associative Property The way you group three or more numbers when adding or multiplying does not change their sum or product.

FOLDABLES List the properties on the tab for Lesson 1–6.

EXAMPLE Multiplication Properties

1 Evaluate $2 \cdot 8 \cdot 5 \cdot 7$.

You can rearrange and group the factors to make mental calculations easier.

$2 \cdot 8 \cdot 5 \cdot 7 = 2 \cdot 5 \cdot 8 \cdot 7$ ____ Property (×)

$= $ ____ $\cdot (8 \cdot 7)$ ____ Property (×)

$= $ ____ $\cdot$ ____ Multiply.

$= $ ____ Multiply.

EXAMPLE Use Addition Properties

2 TRANSPORTATION **Refer to Example 2 in Lesson 1–6 of your book. Find the distance between Lakewood/Ft. McPherson and Five Points. Explain how the Commutative Property makes calculating the answer unnecessary.**

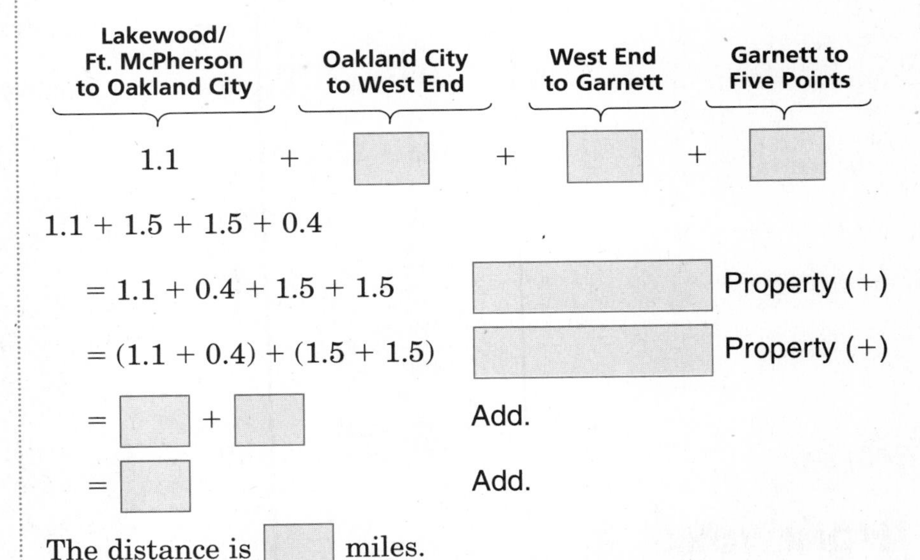

$1.1 + 1.5 + 1.5 + 0.4$

$= 1.1 + 0.4 + 1.5 + 1.5$ ____ Property (+)

$= (1.1 + 0.4) + (1.5 + 1.5)$ ____ Property (+)

$= $ ____ $+$ ____ Add.

$= $ ____ Add.

The distance is ____ miles.

Calculating the answer is actually unnecessary because the route is the opposite of the one in Example 2. The ____ Property states that the order in which numbers are added does not matter.

Your Turn

a. Evaluate $3 \cdot 5 \cdot 3 \cdot 4$.

b. The distance from Five Points to Garnett is 0.4 mile. From Garnett, West End is 1.5 miles. From West End, Oakland City is 1.5 miles. Write an expression to find the distance from Five Points to Oakland City, then write an expression to find the distance from Oakland City to Five Points.

EXAMPLE Simplify an Expression

3 **Simplify $8(2b + 4) + 7b$.**

$8(2b + 4) + 7b =$

$= 8(\quad) + 8(\quad) + 7\quad$	Distributive Property.
$= \quad + \quad + \quad$	Multiply.
$= 16b + 7b + 32$	______ Property (+)
$= (16b + 7b) + 32$	______ Property (+)
$= \quad b + 32$	______ Property
$= \quad + 32$	______

Your Turn Simplify $5(3c + 4) + 6c$.

HOMEWORK ASSIGNMENT

Page(s):

Exercises:

1–7 Logical Reasoning

What You'll Learn

- Identify the hypothesis and conclusion in a conditional statement.
- Use a counterexample to show that an assertion is false.

BUILD YOUR VOCABULARY (pages 2–3)

Conditional statements can be written in the form

____ A ____ B.

The part of the statement immediately after ____ is called the **hypothesis**.

The part of the statement immediately after ____ is called the **conclusion.**

EXAMPLE Identify Hyphothesis and Conclusion

1 Identify the hypothesis and conclusion of each statement.

a. If it is raining, then Beau and Chloe will not play softball.

The hypothesis follows the word ____ and the conclusion follows the word ____.

Hypothesis: ____

Conclusion: ____

b. If $7y + 5 \leq 26$, then $y \leq 3$.

Hypothesis: ____

Conclusion: ____

FOLDABLES

Organize It

On the tab for Lesson 1–7, write a conditional sentence and label the hypothesis and conclusion.

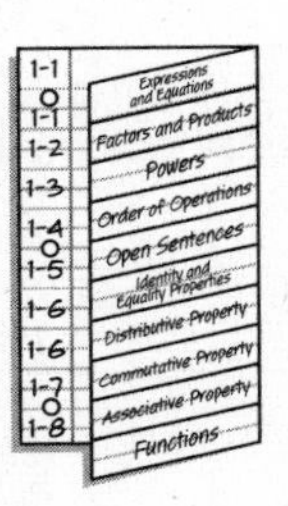

Your Turn **Identify the hypothesis and conclusion of each statement.**

a. If it is above 75°, then you can go swimming.

____

b. If $2x + 3 = 5$, then $x = 1$.

____

EXAMPLE Write a Conditional in If-Then Form

2 Identify the hypothesis and conclusion of the statement. Then write the statement in if-then form.

I eat light meals.

Hypothesis: I eat a meal
Conclusion: it is light

Your Turn **Identify the hypothesis and conclusion of each statement. Then write each statement in if-then form.**

a. We go bowling on Fridays.

b. For a number x such that $11 + 5x < 21$, $x < 2$.

BUILD YOUR VOCABULARY (page 2)

Deductive reasoning is the process of using facts, rules, definitions, or properties to reach a valid .

A **counterexample** is a specific case in which a statement is .

EXAMPLE Find Counterexamples

3 Find a counterexample for the conditional statement. If Joe does not eat lunch, then he must not feel well.

Perhaps Joe was not hungry.

Your Turn Find a counterexample for the conditional statement.

If you are 16, then you have a driver's license.

HOMEWORK ASSIGNMENT

Page(s):
Exercises:

1-8 Graphs and Functions

WHAT YOU'LL LEARN

- Interpret graphs of functions.
- Draw graphs of functions.

BUILD YOUR VOCABULARY (page 2)

A **function** is a relationship between input and output, in which the ______ depends on the ______.

A **coordinate system** is used to graph ______.

In a function, the value of one quantity ______ on the ______ of the other. This ______ is called the **dependent variable.** The other quantity is called the **independent variable.** The set of values for the ______ variable is called the **domain.**

The set of values for the ______ variable is called the **range.**

EXAMPLE Analyze Graphs

1 **The graph represents the temperature in Ms. Ling's classroom on a winter school day. Describe what is happening in the graph.**

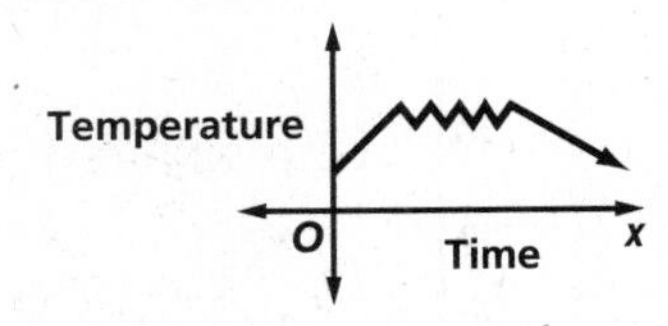

The ______ is low until the heat is turned on. Then the temperature fluctuates ______ and ______ because of the thermostat. Finally the temperature drops when the heat is turned ______.

REMEMBER IT

The x-axis is the horizontal axis and the y-axis is the vertical axis. The independent variable is graphed on the x-axis and the dependent variable is graphed on the y-axis.

Your Turn The graph below represents Macy's speed as she swims laps in a pool. Describe what is happening in the graph.

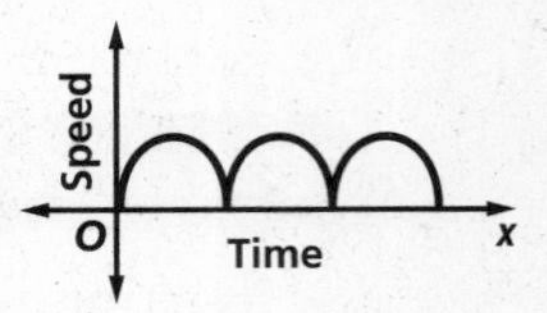

EXAMPLE Draw Graphs

WRITE IT

List three ways data can be represented.

______

______

______

______

2 **There are three lunch periods at a school cafeteria. During the first period, 352 students eat lunch. During the second period, 304 students eat lunch. During the third period, 391 students eat lunch.**

a. Make a table showing the number of students for each of the three lunch periods.

Period	1	2	3
Number of Students			

b. Write the data as a set of ordered pairs.

The period is the ______ variable and the number of students is the ______ variable.

The ordered pairs are (1, ______), (2, ______), and (3, ______)

c. Draw a graph that shows the relationship between the lunch period and the number of students.

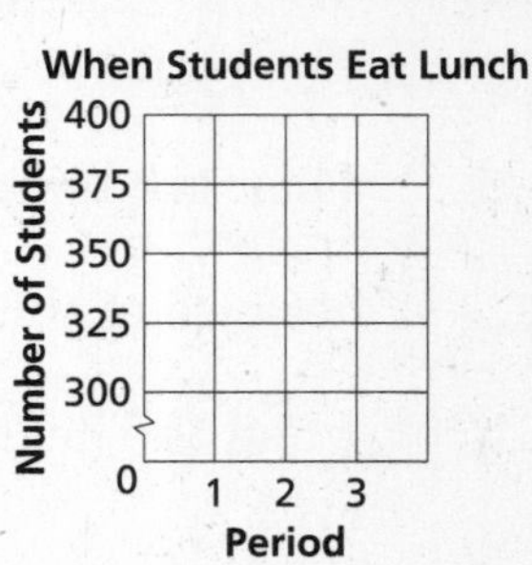

Your Turn **At a car dealership, a salesman worked for three days. On the first day he sold 5 cars. On the second day he sold 3 cars. On the third day he sold 8 cars.**

a. Make a table showing the number of cars sold for each day.

Day			
Number of Cars Sold			

b. Write the data as a set of ordered pairs.

c. Draw a graph that shows the relationship between the day and the number of cars sold.

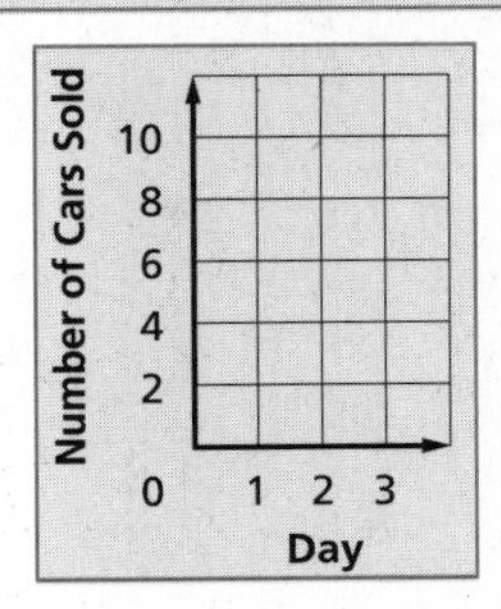

EXAMPLE Domain and Range

3 Mr. Mar is taking his biology classes to the zoo. The zoo admission price is $4 per student, and at most, 120 students will go. Identify a reasonable domain and range for this situation.

The domain contains the number of students on the field trip. Up to ____ students are going on the field trip. Therefore, a reasonable domain would be values from ____ to ____ students. The range contains the total admission price from $0 to 120 × $4 or ____. Thus, a reasonable range is ____ to ____.

Your Turn Prom tickets are on sale at a high school for $25 per person. The banquet room where the prom is being held can hold up to 250 people. Identify a reasonable domain and range for the situation.

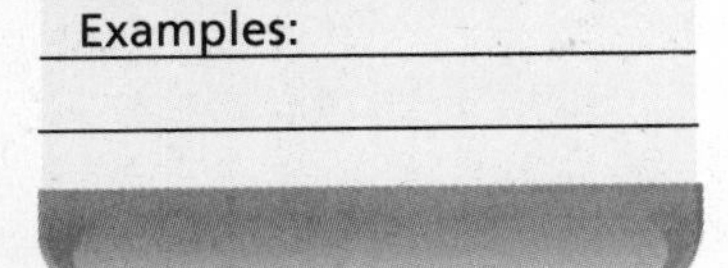

HOMEWORK ASSIGNMENT

Page(s):

Examples:

1–9 Statistics: Analyzing Data by Using Tables and Graphs

EXAMPLE Analyze a Bar Graph

WHAT YOU'LL LEARN

- Analyze data given in tables and graphs (bar, line, and circle).
- Determine whether graphs are misleading.

1 The bar graph shows the number of men and women participating in the NCAA championship sports programs from 1995 to 1999. These same data are displayed in a table.

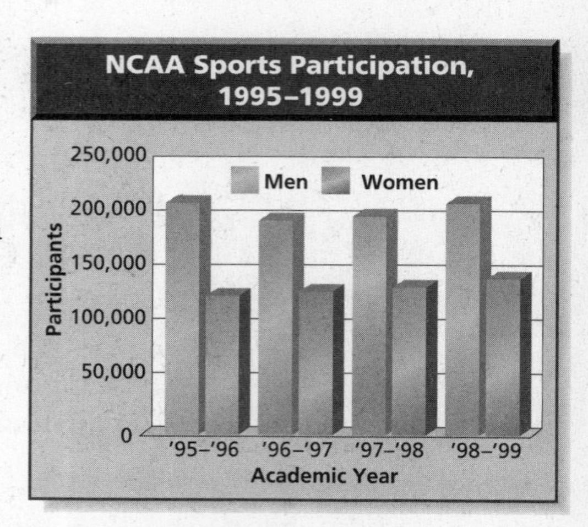

NCAA Championship Sports Participation 1995–1999				
Year	'95–'96	'96–'97	'97–'98	'98–'99
Men	206,366	199,375	200,031	207,592
Women	125,268	129,295	133,376	145,832

Describe how you can tell from the graph that the number of men in NCAA sports remained about the same, while the number of women increased.

Each bar for men is either just above or just below

______. The bars for the women increase each year

from about ______ to ______.

Your Turn **Refer to Example 1.**

a. Approximately how many more men than women participated in sports during the 1996–1997 school year?

b. What was the total participation among men and women in the 1995–1996 academic year?

EXAMPLE Analyze a Circle Graph

2 **A recent poll in New York asked residents whether cell phone use while driving should be banned. The results are shown in the circle graph.**

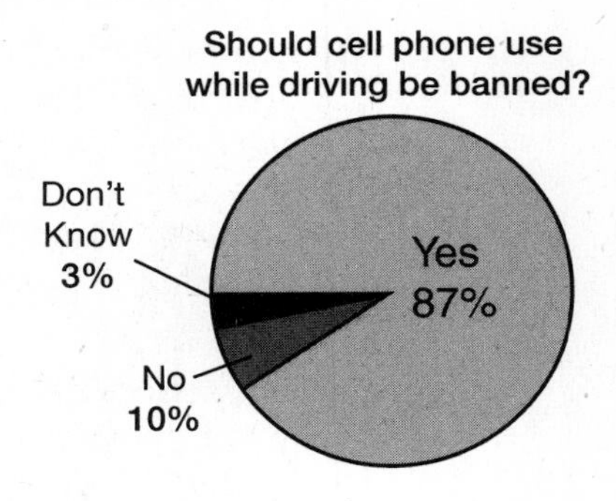

a. If 250 people in New York were surveyed, about how many thought that cell phone use while driving should be banned?

The section of the graph representing people who said cell phone use should be banned while driving is ______ of the circle, so find ______ of 250.

______ of 250 → ______ × 250 or ______

About ______ people said cell phone use while driving should be banned.

b. If a city of 516,000 is representative of those surveyed, how many people could be expected not to know whether cell phone use while driving should be banned?

Of those surveyed, 3% said they did not know if cell phone use while driving should be banned, so find ______ of ______.

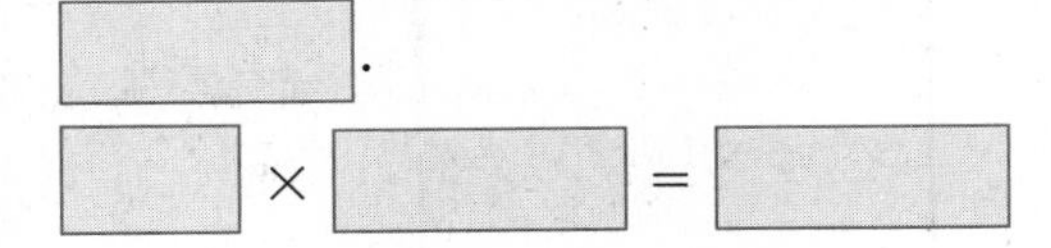

So, ______ people don't know if cell phone use while driving should be banned.

FOLDABLES™

ORGANIZE IT

Under the tab for Lesson 1–9, write a summary of when a bar graph and a circle graph might be used to display data.

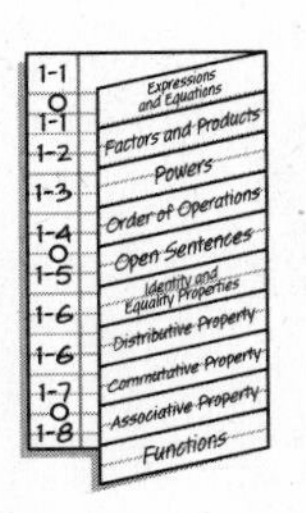

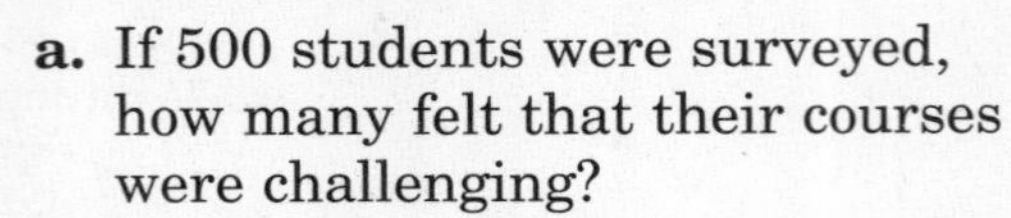

Your Turn **A recent survey asked high school students if they thought their courses were challenging. The results are shown in the circle graph.**

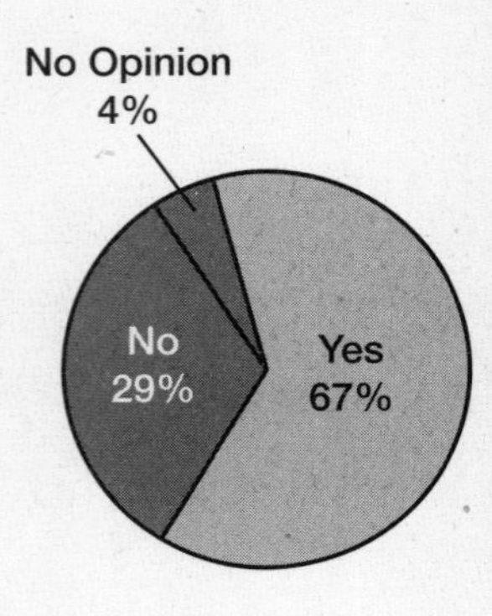

a. If 500 students were surveyed, how many felt that their courses were challenging?

b. If a school of 2350 is representative of those surveyed, how many had no opinion about whether their courses were challenging?

EXAMPLE Misleading Graphs

3 **Joel used the graph below to show his Algebra grade for the first four reporting periods of the year. Does the graph misrepresent the date? Explain.**

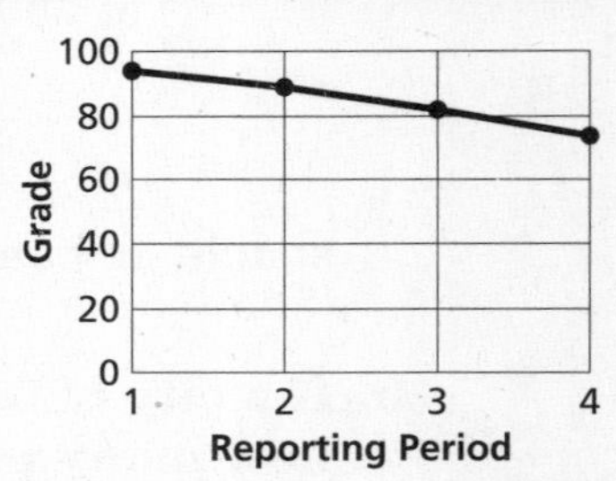

Yes, the scale on the x-axis is too ______ and minimizes the amount that Joel's grade ______.

Your Turn Explain how you could draw a graph that better represents this data.

HOMEWORK ASSIGNMENT

Page(s):

Exercises:

CHAPTER 1

BRINGING IT ALL TOGETHER

STUDY GUIDE

FOLDABLES™	VOCABULARY PUZZLEMAKER	BUILD YOUR VOCABULARY
Use your **Chapter 1 Foldable** to help you study for your chapter test.	To make a crossword puzzle, word search, or jumble puzzle of the vocabulary words in Chapter 1, go to www.glencoe.com/sec/math/t_resources/free/index.php	You can use your completed **Vocabulary Builder** (pages 2–3) to help you solve the puzzle.

1-1 Variables and Expressions

Write the letter of the algebraic expression that best matches each phrase.

1. three more than a number n
2. five times the difference of x and 4
3. one half the number r
4. the product of x and y divided by 2
5. x to the fourth power

a. $5(x - 4)$
b. x^4
c. $\frac{1}{2}r$
d. $n + 3$
e. $\frac{xy}{2}$

1-2 Order of Operations

For each of the following expressions, write *addition, subtraction, multiplication, division,* or *evaluate powers* to tell what operation to use first when evaluating the expression.

6. $400 - 5[12 + 9]$
7. $26 - 8 + 14$
8. $17 + 3 \cdot 6$
9. $69 + 57 \div 3 + 16 \cdot 4$
10. $\frac{51 \div 729}{9^2}$

1-3

Open Sentences

11. How would you read each inequality symbol in words?

Inequality symbol	Words
$<$	
$>$	
$\leq$	
$\geq$	

1-4

Identity and Equality Properties

Write the letter of the sentence that best matches each term.

12. additive identity

13. mutliplicative identity

14. Multiplicative Property of Zero

15. Multiplicative Inverse Property

16. Reflexive Property

17. Symmetric Property

18. Transitive Property

19. Substitution Property

a. $\frac{5}{7} \cdot \frac{7}{5} = 1$

b. $18 = 18$

c. $3 \cdot 1 = 3$

d. If $12 = 8 + 4$, then $8 + 4 = 12$.

e. $6 + 0 = 6$

f. If $2 + 4 = 5 + 1$ and $5 + 1 = 6$, then $2 + 4 = 6$.

g. If $n = 2$, then $5n = 5 \cdot 2$.

h. $4 \cdot 0 = 0$

i. If $n = 2$, then $2n = 4$.

1-5

The Distributive Property

Rewrite using the distributive property.

20. $3(1 + 5)$

21. $5(6 - 4)$

22. $12m + 8m$

1-6

Communitative and Associative Properties

Write the letter of the term that best matches each equation.

23. $3+6 = 6+3$

24. $2 + (3+4) = (2+3) + 4$

25. $2 \cdot (3 \cdot 4) = (2 \cdot 3) \cdot 4$

26. $2 \cdot (3 \cdot 4) = 2 \cdot (4 \cdot 3)$

a. Associative Property of Addition

b. Associative Property of Multiplication

c. Commutative Property of Addition

d. Commutative Property of Multiplication

1-7

Logical Reasoning

Write *hypothesis* or *conclusion* to tell which part of the if-then statement is underlined.

27. If it is Tuesday, then it is raining.

28. If $3x + 7 = 13$, then $x = 2$.

1-8

Graphs and Functions

29. Identify each part of the coordinate system.

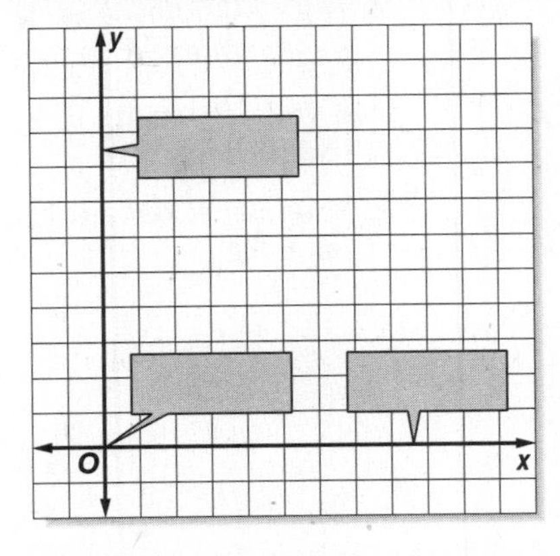

1-9

Analyzing Data by Using Tables and Graphs

Choose *bar graph*, *circle graph*, or *line graph* to complete each statement.

30. A ________ compares parts of a set of data as a percent of the whole set.

31. ________ are useful when showing how a set of data changes over time.

ARE YOU READY FOR THE CHAPTER TEST?

Visit **algebra1.com** to access your textbook, more examples, self-check quizzes, and practice tests to help you study the concepts in Chapter 1.

Check the one that applies. Suggestions to help you study are given with each item.

☐ **I completed the review of all or most lessons without using my notes or asking for help.**

- You are probably ready for the Chapter Test.
- You may want to take the Chapter 1 Practice Test on page 63 of your textbook as a final check.

☐ **I used my Foldable or Study Notebook to complete the review of all or most lessons.**

- You should complete the Chapter 1 Study Guide and Review on pages 57–62 of your textbook.
- If you are unsure of any concepts or skills, refer back to the specific lesson(s).
- You may also want to take the Chapter 1 Practice on page 63.

☐ **I asked for help from someone else to complete the review of all or most lessons.**

- You should review the examples and concepts in your Study Notebook and Chapter 1 Foldable.
- Then complete the Chapter 1 Study Guide and Review on pages 57–62 of your textbook.
- If you are unsure of any concepts or skills, refer back to the specific lesson(s).
- You may also want to take the Chapter 1 Practice Test on page 63.

Student Signature

Parent/Guardian Signature

Teacher Signature

Real Numbers

Use the instructions below to make a Foldable to help you organize your notes as you study the chapter. You will see Foldable reminders in the margin of this Interactive Study Notebook to help you in taking notes.

Begin with a sheet of grid paper.

STEP 1 **Fold**
Fold the short sides to meet in the middle.

STEP 2 **Fold Again**
Fold the top to the bottom.

STEP 3 **Cut**
Open. Cut along second fold to make four tabs.

STEP 4 **Label**
Add a number line and label the tabs as shown.

NOTE-TAKING TIP: Write your notes in different color inks or pencils. For example; use red to write definitions, blue for rules, and pencil to write problems and explanations.

BUILD YOUR VOCABULARY

This is an alphabetical list of new vocabulary terms you will learn in Chapter 2. As you complete the study notes for the chapter, you will see Build Your Vocabulary reminders to complete each term's definition or description on these pages. Remember to add the textbook page number in the second column for reference when you study.

Vocabulary Term	Found on Page	Definition	Description or Example
absolute value			
additive inverses [A-duh-tihv]			
equally likely			
frequency			
integers			
irrational number [ih-RA-shuh-nuhl]			
line plot			
measures of central tendency			

Vocabulary Term	Found on Page	Definition	Description or Example
natural number			
odds			
opposites			
perfect square			
principal square root			
probability [PRAH-buh-BIH-luh-tee]			
rational number [RA-shuh-nuhl]			
real number			
sample space			
simple event			
square root			
stem-and-leaf plot			
whole number			

2–1 Rational Numbers on the Number Line

WHAT YOU'LL LEARN

- Graph rational numbers on a number line.
- Find absolute values of rational numbers.

BUILD YOUR VOCABULARY (pages 30–31)

A ______ line can be used to show the ______ of **natural numbers, whole numbers,** and **integers.**

A **rational number** is any number that can be written in the form $\frac{a}{b}$, where a and b are ______ and b ≠ ______.

The **absolute value** of any number n is its distance from ______ on a number line.

EXAMPLE Identifying Coordinates on a Number Line

1 Name the coordinates of the points graphed on each number line.

REMEMBER IT

Arrowheads on a number line mean that the line and the number set continue to infinity, or never end.

a.

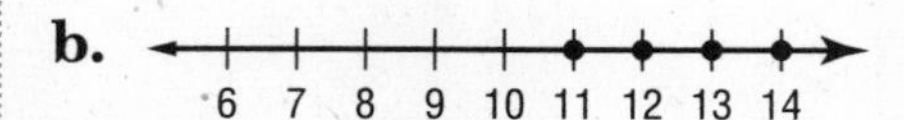

The dots indicate each point on the graph.

The coordinates are ______.

b. 6 7 8 9 10 11 12 13 14

The bold arrow on the graph indicates that the graph continues infinitely in that direction. The coordinates are ______.

Your Turn **Name the coordinates of the points graphed on each number line.**

a.

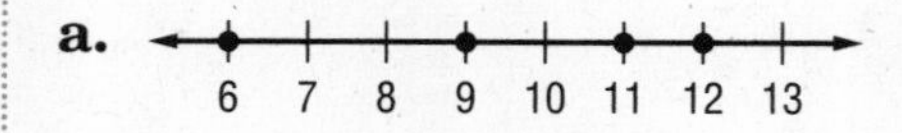

______

b.

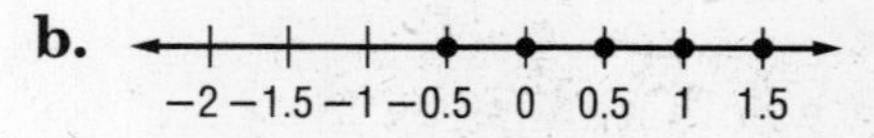

______

EXAMPLE Graph Numbers on a Number Line

2 **Graph each set of numbers.**

a. $\left\{-\frac{1}{2}, 0, \frac{1}{2}, 1\right\}$ **b.** $\left\{-1.5, 0, 1.5, \ldots\right\}$

c. {integers less than −6 or greater than or equal to 1}

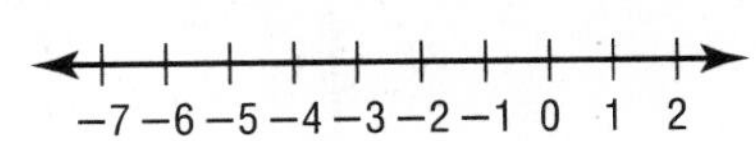

Your Turn **Graph each set of numbers.**

a. {−5, 2, 3, 5}

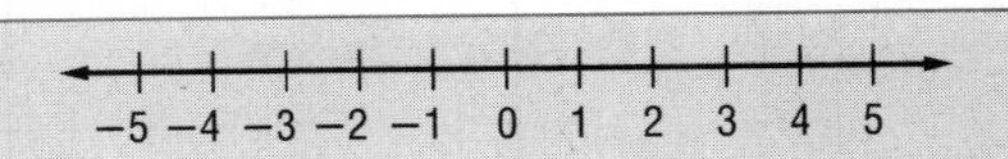

b. $\left\{-\frac{1}{4}, \frac{1}{2}, \frac{7}{4}, 2\right\}$

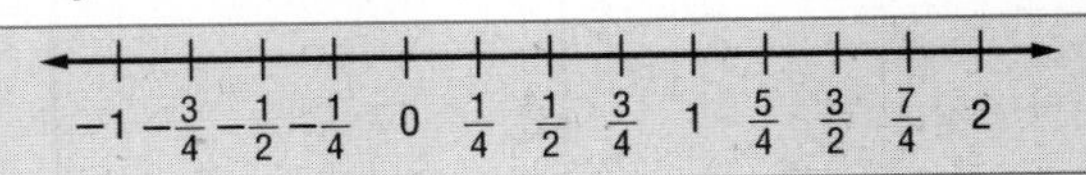

c. {integers less than or equal to −2 or greater than 4}

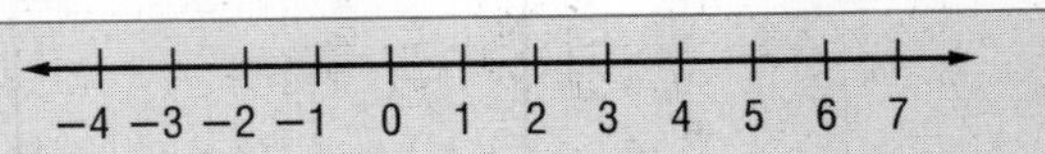

KEY CONCEPT

Absolute Value The absolute value of any number *n* is its distance from zero on a number line and is written as $|n|$.

EXAMPLE Absolute Value of Rational Numbers

3 **Find each absolute value.**

a. $\left|\frac{-5}{8}\right|$

$-\frac{5}{8}$ is $\frac{5}{8}$ unit from 0 in the negative direction.

$\left|\frac{-5}{8}\right| =$ ______

b. $|0.25|$

0.25 is 0.25 unit from 0 in the positive direction.

$|0.25| =$ ______

HOMEWORK ASSIGNMENT

Page(s): ______

Exercises: ______

Your Turn **Find each absolute value.**

a. $\left|\frac{3}{5}\right|$ ______ **b.** $|-6.3|$ ______

2–2 Adding and Subtracting Rational Numbers

What You'll Learn

- Add integers and rational numbers.
- Subtract integers and rational numbers.

Build Your Vocabulary (pages 30–31)

Every ____ rational number can be paired with a ____ rational number. These pairs are called **opposites**.

A number and its ____ are **additive inverses** of each other.

EXAMPLE Use a Number Line to Add Rational Numbers

1 Use a number line to find each sum.

a. $8 + (-5)$

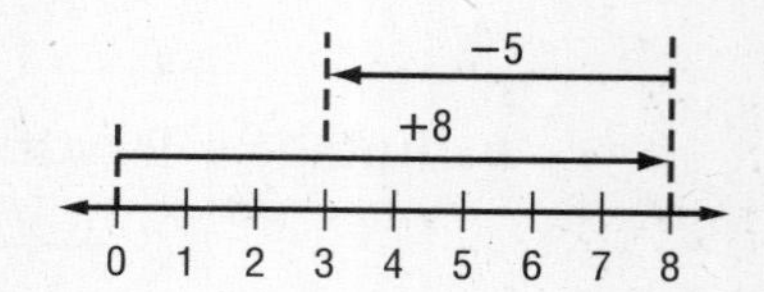

Step 1 Draw an arrow from ____ to ____.

Step 2 Then draw a second arrow 5 units to the left to represent adding -5.

Step 3 The second arrow ends at the sum ____.

b. $-1 + (-4)$

Step 1 Draw an arrow from ____ to ____.

Step 2 Draw a second arrow 4 units to the left.

Step 3 The second arrow ends at the sum ____.

Your Turn **Use a number line to find each sum.**

a. $-6 + 3$

____

b. $-2 + (-7)$

____

KEY CONCEPTS

Addition of Rational Numbers

- To add rational numbers with the *same sign*, add their absolute values. The sum has the same sign as the addends.
- To add rational numbers with *different signs*, subtract the lesser absolute value from the greater absolute value. The sum has the same sign as the number with the greater absolute value.

Additive Inverse Property The sum of a number and its additive inverse is 0.

Subtraction of Rational Numbers To subtract a rational number, add its additive inverse.

FOLDABLES

Write an example of each of these concepts in your notes.

EXAMPLE Add Rational Numbers

2 Find 6 + (−14).

$6 + (-14) = -(|-14| - |6|)$ Subtract the lesser absolute value from the greater absolute value.

$= \square$ Since the number with the greater absolute value is −14, the sum is negative.

$= \square$

Your Turn **Find each sum.**

a. $-12 + (-5)$

b. $\frac{5}{18} + \left(-\frac{2}{9}\right)$

EXAMPLE Subtract Rational Numbers to Solve a Problem

3 STOCKS **In the past year, a publishing company's stock went from $52.08 per share to $70.87 per share. Find the change in the price of the stock.**

$70.87 - 52.08 = 70.87 + (-52.08)$ To subtract 52.08, add its ______.

$= (|70.87| - |52.08|)$ Subtract the absolute values.

$= (70.87 - 52.08)$ The absolute value of 70.87 is greater so the result is ______.

$= \square$

The price of the stock changed by ______.

Your Turn The stock in a company went from $46.98 to $35.09 over a one-month period. Find the change in price for the stock.

HOMEWORK ASSIGNMENT

Page(s):

Exercises:

2–3 Multiplying Rational Fractions

WHAT YOU'LL LEARN

- Multiply integers.
- Multiply rational numbers.

KEY CONCEPTS

Multiplication of Integers The product of two numbers having the *same sign* is positive. The product of two numbers having *different signs* is negative.

Multiplicative Property of −1 The product of any number and −1 is its additive inverse.

EXAMPLE Multiply Integers

1 Find each product.

a. (−8)(−6)

$(-8)(-6) =$ ______ same signs → ______ product

b. (10)(−11)

$(10)(-11) =$ ______ different signs → ______ product

Your Turn **Find each product.**

a. $(-4)(8)$

b. $(-6)(-12)$

EXAMPLE Simplify Expressions

2 Simplify the expression $13x + (-6)(4x)$.

$13x + (-6)(4x) = 13x + (-6)(4)x$ ______ Property (×)

$= 13x +$ ______ Substitution

$= (13 - 24)x$ ______ Property

$=$ ______ Simplify.

Your Turn Simplify the expression $-4(3x) + 15x$.

EXAMPLE Multiply Rational Numbers

3 Find $\left(-\frac{2}{3}\right)\left(-\frac{3}{4}\right)$.

$\left(-\frac{2}{3}\right)\left(-\frac{3}{4}\right) =$ ______ or ______ same signs → ______ product

FOLDABLES™

ORGANIZE IT

Under the tab for multiplying rational numbers, write an example that involves multiplying an odd number of negative integers. Then find the product.

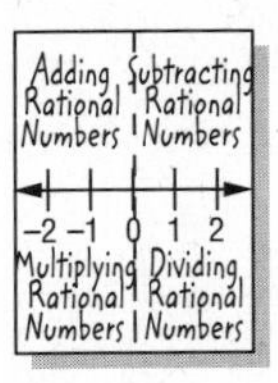

Your Turn Find $\left(-\frac{1}{2}\right)\left(-\frac{2}{5}\right)$.

______

EXAMPLE Multiply Rational Numbers to Solve a Problem

4 STOCKS **The value of a company's stock dropped by \$1.25 per share. By what amount did the total value of the company's stock change if the company has issued 500,000 shares of stock?**

To find the change in the total value of the company's stock, multiply the price lost per share by the number of ______.

______ · ______ = ______ different signs → ______ product

The total value of the company's stock changed by ______.

Your Turn A construction project is stopped by a winter storm. For every day that they are unable to work, the company loses \$35,000. If the storm keeps them from working for 4 days, how much money do they lose?

______

EXAMPLE Evaluate Expressions

5 **Evaluate $\left(-\frac{3}{7}\right)x^3$ if $x = \left(-\frac{1}{2}\right)$.**

$\left(-\frac{3}{7}\right)x^3 = \left(-\frac{3}{7}\right)\left(-\frac{1}{2}\right)^3$ Substitution

$= \left(-\frac{3}{7}\right)(\text{______})$ $\left(-\frac{1}{2}\right)^3 = \left(-\frac{1}{2}\right)\left(-\frac{1}{2}\right)\left(-\frac{1}{2}\right)$ or ______

$=$ 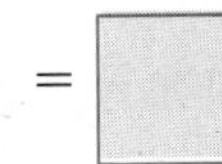 same signs → ______ product

HOMEWORK ASSIGNMENT

Page(s):

Exercises:

Your Turn Evaluate $\left(-\frac{3}{4}\right)m^2$ if $m = \left(-\frac{2}{3}\right)$.

______

2–4 Dividing Rational Numbers

What You'll Learn

- Divide integers.
- Divide rational numbers.

Key Concept

Division of Integers The quotient of two numbers having the *same sign* is positive. The quotient of two numbers having *different signs* is negative.

Foldables Under the tab for dividing rational numbers, write the rule for dividing integers in your own words.

EXAMPLE Divide Integers

1 **Find each quotient.**

a. $\mathbf{-60 \div (-5)}$

$-60 \div (-5) = \square$ positive quotient

b. $\mathbf{-\frac{108}{18}}$

$\frac{-108}{18} = \square \div \square$ Divide.

$= \square$ negative quotient

Your Turn **Find each quotient.**

a. $-80 \div -4$

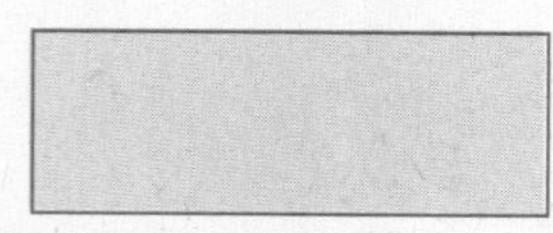

b. $\frac{-105}{7}$

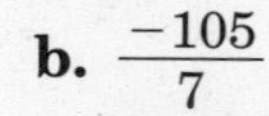

EXAMPLE Simplify Before Dividing

2 **Simplify** $\mathbf{\frac{2(1-5)}{17+(-13)}}$.

$\frac{2(1-5)}{17+(-13)} = \frac{\square}{17+(-13)}$ Simplify the numerator first.

$= \frac{\square}{17+(-13)}$ Multiply.

$= \square$ or $\square$ different signs → negative quotient

Your Turn Simplify $\frac{3(-7-6)}{-16+3}$.

EXAMPLE Divide Rational Numbers

3 Find $-\frac{3}{8} \div \left(-\frac{1}{3}\right)$.

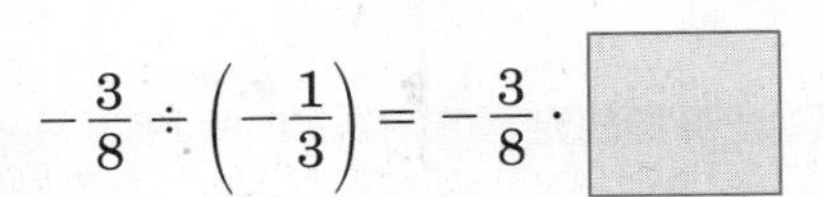

$-\frac{3}{8} \div \left(-\frac{1}{3}\right) = -\frac{3}{8} \cdot$ ______ Multiply by the reciprocal of $-\frac{1}{3}$.

$=$ ______ or ______ same signs → ______ quotient

REVIEW IT

By what are you multiplying when you divide by any nonzero number?

______

______

______

______

______

Your Turn **Find each quotient.**

a. $-39.78 \div (-2.6)$

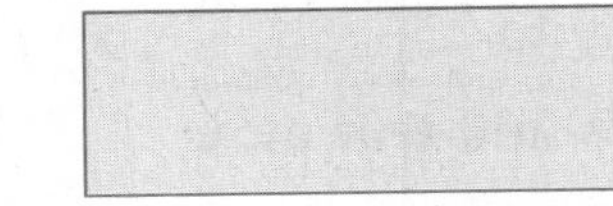

b. $-\frac{3}{8} \div \frac{2}{5}$

EXAMPLE Evaluate Algebraic Expressions

4 **Evaluate $\frac{wx}{y^2}$ if $w = 2$, $x = -9.1$ and $y = 4$.**

$\frac{wx}{y^2} =$ ______ Replace the variables.

$=$ ______ Find the numerator and the denominator.

$=$ ______ Use a calculator.

Your Turn

a. Simplify $\frac{-51 + 34c}{-17}$.

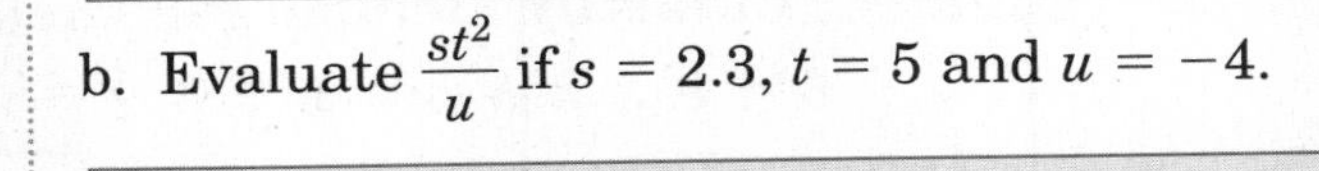

b. Evaluate $\frac{st^2}{u}$ if $s = 2.3$, $t = 5$ and $u = -4$.

HOMEWORK ASSIGNMENT

Page(s): ______

Exercises: ______

2–5 Statistics: Displaying and Analyzing Data

What You'll Learn

- Interpret and create line plots and stem-and-leaf plots.
- Analyze data using mean, median, and mode.

Build Your Vocabulary (pages 30–31)

Most **line plots** have a number line labeled with a scale to include all the data.

An × above the ______ point represents the **frequency** of the data.

Another way to ______ and ______ data is by using a **stem-and-leaf plot.**

EXAMPLE Use a Line Plot to Solve a Problem

1. **TRAFFIC The speeds (in miles per hour) of 20 cars that passed a radar survey are listed.**

72 70 72 74 68 69 70 72 74 75
79 75 74 72 70 64 69 66 68 67

a. Make a line plot of the data.

The lowest value is ______ and the highest value is ______, so use a scale that includes those values. Place an × above each value for each occurrence.

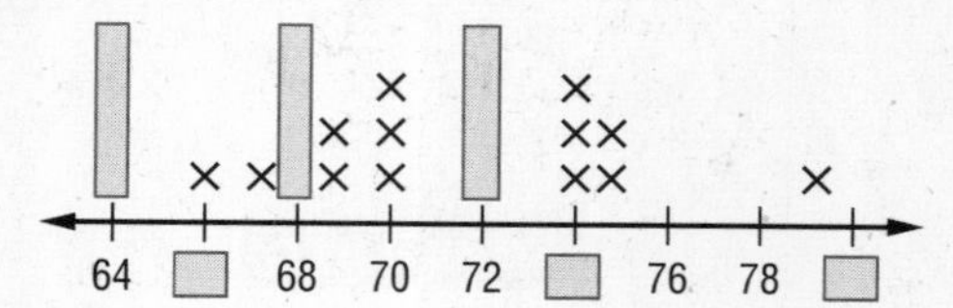

b. Which speed occurs the most frequently?

The line plot shows that ______ miles per hour occurs most frequently.

Your Turn **Students in Mrs. Barrett's class listed the number of family members in their households.**

6 4 8 3 3 5 4 4 3 5 5 2 5 6 3 5 6 2 4 4

a. Make a line plot of the data.

b. Which family size occurs the most frequently?

EXAMPLE Create a Stem-and-Leaf Plot

2 Use the data below to make a stem-and-leaf plot.

85	115	126	92	104	107	78	131	114	92
85	116	100	121	123	131	88	97	99	116
79	90	110	129	108	93	84	75	70	132

The greatest common place value is tens, so the digits in the items place are the stems.

Stem	Leaf
7	0 5 8 9
8	
	0 2 2 3 7 9
10	0 4 7 8
11	
12	1 3 6 9
	1 1 2

11|5 = 115

REVIEW IT

List three other ways that data can be represented. *(Lesson 1–8)*

Your Turn Use the data to make a stem-and-leaf plot.

3	5	7	11	10
15	21	11	13	25
32	37	21	10	12

BUILD YOUR VOCABULARY (page 30)

Numbers known as **measures of central tendency** are often used to describe ______ of ______ because they represent a centralized or middle value.

EXAMPLE Analyze Data

3 Which measure of central tendency best represents the data?

Stem	Leaf
4	1 1 2 4 4 4 5 8
5	0
6	2 5 7
7	3 9
8	1

6|2 = 6.2

The mean is about [ ].

The median is [ ].

The mode is [ ]

Either the [ ] or the [ ] best represent the set of data since both measures are located in the center of the majority of the data. In this instance, the mean is too [ ].

Your Turn Which measure of central tendency best represents the data?

Stem	Leaf
1	0 1 1 5 6 8
2	3 7 8
3	2
4	6
5	4 5 9

3|2 = 3.2

HOMEWORK ASSIGNMENT

Page(s):

Exercises:

2–6 Probability: Simple Probability and Odds

WHAT YOU'LL LEARN

- Find the probability of a simple event.
- Find the odds of a simple event.

BUILD YOUR VOCABULARY (page 31)

The **probability** of a **simple event** like a coin landing heads up when it is tossed, is a ______ of the number of ______ outcomes for the event to the ______ number of possible outcomes of the event.

A list of all possible ______ is called a **sample space**.

EXAMPLE Find Probabilities of Simple Events

KEY CONCEPT

Probability The probability of an event *a* can be expressed as $P(a) = \frac{\text{number of favorable outcomes}}{\text{total number of possible outcomes}}$

1 a. Find the probability of rolling a number greater than 2 on a die.

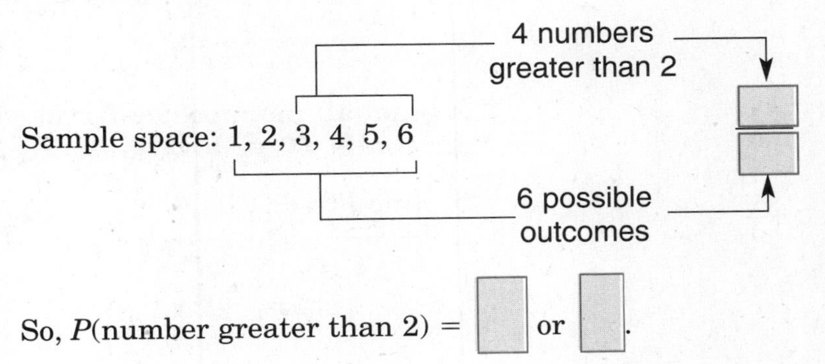

So, P(number greater than 2) = ______ or ______.

A class contains 6 students with black hair, 8 with brown hair, 4 with blonde hair, and 2 with red hair.

b. Find P(black).

There are 6 students with black hair and 20 total students.

P(black) = ______ ← number of favorable outcomes / ← number of possible outcomes

= ______ or ______ Simplify.

So, P(black) = ______ or ______.

c. Find *P*(red or brown).

There are ______ students with red hair and ______ students with brown hair. So there are ______ + ______ or ______ students with red or brown hair.

P(red or brown) = ______ ← number of favorable outcomes / ← number of possible outcomes

= ______ or ______ Simplify.

So, *P*(red or brown) = ______ or ______.

Your Turn

a. Find the probability of rolling a number less than 3 on a die.

A gumball machine contains 40 red gumballs, 30 green gumballs, 50 yellow gumballs, and 40 blue gumballs.

b. Find *P*(red).

c. Find *P*(green or yellow).

BUILD YOUR VOCABULARY (pages 30–31)

When there are *n* ______ and the ______ of each one is $\frac{1}{n}$, we say that the outcomes are **equally likely**.

Another way to express the chance of an event ______ is with **odds**.

EXAMPLE Odds of an Event

KEY CONCEPT

Odds The odds of an event occurring is the ratio that compares the number of ways an event can occur (successes) to the number of ways it cannot occur (failures).

2 **Find the odds of rolling a number greater than 2.**

There are six possible outcomes, 4 are successes and 2 are failures.

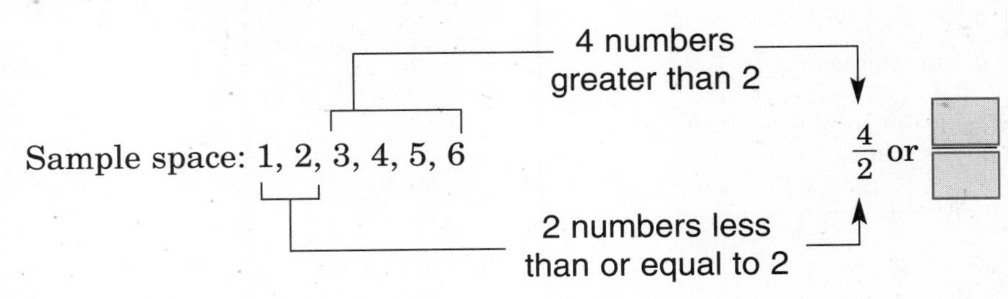

The odds of rolling a number greater than 2 are

____ or ____.

Your Turn Find the odds of rolling a number less than 4.

EXAMPLE Odds Against an Event

REMEMBER IT

A standard deck of playing cards has 52 cards and a die has six sides.

3 **A card is selected at random from a standard deck of 52 cards. What are the odds against selecting a 2 or 3?**

There are four 2s and four 3s in a deck of cards, and there are $52 - 4 - 4$ or ____ cards that are not a 2 or a 3.

odds against a 2 or 3 = ____ ← number of ways *not* to pick a 2 or 3
____ ← number of ways to pick a 2 or 3

The odds against selecting a 2 or 3 are ____.

Your Turn A card is selected at random from a standard deck of 52 cards. What are the odds against selecting a 5, 6, or 7?

HOMEWORK ASSIGNMENT

Page(s):

Exercises:

2–7 Square Roots and Real Numbers

WHAT YOU'LL LEARN

- Find square roots.
- Classify and order real numbers.

BUILD YOUR VOCABULARY (page 31)

A **square root** is one of two square ______ of a number.

A number whose square root is a ______ number is called a **perfect square**.

A **radical sign** is used to indicate the ______ or **principal square root** of the expression under the radical sign.

EXAMPLE Find Square Roots

1 **Find each square root.**

a. $\pm\sqrt{\frac{16}{9}}$

$\pm\sqrt{\frac{16}{9}}$ represents the ______ and ______ square roots of $\frac{16}{9}$.

$\frac{16}{9} =$ ______ and $\frac{16}{9} =$ ______

$\pm\sqrt{\frac{16}{9}} =$ ______

b. $\sqrt{0.0144}$

$\sqrt{0.0144}$ represents the ______ square root of 0.0144.

$0.0144 =$ ______

$\sqrt{0.0144} =$ ______

Your Turn **Find each square root.**

a. $\sqrt{\frac{64}{25}}$ ______ **b.** $\sqrt{0.36}$ ______

WRITE IT

Are the square roots for $\sqrt{-81}$ and $\sqrt{81}$ the same? Explain.

______

______

______

______

BUILD YOUR VOCABULARY (pages 30–31)

Numbers that cannot be expressed as ________ or ________ decimals, or in the form $\frac{a}{b}$, where a and b are integers and $b \neq 0$, are called **irrational numbers.**

________ numbers and rational numbers together form the set of **real numbers.**

KEY CONCEPT

Real Numbers The set of real numbers consists of the set of rational numbers and the set of irrational numbers.

EXAMPLE Classify Real Numbers

2 **Name the set or sets of numbers to which each number belongs.**

a. $\sqrt{17}$

Because $\sqrt{17}$ = ________, which is neither a repeating nor terminating decimal, this number is ________.

b. $\sqrt{169}$

Because $\sqrt{169}$ = ____, this number is a ________ number, a ________ number, an ________ and a ________ number.

Your Turn **Name the set or sets of numbers to which each real number belongs.**

a. $\frac{7}{9}$

b. $\sqrt{36}$

c. $\sqrt{45}$

d. $-\frac{56}{7}$

EXAMPLE Compare Real Numbers

3 Replace each ● with <, >, or = to make each sentence true.

a. 14 ● $\sqrt{196}$

Since $\sqrt{196} = 14$, the numbers are ____.

So, 14 ____ $\sqrt{196}$.

b. $\sqrt{48}$ ● $6.\overline{9}$

$\sqrt{48} = 6.9282032\ldots$

$6.\overline{9} = 6.999999\ldots$

So, $\sqrt{48}$ ____ $6.\overline{9}$.

Your Turn **Replace each ● with <, >, or = to make each sentence true.**

a. 12 ● $\sqrt{169}$

b. $\sqrt{75}$ ● $8.\overline{6}$

EXAMPLE Order Real Numbers

4 Write $\frac{12}{5}$, $\sqrt{6}$, $2.\overline{4}$, and $\frac{61}{25}$ in order from least to greatest.

Write each number as a decimal.

$\frac{12}{5} =$ ____

$\sqrt{6} = 2.4494897\ldots$ or about 2.4495.

$2.\overline{4} = 2.444444\ldots$ or about 2.4444.

$\frac{61}{25} =$ ____

Since $2.4 < 2.44 < 2.4444 < 2.4495$, the numbers arranged in order from least to greatest are ____.

Your Turn Write $\frac{5}{2}$, $\sqrt{5}$, $2.\overline{2}$, and $\frac{51}{20}$ in order from least to greatest.

HOMEWORK ASSIGNMENT

Page(s):

Examples:

CHAPTER 2

BRINGING IT ALL TOGETHER

STUDY GUIDE

FOLDABLES™	VOCABULARY PUZZLEMAKER	BUILD YOUR VOCABULARY
Use your **Chapter 2 Foldable** to help you study for your chapter test.	To make a crossword puzzle, word search, or jumble puzzle of the vocabulary words in Chapter 2, go to www.glencoe.com/sec/math/t_resources/free/index.php	You can use your completed **Vocabulary Builder** (pages 30–31) to help you solve the puzzle.

2-1 Rational Numbers on the Number Line

Write *true* or *false* for each of the following statements.

1. All whole numbers are integers.
2. All natural numbers are integers.
3. All whole numbers are natural numbers.
4. All natural numbers are whole numbers.
5. Explain why $\frac{-3}{7}$, $0.\overline{6}$, and 15 are rational numbers.

2-2 Adding and Subtracting Rational Numbers

Find each sum or difference.

6. $-8 - (-2.4)$
7. $\frac{1}{2} + \left(-\frac{5}{4}\right)$
8. $-21 + 32$
9. $-1.9 - 4$
10. If two numbers are additive inverses, what must be true about their absolute values?

2-3 Multiplying Rational Numbers

Find each product.

11. $(-4)(9)$

12. $(-2)(-13)$

13. $5(-8)$

14. $6(3)$

15. A staircase in a school starts at ground level. Each step up raises you by 8.25 inches. What is your height in relation to ground level after ascending 16 steps?

2-4 Dividing Rational Numbers

Find each quotient.

16. $-\frac{35}{7}$

17. $\frac{-78}{-13}$

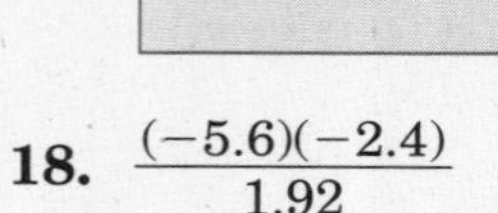

18. $\frac{(-5.6)(-2.4)}{1.92}$

19. Explain what the term *inverse operations* means to you.

2-5 Statistics: Displaying and Analyzing Data

Use the stem-and-leaf plot shown at the right.

20. How is the number 758 represented on the plot?

21. Explain how you know there are 23 numbers in the data.

Stem	Leaf
72	0 1 1 2 5
73	2 2 2 7 9 9
74	1 3 3
75	6 6 8 9
76	0 1 8 8 8
	74|2 = 742

2-6

Probability: Simple Probability and Odds

Write whether each statement is *true* or *false*. If false, replace the underlined word or number to make a true statement.

22. Probability can be written as a fraction, a decimal, or a percent.

23. The sample space of flipping one coin is heads or tails.

24. The probability of an impossible event is 1.

25. The odds against an event occurring are the odds that the event will occur.

Two dice are rolled, and their sum is recorded. Find each probability.

26. *P*(even sum)

27. *P*(sum less than 5)

28. *P*(sum greater than 10)

29. *P*(sum between 2 and 7)

2-7

Square Roots and Real Numbers

Complete each statement.

30. The positive square root of a number is called the ______ square root of the number.

31. A number whose positive square root is a rational number is a ______.

Write each of the following as a mathematical expression that uses the $\sqrt{\ }$ symbol. Then find each square root.

32. the positive square root of 1600

33. the negative square root of 729

34. the principal square root of 3025

35. the irrational numbers and rational numbers together form the set of ______ numbers

ARE YOU READY FOR THE CHAPTER TEST?

Visit **www.algebra1.com** to access your textbook, more examples, self-check quizzes, and practice tests to help you study the concepts in Chapter 2.

Check the one that applies. Suggestions to help you study are given with each item.

☐ **I completed the review of all or most lessons without using my notes or asking for help.**

- You are probably ready for the Chapter Test.
- You may want to take the Chapter 2 Practice Test on page 115 of your textbook as a final check.

☐ **I used my Foldable or Study Notebook to complete the review of all or most lessons.**

- You should complete the complete the Chapter 2 Study Guide and Review on pages 110–114 of your textbook.
- If you are unsure of any concepts or skills, refer back to the specific lesson(s).
- You may also want to take the Chapter 2 Practice Test on page 115.

☐ **I asked for help from someone else to complete the review of all or most lessons.**

- You should review the examples and concepts in your Study Notebook and Chapter 2 Foldable.
- Then complete the Chapter 2 Study Guide and Review on pages 110–114 of your textbook.
- If you are unsure of any concepts or skills, refer back to the specific lesson(s).
- You may also want to take the Chapter 2 Practice Test on page 115.

Student Signature

Parent/Guardian Signature

Teacher Signature

Solving Linear Equations

Use the instructions below to make a Foldable to help you organize your notes as you study the chapter. You will see Foldable reminders in the margin of this Interactive Study Notebook to help you in taking notes.

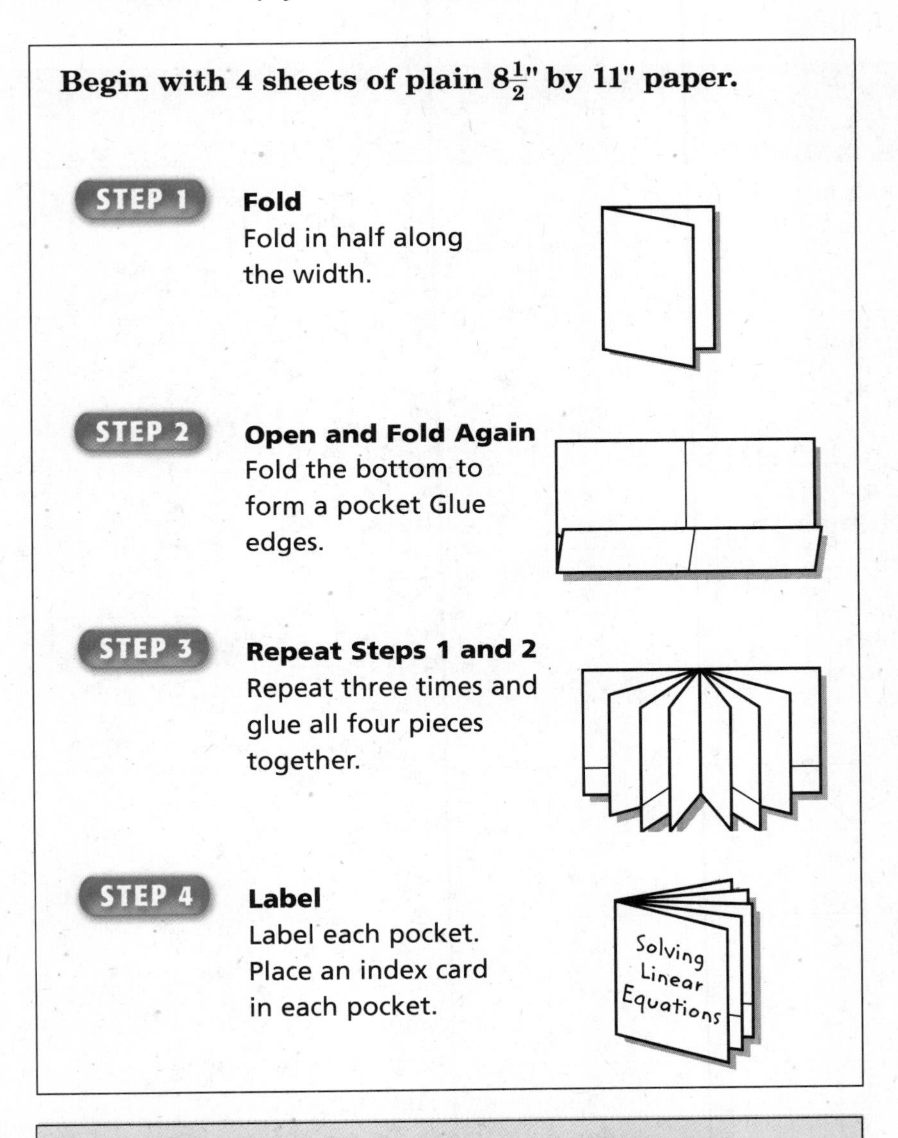

NOTE-TAKING TIP: When taking notes, write down a question mark next to anything you do not understand. Before your next quiz, ask your instructor to explain these sections.

BUILD YOUR VOCABULARY

This is an alphabetical list of new vocabulary terms you will learn in Chapter 3. As you complete the study notes for the chapter, you will see Build Your Vocabulary reminders to complete each term's definition or description on these pages. Remember to add the textbook page number in the second column for reference when you study.

Vocabulary Term	Found on Page	Definition	Description or Example
consecutive integers [kuhn-SEH-kyuh-tihv]			
defining a variable			
dimensional analysis [duh-MEHNCH-nuhl]			
equivalent equation [ih-KWIHV-luhnt]			
extremes			
formula			
identity			
means			

Vocabulary Term	Found on Page	Definition	Description or Example
multi-step equations			
number theory			
percent of change			
percent of decrease			
percent of increase			
proportion [pruh-POHR-shun]			
ratio			
rate			
scale			
solve an equation			
weighted average			
work backward			

3–1 Writing Equations

WHAT YOU'LL LEARN

- Translate verbal sentences into equations.
- Translate equations into verbal sentences.

BUILD YOUR VOCABULARY (page 54)

Choosing a variable to represent an unspecific ______ in a problem is called **defining a variable.**

EXAMPLE Translate Sentences into Equations

1 Translate the sentence into an equation.

A number b divided by three is equal to six less than c.

b divided by three	is equal to	six less than c.
$\frac{b}{3}$	$=$	______

The equation is ______.

Your Turn **Translate the sentence into an equation.**

A number c multiplied by six is equal to two more than d.

______

EXAMPLE Use the Four-Step Plan

2 JELLYBEANS **A popular jellybean manufacturer produces 1,250,000 jellybeans per hour. How many hours does it take them to produce 10,000,000 jellybeans?**

Write an equation. Let h represent the number of hours needed to produce the jellybeans.

1,250,000	times	hours	equals	10,000,000.
1,250,000	______	h	______	10,000,000

$1{,}25\cancel{0{,}000}h = 10{,}00\cancel{0{,}000}$

Find h mentally by asking, "What number times 125 equals 1,000?"

$h =$ ______

It will take ______ hours to produce 10,000,000 jellybeans.

KEY CONCEPT

Four-Step Problem-Solving Plan

Step 1 Explore the problem.

Step 2 Plan the solution.

Step 3 Solve the problem.

Step 4 Examine the solution.

FOLDABLES Write the four-step problem-solving plan on an index card.

Your Turn A person at the KeyTronic World Invitational Type-Off typed 148 words per minute. How many minutes would it take to type 3552 words?

BUILD YOUR VOCABULARY (page 54)

A **formula** is an ______ that states a ______ for the relationship between certain quantities.

EXAMPLE Write a Formula

3 **Translate the sentence into a formula.**

The perimeter of a square equals four times the length of the side.

Words	Perimeter equals four times the length of the side.
Variable	Let P = perimeter and s = length of a side.
Expression	Perimeter → P; equals → $=$; four times the length of a side. → ______

The formula is ______ = ______

Your Turn Translate the sentence. *The area of a circle equals the product of* π *and the square of the radius* r into a formula.

EXAMPLE Translate Equations into Sentences

4 **Translate each equation into a verbal sentence.**

a. $\mathbf{12 - 2x = -5}$

12 $-$ $2x$ $=$ -5

_______ minus _______ equals _______.

b. $\mathbf{a^2 + 3b = \frac{c}{6}}$

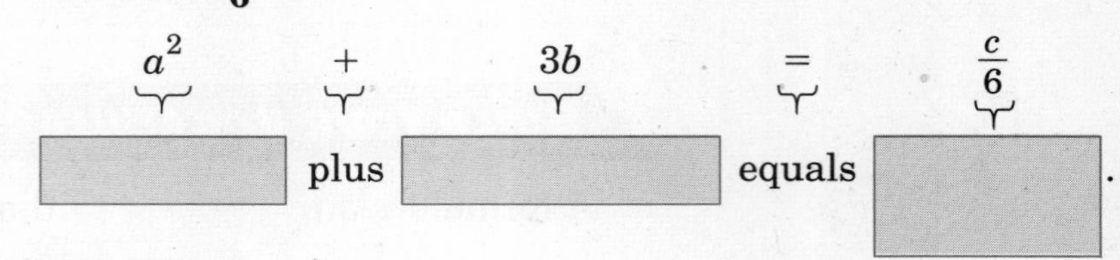

Your Turn **Translate each equation into a verbal sentence.**

a. $\frac{12}{b} - 4 = -1$

b. $5a = b^2 + 1$

HOMEWORK ASSIGNMENT

Pages:

Exercises:

3–2 Solving Equations by Using Addition and Subtraction

What You'll Learn

- Solve equations by using addition.
- Solve equations by using subtraction.

BUILD YOUR VOCABULARY (pages 54–55)

Equivalent equations have the ______ solution.

To **solve an equation** means to find all values of the ______ that make the equation a ______ statement.

Key Concepts

Addition Property of Equality If an equation is true and the same number is added to each side, the resulting equation is true.

Subtraction Property of Equality If an equation is true and the same number is subtracted from each side, the resulting equation is true.

EXAMPLE Solve by Adding a Positive Number

1 **Solve $h - 12 = -27$.**

$h - 12 = -27$ Original equation

$h - 12 + \_\_\_ = -27 + \_\_\_$ Add ______ to each side.

$h = \_\_\_$ $-12 + 12 = 0$ and $-27 + 12 = \_\_\_$.

EXAMPLE Solve by Adding a Negative Number

2 **Solve $k + 63 = 92$.**

$k + 63 = 92$ Original equation

$k + 63 + \_\_\_ = 92 + \_\_\_$ Add ______ to each side.

$k = \_\_\_$ $63 + (-63) = 0$ and $92 + (-63) = \_\_\_$.

EXAMPLE Solve by Subtracting

3 **Solve $c + 102 = 36$.**

$c + 102 = 36$ Original equation

$c + 102 - \_\_\_ = 36 - \_\_\_$ Subtract ______ from each side.

$c = \_\_\_$ $102 - 102 = 0$ and $36 - 102 = \_\_\_$

EXAMPLE Solve by Adding or Subtracting

4 **Solve $y + \frac{4}{5} = \frac{2}{3}$ in two ways.**

Method 1 Use the Subtraction Property of Equality.

$y + \frac{4}{5} = \frac{2}{3}$	Original equation
$y + \frac{4}{5} - \square = \frac{2}{3} - \square$	Subtract $\square$ from each side.
$y = -\frac{2}{15}$	Simplify.

Method 2 Use the Addition Property of Equality.

$y + \frac{4}{5} = \frac{2}{3}$	Original equation
$y + \frac{4}{5} + \square = \frac{2}{3} + \square$	Add $\square$ to each side.
$y = -\frac{2}{15}$	Simplify.

Your Turn **Solve each equation.**

a. $a - 24 = 16$

b. $t + 22 = -39$

c. $129 + k = -42$

d. $\frac{2}{3} + y = \frac{5}{6}$

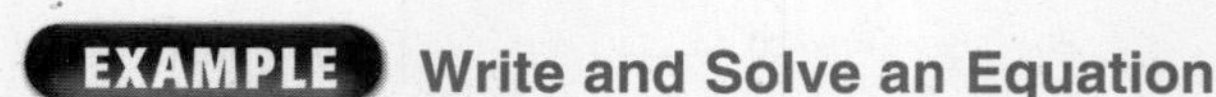

EXAMPLE Write and Solve an Equation

5 **Write and solve an equation for the problem.**

Fourteen more than a number is equal to twenty-seven. Find this number.

Fourteen	more than	a number	is equal to	twenty-seven.
14	+	n	=	27

$14 + n = 27$	Original equation
$14 + n - \square = 27 - \square$	Subtract $\square$ from each side.
$n = \square$	14 – 14 = 0 and 27 – 14 = 13

Your Turn Twelve less than a number is equal to negative twenty-five. Find the number.

HOMEWORK ASSIGNMENT

Pages:

Exercises:

3–3 Solving Equations by Using Multiplication and Division

What You'll Learn

- Solve equations by using multiplication.
- Solve equations by using division.

Key Concept

Multiplication Property of Equality If an equation is true, and each side is multiplied by the same number, the resulting equation is true.

Foldables On an index card write and solve an equation that uses the Multiplication Property of Equality.

EXAMPLE Solve Using Multiplication by a Positive Number

1 Solve $\frac{s}{12} = \frac{3}{4}$.

$\frac{s}{12} = \frac{3}{4}$ Original equation

$\square\left(\frac{s}{12}\right) = \square\left(\frac{3}{4}\right)$ Multiply each side by  .

$s = \square$ Simplify.

The solution is ____.

EXAMPLE Solve Using Multiplication by a Fraction

2 Solve $-3\frac{3}{8}k = 1\frac{4}{5}$.

$\left(-3\frac{3}{8}\right)k = 1\frac{4}{5}$ Original equation

$\left(-\frac{27}{8}\right)k = \square$ Rewrite $1\frac{4}{5}$ as an improper fraction.

$\square\left(-\frac{27}{8}\right)k = \square\left(\frac{9}{5}\right)$ Multiply by the reciprocal of $-\frac{27}{8}$.

$k = -\frac{72}{135}$ or $\square$ Simplify.

The solution is ____.

EXAMPLE Solve Using Multiplication by a Negative Number

3 Solve $-75 = -15b$.

$-75 = -15b$ Original equation

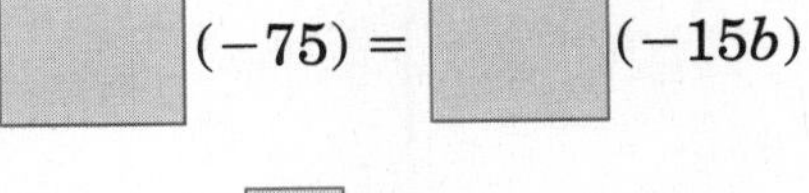

$\square(-75) = \square(-15b)$ Multiply each side by the reciprocal of -15.

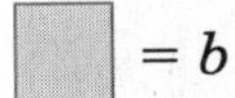

$\square = b$ Simplify.

The solution is ____.

Your Turn **Solve each equation.**

a. $\frac{a}{18} = \frac{2}{3}$

b. $\left(4\frac{1}{3}\right)m = 5\frac{3}{7}$

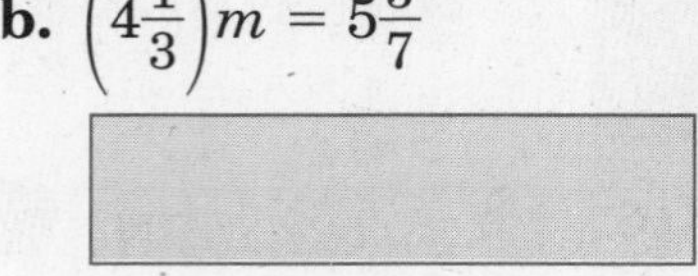

c. $32 = -14c$

EXAMPLE Write and Solve an Equation Using Multiplication

4 SPACE TRAVEL **Using information from Example 4 in the Student Edition, what would be the weight of Neil Armstrong's suit and life-support backpack on Mars if three times the Mars weight equals the Earth weight?**

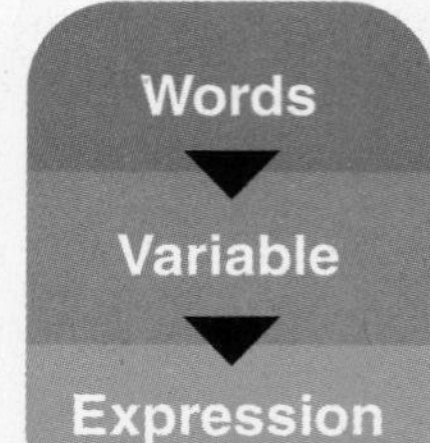

Three times the weight on Mars equals the weight on Earth.

Let $w =$ ______.

Three	times	the weight on Mars	equals	the weight on Earth.
3	______	w	=	______

$3w = 198$ — Original equation

______ $(3w) =$ ______ (198) — Multiply each side by ______.

$w =$ ______ — $\frac{1}{3}(3) = 1$ and $\frac{1}{3}(198) = 66$

The weight of Neil Armstrong's suit and life-support backpacks on Mars would be ______ pounds.

Your Turn Refer to Example 4. If Neil Armstrong weighed 216 pounds on Earth, how much would he weigh on Mars?

KEY CONCEPT

Division Property of Equality If an equation is true, and each side is divided by the same nonzero number, the resulting equation is true.

EXAMPLE Solve Using Division by a Positive Number

5 **Solve $11w = 143$.**

$11w = 143$ Original equation

$\frac{11w}{\square} = \frac{143}{\square}$ Divide each side by $\square$.

$w = \square$ $\frac{11w}{11} = w$ and $\frac{143}{11} = 13$

EXAMPLE Solve Using Division by a Negative Number

6 **Solve $-8x = 96$.**

$-8x = 96$ Original equation

$\frac{-8x}{\square} = \frac{96}{\square}$ Divide each side by $\square$.

$x = \square$ $\frac{-8x}{-8} = x$ and $\frac{96}{-8} = -12$

Your Turn **Solve each equation.**

a. $35t = 595$

b. $-12b = 276$

HOMEWORK ASSIGNMENT

Pages: ____

Exercises: ____

3–4 Solving Multi-Step Equations

BUILD YOUR VOCABULARY (page 55)

Work backward is one of the many problem-solving strategies that you can use to solve multi-step equations.

To solve equations with more than one operation, often called **multi-step equations,** ______ operations by working backward.

WHAT YOU'LL LEARN

- Solve problems by working backward.
- Solve equations involving more than one operation.

FOLDABLES™

ORGANIZE IT

Explain how to solve multi-step equations on an index card. Include an example.

EXAMPLE Work Backward to Solve a Problem

1 **Danny took some rope with him on his camping trip. He used 32 feet of rope to tie his canoe to a log on the shore. The next night, he used half of the remaining rope to secure his tent during a thunderstorm. On the last day, he used 7 feet as a fish stringer to keep the fish that he caught. After the camping trip, he had 9 feet left. How much rope did he have at the beginning of the camping trip?**

Start at the end of the problem and undo each step.

Statement	Undo the Statement
He had ______ feet left.	______
He used ______ feet as a fish stringer.	$9 + 7 = 16$
He used half of the remaining rope to secure his tent.	$16 \times 2 =$ ______
He used 32 feet to tie his canoe.	$32 + 32 =$ ______

He had ______ feet of rope.

Your Turn Olivia went to the mall to spend some of her monthly allowance. She put $10 away so it could be deposited in the savings account at a later date. The first thing she bought was a CD for $15.99. The next stop was to buy hand lotion and a candle, which set her back $9.59. For lunch, she spent half of the remaining cash. She went to the arcade room and spent $5.00 and took home $1.21. How much was Olivia's monthly allowance?

______

EXAMPLE Solve Using Addition and Division

2 Solve $5q - 13 = 37$.

$5q - 13 = 37$	Original equation
$5q - 13 + \square = 37 + \square$	Add ____ to each side.
$\square = \square$	Simplify.
$\frac{\square}{\square} = \frac{\square}{\square}$	Divide each side by ____.
$q = \square$	Simplify.

EXAMPLE Solve Using Subtraction and Multiplication

3 Solve $\frac{s}{12} + 9 = -11$.

$\frac{s}{12} + 9 = -11$	Original equation
$\frac{s}{12} + 9 - \square = -11 - \square$	Subtract ____ from each side.
$\square = \square$	____.
$\square\left(\frac{s}{12}\right) = \square(-20)$	Multiply each side by ____.
$s = \square$	Simplify.

Your Turn **Solve each equation.**

a. $4 + \frac{a}{11} = 37$

b. $4a - 42 = 14$

BUILD YOUR VOCABULARY (pages 54–55)

Consecutive integers are integers in ____ order, such as 7, , and 9.

The study of ____ and the relationships between them is called **number theory**.

WRITE IT

What is meant by undoing an equation?

____

____

____

____

EXAMPLE Solve Using Multiplication and Division

Solve $\frac{r-8}{-3} = -2$.

$\frac{r-8}{-3} = -2$ — Original equation

$\square\left(\frac{r-8}{-3}\right) = \square(-2)$ — Multiply each side by $\square$.

$r - 8 = 6$ — .

$r - 8 + \square = 6 + \square$ — Add $\square$ to each side.

$r = \square$ — Simplify.

EXAMPLE Write and Solve a Multi-Step Equation

5 **Write an equation for the problem below. Then solve the equation.**

Eight more than five times a number is negative 62.

Eight	more than	five	times	a number	is	negative 62.
8	+	5	×	n	=	−62

$5n + 8 = -62$ — Original equation

$5n + 8 \square = -62 \square$ — Subtract $\square$ from each side.

$\square = -70$ — Simplify.

$\square(5n) = \square(-70)$ — Multiply each side by $\square$.

$n = \square$ — Simplify.

Your Turn

a. Solve $\frac{a-9}{2} = 6$.

b. Three-fourths of seven subtracted from a number is negative fifteen. What is the number?

HOMEWORK ASSIGNMENT

Page(s):

Exercises:

3–5 Solving Equations with the Variable on Each Side

WHAT YOU'LL LEARN

- Solve equations with the variable on each side.
- Solve equations involving grouping symbols.

REVIEW IT

Give an example of the Distributive Property. (*Lesson 1–5*)

____

____

____

____

____

____

EXAMPLE Solve an Equation with Variables on Each Side

1 Solve $8 + 5s = 7s - 2$.

$8 + 5s = 7s - 2$	Original equation
$8 + 5s - \square = 7s - 2 - \square$	Subtract $\square$ from each side.
$\square = -2$	Simplify.
$8 - 2s - \square = -2 - \square$	Subtract $\square$ from each side.
$-2s = \square$	Simplify.
$\frac{-2s}{\square} = \frac{-10}{\square}$	Divide each side by $\square$.
$s = 5$	Simplify.

EXAMPLE Solve an Equation with Grouping Symbols

2 Solve $\frac{1}{3}(18 + 12q) = 6(2q - 7)$.

$\frac{1}{3}(18 + 12q) = 6(2q - 7)$	Original equation
$6 + 4q = \square$	Distributive Property
$6 + 4q - \square = 12q - 42 - \square$	Subtract $\square$ from each side.
$\square = -42$	Simplify.
$6 - 8q - \square = -42 - \square$	Subtract $\square$ from each side.
$\square = -48$	Simplify.
$\frac{-8q}{\square} = \frac{-48}{\square}$	Divide each side by $\square$.
$q = \square$	Simplify.

EXAMPLE No Solutions

3 Solve $8(5c - 2) = 10(32 + 4c)$.

$8(5c - 2) = 10(32 + 4c)$	Original equation
______ $= 320 + 40c$	Distributive Property
$40c - 16 -$ ______ $= 320 + 40c -$ ______	Subtract.
$0c - 16 = 320$	This statement is false.

There must be at least one c to represent the variable. This equation has no solution.

FOLDABLES™

ORGANIZE IT

On an index card, write an equation that has no solution.

BUILD YOUR VOCABULARY (page 54)

An equation that is ______ for every value of the variable is called an **identity**.

EXAMPLE An Identity

4 Solve $4(t + 20) = \frac{1}{5}(20t + 400)$.

$4(t + 20) = \frac{1}{5}(20t + 400)$	Original equation
______ = ______	Distributive Property

Since the expression on each side of the equation is the ______, this equation is an identity. The statement $4t + 80 = 4t + 80$ is ______ for all values of t.

Your Turn **Solve each equation.**

a. $9f - 6 = 3f + 7$

b. $6(3r - 4) = \frac{3}{8}(46r + 8)$

______

c. $2(4a + 8) = 3\left(\frac{8a}{3} - 10\right)$

______

d. $\frac{1}{7}(21c - 56) = 3\left(c - \frac{8}{3}\right)$

______

HOMEWORK ASSIGNMENT

Page(s):

Exercises:

3–6 Ratios and Proportions

WHAT YOU'LL LEARN

- Determine whether two ratios form a proportion.
- Solve proportions.

BUILD YOUR VOCABULARY (pages 54–55)

A **ratio** is a comparison of two numbers by ______.

An equation stating that two ratios are ______ is called a **proportion**.

In the proportion $\frac{0.4}{0.8} = \frac{0.7}{1.4}$, 0.4 and 1.4 are called the **extremes** and 0.8 and 0.7 are called the **means**.

REMEMBER IT

A whole number can be written as a ratio with a denominator of 1.

EXAMPLE Determine Whether Ratios Form a Proportion

1 Determine whether the ratios $\frac{7}{8}$ and $\frac{49}{56}$ form a proportion.

$\frac{7}{8} = \frac{7}{8}$ ______ ______

$\frac{49}{56} = \frac{7}{8}$ ______ ______

The ratios are ______. Therefore, they form a proportion.

Your Turn Do the ratios $\frac{5}{6}$ and $\frac{40}{49}$ form a proportion?

______

FOLDABLES

ORGANIZE IT

On an index card, explain the difference between a ratio and proportion.

EXAMPLE Use Cross Products

2 Use cross products to determine whether the pair of ratios below forms a proportion.

a. $\frac{0.25}{0.6}, \frac{1.25}{2}$

$\frac{0.25}{0.6} \times \frac{1.25}{2}$	Write the equation.
______ $(2) = 0.6$ ______	Find the cross products.
$0.5 \neq 0.75$	Simplify.

The cross products are not equal, so $\frac{0.25}{0.6} \neq \frac{1.25}{2}$.

The ratios ______ form a proportion.

b. $\frac{4}{5}, \frac{16}{20}$

$\frac{4}{5} \times \frac{16}{20}$ Write the equation.

$\square = 5(16)$ Find the cross products.

$\square = \square$ Simplify.

The cross products are ____, so $\frac{4}{5} = \frac{16}{20}$.

Since the ratios are equal, they form a ____.

Your Turn **Use cross products to determine whether each pair of ratios below forms a proportion.**

a. $\frac{0.5}{1.3}, \frac{0.45}{1.17}$

b. $\frac{5}{6}, \frac{12}{15}$

EXAMPLE Solve a Proportion

3 **Solve the proportion $\frac{n}{12} = \frac{3}{8}$.**

$\frac{n}{12} \times \frac{3}{8}$ Original equation

$8(n) = 12(3)$ Find the cross products.

$\square = \square$ Simplify.

$\frac{\square}{\square} = \frac{\square}{\square}$ Divide each side by $\square$.

$n =$ 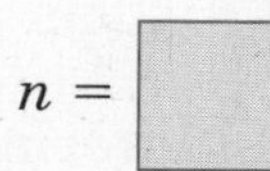 .

Key Concept

Means-Extremes Property of Proportion
In a proportion, the product of the extremes is equal to the product of the means.

Your Turn Solve the proportion $\frac{r}{9} = \frac{7}{10}$.

BUILD YOUR VOCABULARY (page 55)

The ______ of two measurements having ______ units of measure is called a **rate.**

EXAMPLE Use Rates

4 BICYCLING **The gear on a bicycle is 8:5. This means that for every eight turns of the pedals, the wheel turns five times. Suppose the bicycle wheel turns about 2435 times during a trip. How many times would you have to turn the pedals during the trip?**

$\frac{8}{5} = \frac{p}{2435}$ Original proportion

$\_\_\_\_ = 5(p)$ Find the ______.

$\_\_\_\_ = 5p$ ______.

$\frac{\_\_\_\_}{\_\_\_\_} = \frac{\_\_\_\_}{\_\_\_\_}$ Divide each side by ___.

$\_\_\_\_ = p$ Simplify.

Your Turn Before 1980, Disney created animated movies using cels. These hand drawn cels (pictures) of the characters and scenery represented the action taking place, one step at a time. For the movie *Snow White*, it took 24 cels per second to have the characters move smoothly. The movie is around 42 minutes long. About how many cels were drawn to produce *Snow White*?

HOMEWORK ASSIGNMENT

Page(s):

Exercises:

3–7 Percent of Change

WHAT YOU'LL LEARN

- Find percents of increase and decrease.
- Solve problems involving percents of change.

BUILD YOUR VOCABULARY (page 55)

When an ______ or ______ is expressed as a percent, the percent is called the **percent of change**.

If the new number is ______ than the original number, the percent of change is a **percent of increase**.

If the new number is ______ than the original, the percent of change is a **percent of decrease**.

EXAMPLE Find a Percent of Change

1 State whether the percent of change is a percent of increase or a percent of decrease. Then find the percent of change.

original: 32
new: 40

Find the *amount* of change. Since the new amount is greater than the original, the percent of change is a percent of ______.

______ $= 8$

Find the percent using the original number, 32, as the base.

change → original amount → $\frac{8}{32} = \frac{r}{100}$

______ $= 32(r)$ Cross products

$800 =$ ______ Simplify.

$\frac{800}{\square} = \frac{32r}{\square}$ Divide each side by ______.

______ $= r$ Simplify.

The percent of increase is ______.

Your Turn **State whether each percent of change is a percent of increase or a percent of decrease. Then find the percent of change.**

a. original: 20
new: 18

b. original: 12
new: 48

EXAMPLE Find the Missing Value

2 SALES **The price a used-book store pays to buy a book is $5. The store sells the book for 28% above the price that it pays for the book. What is the selling price of the $5 book?**

Let s = the selling price of the book. Since 28% is the percent of increase, the amount the used-book store pays to buy a book is less than the selling price. Therefore, $s - 5$ represents the amount of change.

change → $s - 5$
book store cost → 5

$$\frac{s-5}{5} = \frac{28}{100}$$

$(s - 5)(100) = 5(28)$ Cross products

$100s - 500 = 140$ Distributive Property

$100s - 500 + \square = 140 + \square$ Add $\square$ to each side.

$100s = 640$ Simplify.

$s = \square$ Divide each side by 100.

The selling price of the $5 book is $\square$.

Your Turn At one store the price of a pair of jeans is $26.00. At another store the same pair of jeans has a price that is 22% higher. What is the price of jeans at the second store?

HOMEWORK ASSIGNMENT

Page(s):

Exercises:

3–8 Solving Equations and Formulas

WHAT YOU'LL LEARN

- Solve equations for given variables.
- Use formulas to solve real-world problems.

EXAMPLES Solve an Equation for a Specific Variable

1 Solve $5b + 12c = 9$ for b.

$5b + 12c = 9$ Original equation

$5b + 12c - \square = 9 - \square$ Subtract.

$5b = 9 - 12c$ Simplify.

$\frac{5b}{\square} = \frac{9 - 12c}{\square}$ Divide each side by $\square$.

$b = \frac{9 - 12c}{5}$ Simplify.

or $\frac{-12c + 9}{5}$

The value of b is $\square$.

2 Solve $7x - 2z = 4 - xy$ for x.

$7x - 2z = 4 - xy$ Original equation

$7x - 2z + \square = 4 - xy + \square$ Add $\square$ to each side.

$7x - 2z + xy = 4$ Simplify.

$7x - 2z + xy + \square = 4 + \square$ Add $\square$ to each side.

$\square = 4 + 2z$ Simplify.

$\square = 4 + 2z$ Use the Distributive Property.

$\square = \frac{4 + 2z}{7 + y}$ Divide each side by $\square$.

The value of x is $\square$. Since division by 0 is undefined, $\square \neq 0$, or $y \neq \square$.

Your Turn

a. Solve $2x - 17y = 13$ for y.

b. Solve $12a + 3c = 2ab + 6$ for a.

EXAMPLE Use a Formula to Solve Problems

3 **a.** **FUEL ECONOMY A car's fuel economy E (miles per gallon) is given by the formula $E = \frac{m}{g}$, where m is the number of miles driven and g is the number of gallons of fuel used. Solve the formula for m.**

$E = \frac{m}{g}$ Formula for fuel economy

$E\square = \frac{m}{g}\square$ Multiply each side by $\square$.

$\square = \square$ Simplify.

b. **FUEL ECONOMY If Claudia's car has an average fuel consumption of 30 miles per gallon and she used 9.5 gallons, how far did she drive?**

$Eg = m$ Formula for how many miles driven

$30\square = m$ $E = 30$ mpg and $g = 9.5$ gallons

$\square = m$ Multiply.

She drove $\square$ miles.

Your Turn

a. Refer to Example 3. Solve the formula for g.

b. If Claudia drove 1477 miles and her pickup has an average fuel consumption of 19 miles per gallon, how many gallons of fuel did she use?

BUILD YOUR VOCABULARY (page 54)

Dimensional analysis is the process of carrying ______ throughout a ______.

EXAMPLE Use Dimensional Analysis

4 **a. GEOMETRY The formula for the volume of a cylinder is $V = \pi r^2 h$, where r is the radius of the cylinder and h is the height. Solve the formula for h.**

$V = \pi r^2 h$ Original formula

$\frac{V}{\square} = \frac{\pi r^2 h}{\square}$ Divide each side by ______

$\square = h$

b. What is the height of a cylindrical swimming pool that has a radius of 12 feet and a volume of 1810 cubic feet?

$\square = h$ Formula for h

$\frac{1810}{\pi 12^2} = h$ $V =$ ______ and $r =$ ______

$\square = h$ Use a calculator.

The height of the cylindrical swimming pool is about ______.

Your Turn

a. The formula for the volume of a cylinder is $V = \pi r^2 h$, where r is the radius of the cylinder and h is the height. Solve the formula for r.

b. What is the radius of a cylindrical swimming pool if the volume is 2010 cubic feet and a height of 6 feet?

HOMEWORK ASSIGNMENT

Page(s):

Exercises:

3–9 Weighted Averages

What You'll Learn

- Solve mixture problems.
- Solve uniform motion problems.

BUILD YOUR VOCABULARY (page 55)

The **weighted average** M of a set of data is the sum of the product of the number of units and the value per unit divided by the sum of the number of units.

EXAMPLE Solve a Mixture Problem with Prices

1 PETS **Jeri likes to feed her cat gourmet cat food that costs $1.75 per pound. However, food at that price is too expensive so she combines it with cheaper cat food that costs $0.50 per pound. How many pounds of cheaper food should Jeri buy to go with 5 pounds of gourmet food, if she wants the price to be $1.00 per pound?**

Let w = the number of pounds of cheaper cat food.

	Units (lb)	Price per Unit	Price
Gourmet cat food			
Cheaper cat food	w	$0.50	$0.5w$
Mixed cat food		$1.00	

Price of gourmet cat food	plus	price of cheaper cat food	equals	price of mixed cat food.
8.75	$+$	$0.5w$	$=$	$1.00(5 + w)$

$8.75 + 0.5w = 1.00(5 + w)$ Original equation

$8.75 + 0.5w =$ ______ Distributive Property

$8.75 + 0.5w -$ ______ $= 5.0 + 1w -$ ______ Subtract.

$8.75 = 5.0 + 0.5w$ Simplify.

$8.75 - 5.0 = 5.0 + 0.5w - 5.0$ Subtract.

$3.75 = 0.5w$ Simplify.

$7.5 = w$ Divide.

Your Turn A recipe calls for mixed nuts with 50% peanuts. $\frac{1}{2}$ pound of 15% peanuts has already been used. How many pounds of 75% peanuts needs to be added to obtain the required 50% mix?

FOLDABLES™

ORGANIZE IT

On an index card, take notes on mixture problems and uniform motion problems.

EXAMPLE Solve for an Average Speed

2 AIR TRAVEL **Mirasol took a non-stop flight from Newark to Austin to visit her grandmother. The 1500-mile trip took three hours and 45 minutes. Because of bad weather, the return trip took four hours and 45 minutes. What was her average speed for the round trip?**

To find the average speed for each leg of the trip, rewrite $d = rt$ as $r = \frac{d}{t}$.

Going

$r = \frac{d}{t} = \frac{1500 \text{ miles}}{\underline{\quad} \text{ hours}}$ or ______ miles per hour

Returning

$r = \frac{d}{t} = \frac{1500 \text{ miles}}{\underline{\quad} \text{ hours}}$ or ______ miles per hour

Round trip

$M = \frac{400(1) + 315.79(2)}{1 + 2}$ Definition of weighted average

= ______ or ______ Simplify.

The average speed was about ______ miles per hour.

HOMEWORK ASSIGNMENT

Page(s):

Exercises:

Your Turn In the morning, when traffic is light, it takes 30 minutes to get to work. The trip is 15 miles through towns. In the afternoon when traffic is a little heavier, it takes 45 minutes. What is the average speed for the round trip?

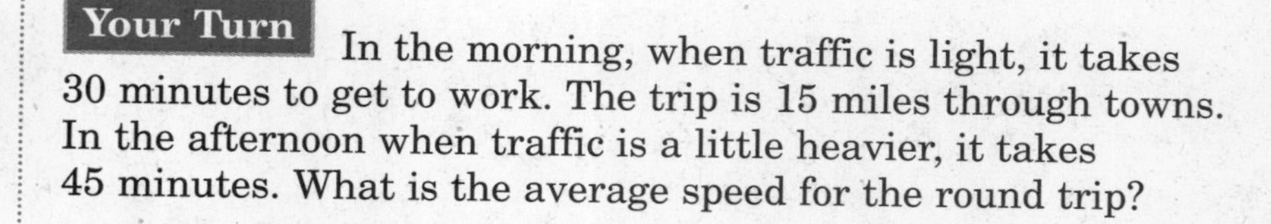

CHAPTER 3

BRINGING IT ALL TOGETHER

STUDY GUIDE

FOLDABLES™	VOCABULARY PUZZLEMAKER	BUILD YOUR VOCABULARY
Use your **Chapter 3 Foldable** to help you study for your chapter test.	To make a crossword puzzle, word search, or jumble puzzle of the vocabulary words in Chapter 3, go to www.glencoe.com/sec/math/t_resources/free/index.php	You can use your completed **Vocabulary Builder** (pages 54–55) to help you solve the puzzle.

3–1 Writing Equations

Translate each sentence into an equation.

1. Two times the sum of x and three minus four equals four times x

2. The difference of k and 3 is two times k divided by five.

3–2 Solving Equations by Using Addition and Subtraction

Complete each sentence.

3. To solve $y - 9 = -30$ using the Addition Property of Equality, you would add ______ to each side.

4. Write an equation that you could solve by subtracting 32 from each side.

3–3 Solving Equations by Using Multiplication or Division

Complete the sentence after each equation to tell how you would solve the equation.

5. $\frac{x}{7} = 16$ ______ each side by ______.

6. $5x = 125$ ______ each side by ______, or multiply each side by ______.

3–4 Solving Multi-Step Equations

Suppose you want to solve $\frac{x + 3}{5} = 6$.

7. What is the grouping symbol in the equation $\frac{x + 3}{5} = 6$?

8. What is the first step in solving the equation?

9. What is the next step in solving the equation?

3–5 Solving Equations with the Variable on Each Side

10. When solving $2(3x - 4) = 3(x + 5)$, why is it helpful first to use the Distributive Property to remove the grouping symbols?

The solutions of three equations are shown in Exercises 11–13. Write a sentence to describe each solution.

11. $x = -4$

12. $6m = 6m$

13. $12 = 37$

3–6 Ratios and Proportions

14. A jet flying at a steady speed traveled 825 miles in 2 hours. If you solved the proportion $\frac{825}{2} = \frac{x}{1.5}$, what would the answer tell you about the jet?

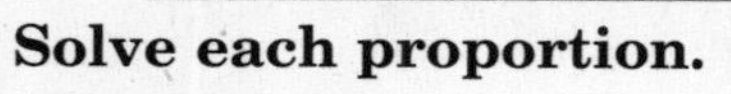

Solve each proportion.

15. $\frac{10}{a} = \frac{60}{108}$

16. $\frac{b}{32} = \frac{12}{8}$

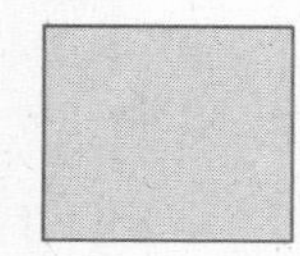

17. $\frac{3}{7} = \frac{x - 2}{6}$

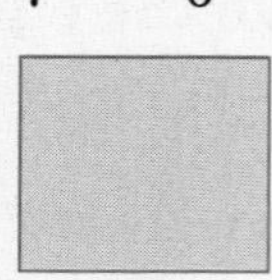

3–7

Percent of Change

Match the problem on the left with its answer on the right.

18. Original Amount = 10
New Amount = 13

19. Original Amount = 10
New Amount = 7

20. Original Amount = 50
New Amount = 42

21. Original Amount = 50
New Amount = 58

a. 30% increase

b. 50% decrease

c. 16% increase

d. 30% decrease

e. 16% decrease

3–8

Solving Equations and Formulas

Solve each equation or formula for the variable specified.

22. $7f + g = 5$ for f

23. 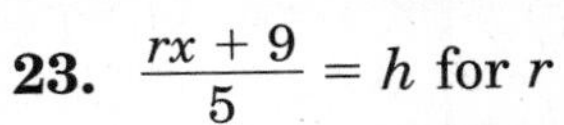 $\frac{rx + 9}{5} = h$ for r

24. $3y + w = 5 + 5y$ for y

3–9

Weighted Averages

25. Suppose Clint drives at 50 miles per hour for 2 hours. Then he drives at 60 miles per hour for 3 hours. Write his speed for each hour of the trip.

Speed					
Hour	1	2	3	4	5

26. What is his average speed?

27. How many grams of sugar must be added to 60 grams of a solution that is 32% sugar to obtain a solution that is 50% sugar?

ARE YOU READY FOR THE CHAPTER TEST?

Math Online

Visit **www.algebra1.com** to access your textbook, more examples, self-check quizzes, and practice tests to help you study the concepts in Chapter 3.

Check the one that applies. Suggestions to help you study are given with each item.

☐ **I completed the review of all or most lessons without using my notes or asking for help.**

- You are probably ready for the Chapter Test.
- You may want to take the Chapter 3 Practice Test on page 185 of your textbook as a final check.

☐ **I used my Foldable or Study Notebook to complete the review of all or most lessons.**

- You should complete the Chapter 3 Study Guide and Review on pages 179–184 of your textbook.
- If you are unsure of any concepts or skills, refer back to the specific lesson(s).
- You may also want to take the Chapter 3 Practice Test on page 185.

☐ **I asked for help from someone else to complete the review of all or most lessons.**

- You should review the examples and concepts in your Study Notebook and Chapter 3 Foldable.
- Then complete the Chapter 3 Study Guide and Review on pages 179–184 of your textbook.
- If you are unsure of any concepts or skills, refer back to the specific lesson(s).
- You may also want to take the Chapter 3 Practice Test on page 185.

Student Signature

Parent/Guardian Signature

Teacher Signature

Graphing Relations and Functions

Use the instructions below to make a Foldable to help you organize your notes as you study the chapter. You will see Foldable reminders in the margin of this Interactive Study Notebook to help you in taking notes.

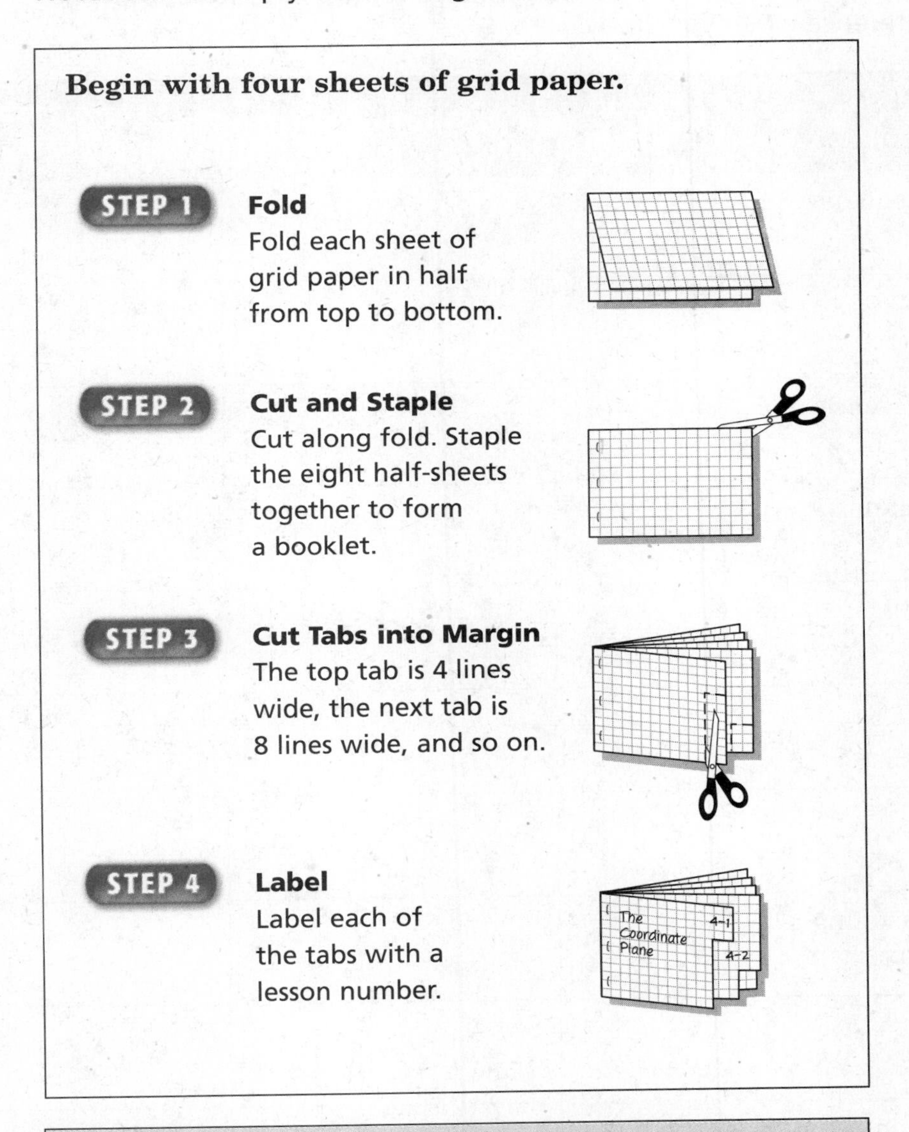

NOTE-TAKING TIP: When you take notes, be sure to listen actively. Always think before you write, but don't get behind in your note-taking. Remember to enter your notes legibly.

BUILD YOUR VOCABULARY

This is an alphabetical list of new vocabulary terms you will learn in Chapter 4. As you complete the study notes for the chapter, you will see Build Your Vocabulary reminders to complete each term's definition or description on these pages. Remember to add the textbook page number in the second column for reference when you study.

Vocabulary Term	Found on Page	Definition	Description or Example
arithmetic sequence			
axes			
common difference			
coordinate plane [koh-AWRD-nuht]			
dilation [dy-LA-shuhn]			
function			
image			
inductive reasoning [ihn-DUHK-tihv]			
inverse			
linear equation			
mapping			

Vocabulary Term	Found on Page	Definition	Description or Example
origin			
quadrant [KWAH-druhnt]			
reflection			
rotation			
sequence			
standard form			
terms			
transformation			
translation			
vertical line test			

4–1 The Coordinate Plane

BUILD YOUR VOCABULARY (pages 84–85)

In mathematics, points are located in reference to the ______ and ______ on a coordinate system or **coordinate plane**.

The x-axis and y-axis separate the coordinate plane into ______ regions, called **quadrants**.

WHAT YOU'LL LEARN

- Locate points on the coordinate plane.
- Graph points on a coordinate plane.

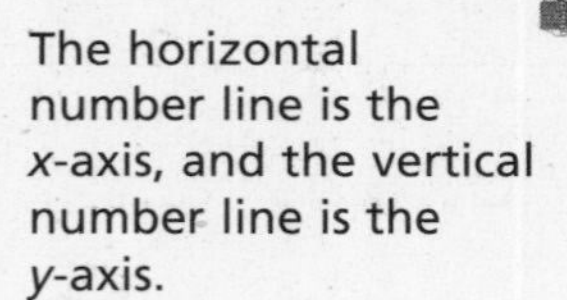

REMEMBER IT

The horizontal number line is the x-axis, and the vertical number line is the y-axis.

FOLDABLES

ORGANIZE IT

Use the tab for Lesson 4–1. Draw a coordinate plane and label each quadrant. Then show whether the x- and y-coordinates are positive or negative in each quadrant.

EXAMPLE Name an Ordered Pair

1 **Write the ordered pair for point *B*.**

- Follow along a vertical line through B to find the x-coordinate on the x-axis. The x-coordinate is 3.
- Follow along a horizontal line through the point to find the y-coordinate on the y-axis. The y-coordinate is -2.
- The ordered pair for point B is ______. This can also be written as ______.

Your Turn Write the ordered pair for point C.

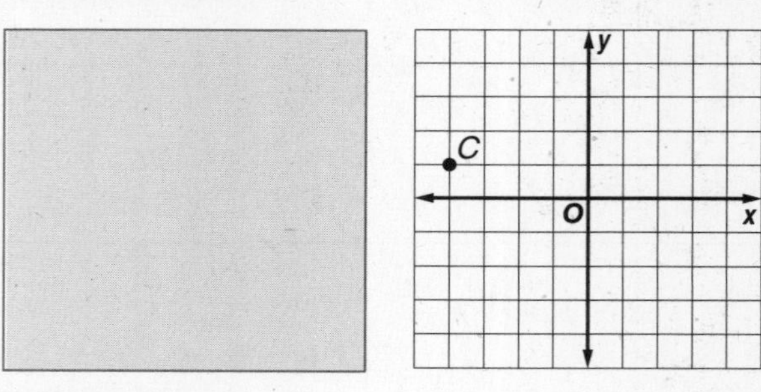

EXAMPLE Identify Quadrants

2 **Write ordered pairs for points *A*, *B*, *C*, and *D*. Name the quadrant in which each point is located.**

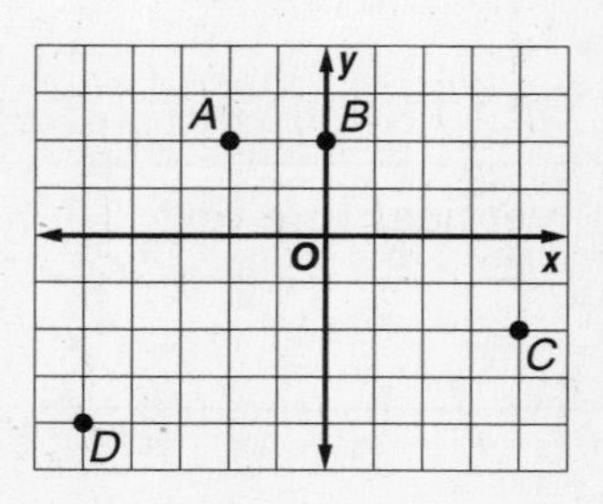

Use a table to help find the coordinates of each point.

Point	*x*-Coordinate	*y*-Coordinate	Ordered Pair	Quadrant
A		2	(−2, 2)	
B	0		(0, 2)	None
C	4	−2	(4, −2)	
D		−4		III

Your Turn Write ordered pairs for points *Q*, *R*, *S*, and *T*. Name the quadrant in which each point is located.

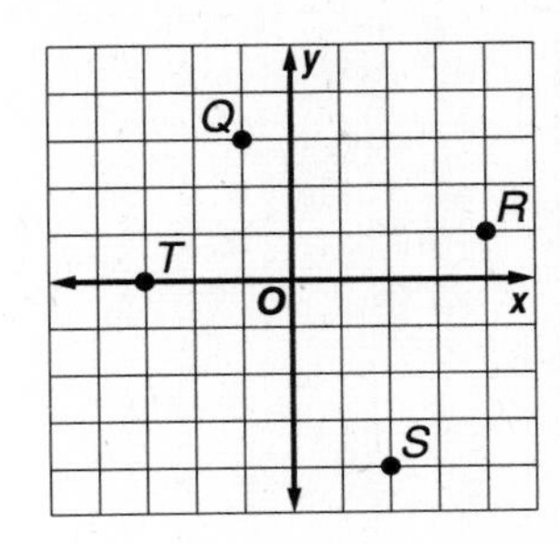

EXAMPLE Graph Points

3 **Plot *A*(3, 1) on the coordinate plane.**

- Start at the origin.
- Move ______ 3 units since the *x*-coordinate is 3.
- Move ______ 1 unit since the *y*-coordinate is 1.
- Draw a dot and label it *A*.

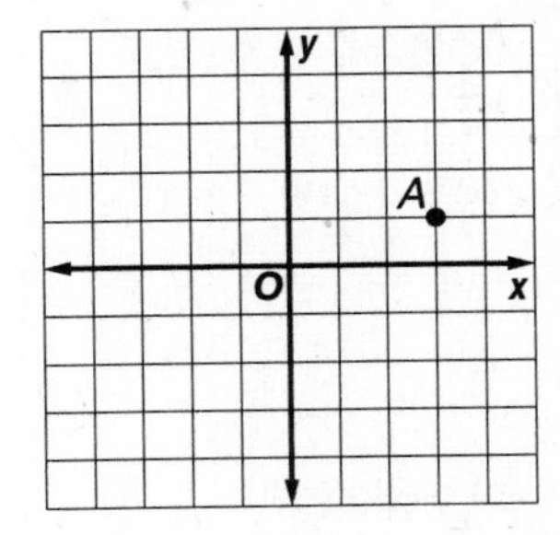

Your Turn Plot each point on a coordinate plane.

a. *H*(3, 5)

b. *J*(0, 4)

c. *K*(6, −2)

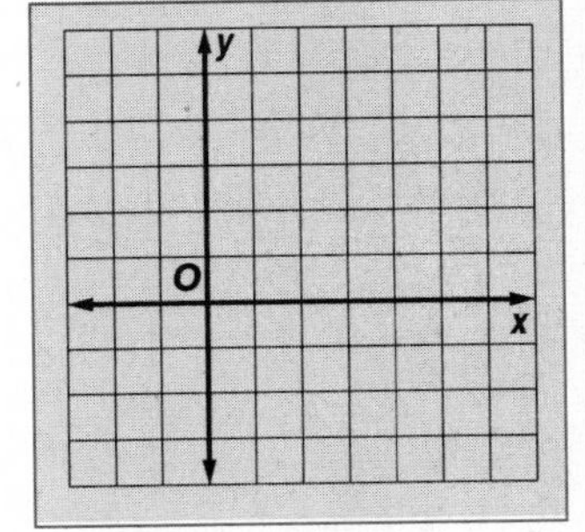

HOMEWORK ASSIGNMENT

Page(s):

Exercises:

4–2 Transformations on the Coordinate Plane

WHAT YOU'LL LEARN

- Transform figures by using reflections, translation, dilations, and rotations.
- Transform figures on a coordinate plane by using reflections, translations, dilations, and rotations.

BUILD YOUR VOCABULARY (pages 84–85)

Transformations are ______ of geometric figures.

The **image** is the position of the figure after the transformation.

When a figure is ______ over a line it is said to be a ______.

A **translation** occurs when a figure is ______ in any direction.

Dilation is when a figure is enlarged or ______.

When a figure is turned around a point it is said to be a **rotation**.

EXAMPLE Identify Transformations

1 **Identify each transformation as a *reflection, translation, dilation,* or *rotation.***

a. The figure has been increased in size. This is a ______.

b. The figure has been shifted horizontally to the right. This is a ______.

Your Turn **Identify each transformation as a *reflection, translation, dilation,* or *rotation.***

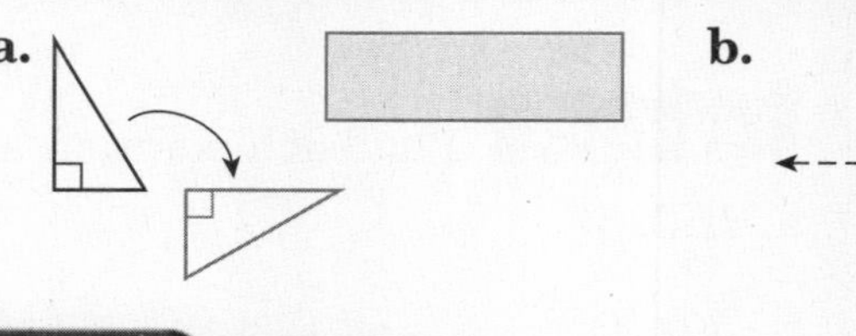

a. ______ **b.** ______

ORGANIZE IT

Under the tab for Lesson 4–2, draw and label each of the four transformations (reflection, translation, dilation, rotation) discussed in the lesson.

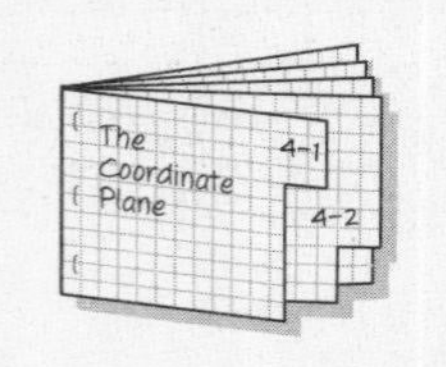

EXAMPLE Reflection

2 **A trapezoid has vertices $W(-1, 4)$, $X(4, 4)$, $Y(4, 1)$ and $Z(-3, 1)$.**

a. Trapezoid *WXYZ* is reflected over the *y*-axis. Find the coordinates of the vertices of the image.

To reflect the figure over the y-axis, multiply each x-coordinate by ______.

Key Concepts

Transformations on the Coordinate Plane

Reflection To reflect a point over the x-axis, multiply the y-coordinate by -1. To reflect a point over the y-axis, multiply the x-coordinate by -1.

Translation To translate a point by an ordered pair (a, b), add a to the x-coordinate and b to the y-coordinate.

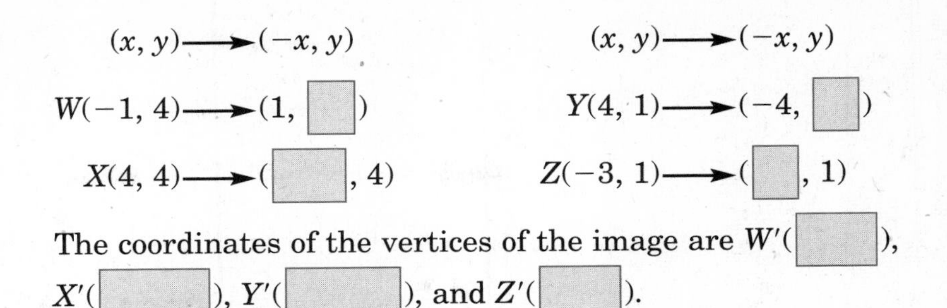

$(x, y) \longrightarrow (-x, y)$ $\qquad$ $(x, y) \longrightarrow (-x, y)$

$W(-1, 4) \longrightarrow (1, \square)$ $\qquad$ $Y(4, 1) \longrightarrow (-4, \square)$

$X(4, 4) \longrightarrow (\square, 4)$ $\qquad$ $Z(-3, 1) \longrightarrow (\square, 1)$

The coordinates of the vertices of the image are $W'(\square)$, $X'(\square)$, $Y'(\square)$, and $Z'(\square)$.

b. Graph trapezoid *WXYZ* and its image *W'X'Y'Z'*.

Graph each vertex of the trapezoid *WXYZ*. Connect the points.

Graph each vertex of the reflected image *W'X'Y'Z'*. Connect the points.

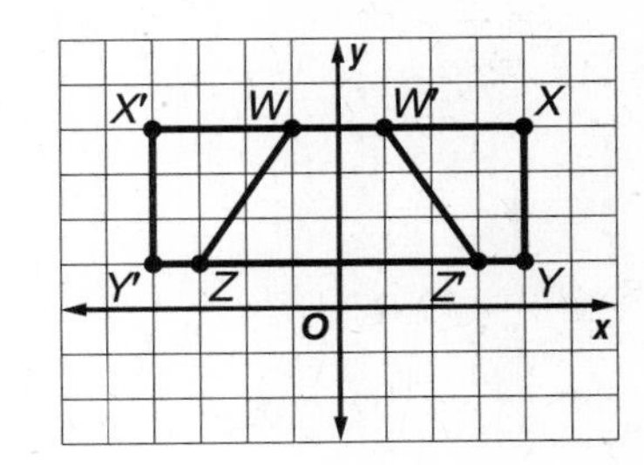

Your Turn A parallelogram has vertices $A(-4, 7)$, $B(2, 7)$, $C(0, 4)$, and $D(-2, 4)$. Parallelogram *ABCD* is reflected over the x-axis. Find the coordinates of the vertices of the image. Then graph parallelogram *ABCD* and its image *A'B'C'D'*.

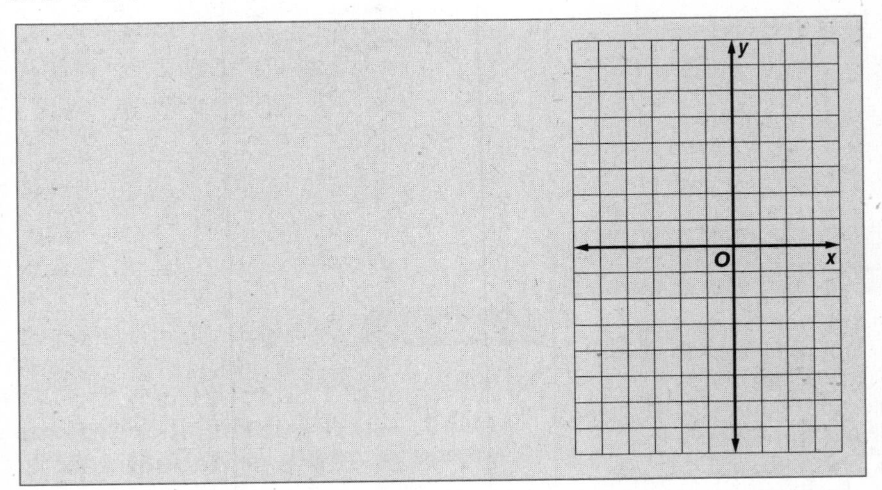

EXAMPLE Translation

3 Triangle *ABC* has vertices $A(-2, 1)$, $B(2, 4)$, and $C(1, 1)$.

a. Find the coordinates of the vertices of the image if it is translated 3 units to the right and 5 units down.

To translate the triangle 3 units to the right, add $\square$ to the x-coordinate of each vertex. To translate the triangle 5 units down, add $\square$ to the y-coordinate of each vertex.

$(x, y) \rightarrow (x + 3, y - 5)$

$A(-2, 1) \rightarrow A'(-2 + 3, 1 - 5) \rightarrow A'(1, \square)$

$B(2, 4) \rightarrow B'(2 + 3, 4 - 5) \rightarrow B'(\square, -1)$

$C(1, 1) \rightarrow C'(1 + 3, 1 - 5) \rightarrow C'(4, \square)$

The coordinates of the vertices of the image are

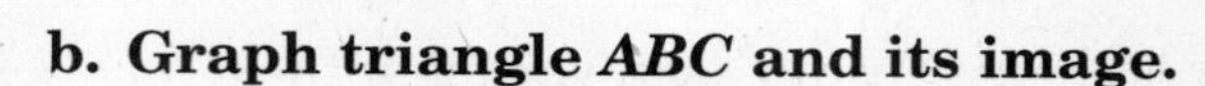
A' ____, B' ____, C' ____.

b. Graph triangle *ABC* and its image.

The preimage is △ ____.

The translated image is △ ____.

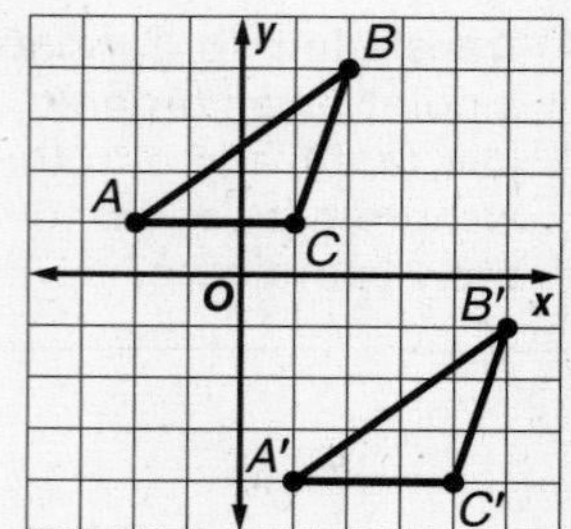

Your Turn Triangle *JKL* has vertices $J(2, -3)$, $K(4, 0)$, and $L(6, -3)$. Find the coordinates of the vertices of the image if it is translated 5 units to the left and 2 units up. Graph triangle *JKL* and its image.

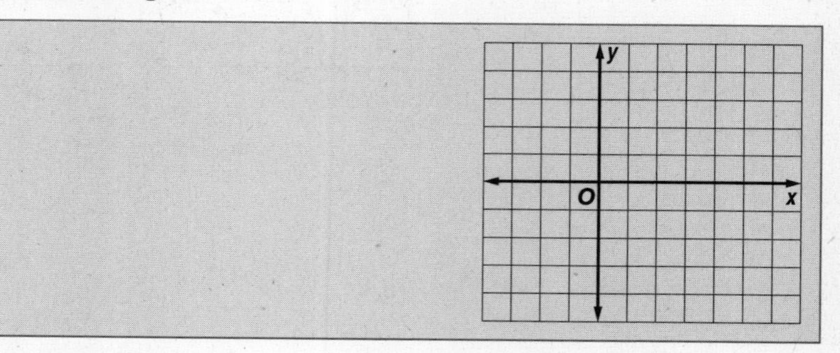

KEY CONCEPTS

Dilation To dilate a figure by a scale factor *k*, multiply both coordinates by *k*. If $k > 1$, the figure is enlarged. If $0 < k < 1$, the figure is reduced.

Rotation To rotate a figure 90° *counterclockwise* about the origin, switch the coordinates of each point and then multiply the new first coordinate by −1. To rotate a figure 180° about the origin, multiply both coordinates of each point by −1.

EXAMPLE Dilation

4 **A trapezoid has vertices $E(-1, 2)$, $F(2, 1)$, $G(2, -1)$, $H(-1, -2)$. Find the coordinates of the dilated trapezoid $E'F'G'H'$ if the scale factor is 2.**

To dilate the figure, multiply the coordinates of each vertex by 2.

$(x, y) \rightarrow (2x, 2y)$

$E(-1, 2) \rightarrow E'(2 \cdot (-1), 2 \cdot 2) \rightarrow E'(-2, \square)$

$F(2, 1) \rightarrow F'(2 \cdot 2, 2 \cdot 1) \rightarrow F'(\square, 2)$

$G(2, -1), \rightarrow G'(2 \cdot 2, 2 \cdot (-1)) \rightarrow G'(4, \square)$

$H(-1, -2) \rightarrow H'(2 \cdot (-1), 2 \cdot (-2)) \rightarrow H'(\square)$

The coordinates of the vertices of the image are

E' ______, F' ______, G' ______, and H' ______.

Your Turn A trapezoid has vertices $E(-4, 7)$, $F(2, 7)$, $G(0, 4)$, and $H(-2, 4)$. Find the coordinates of the dilated trapezoid $E'F'G'H'$ if the scale factor is $\frac{1}{2}$.

EXAMPLE Rotation

5 **Triangle ABC has vertices $A(1, -3)$, $B(3, 1)$, and $C(5, -2)$.**

a. Find the coordinates of the image $\triangle ABC$ after it is rotated 180° about the origin.

To find the coordinates of the image of $\triangle ABC$ after a 180º rotation, multiply both coordinates of each point by -1.

$(x, y) \longrightarrow (-1 \cdot x, -1 \cdot y)$

$A(1, -3) \longrightarrow A'($ ______ $)$

$B(3, 1) \longrightarrow B'($ ______ $)$

$C(5, -2) \longrightarrow C'($ ______ $)$

The coordinates of the vertices of the image are A' ______, B' ______, and C' ______.

b. Graph the preimage and its image.

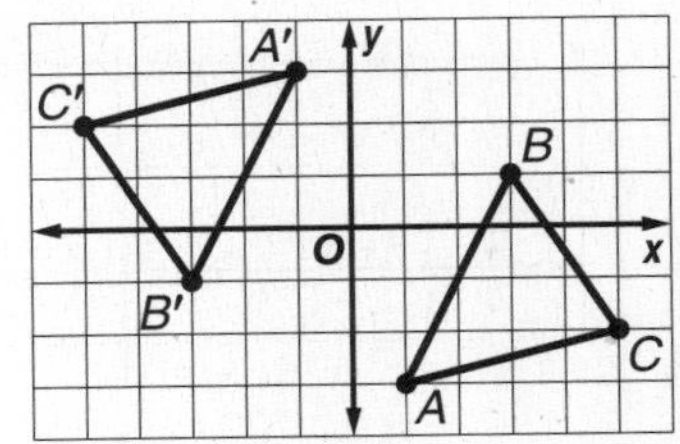

The preimage is $\triangle ABC$.

The translated image is $\triangle A'B'C'$.

Your Turn Triangle RST has vertices $R(4, 0)$, $S(2, -3)$, and $T(6, -3)$. Find the coordinates of the image of $\triangle RST$ after it is rotated 90° counterclockwise about the origin. Graph the preimage and the image.

HOMEWORK ASSIGNMENT

Page(s):

Examples:

4–3 Relations

Build Your Vocabulary (page 84)

A **mapping** illustrates how each element of the ______ is paired with an element in the ______.

The **inverse** of any relation is obtained by switching the ______ in each ______.

What You'll Learn

- Represent relations as sets of ordered pairs, tables, mappings, and graphs.
- Find the inverse of a relation.

Review It

Write an example of a relation.

______
______
______
______
______
______
______

EXAMPLE Represent a Relation

1 Express the relation {(4, 3), (−2, −1), (−3, 2), (2, −4), (0, −4)} as a table, a graph and a mapping.

Table
List the set of *x*-coordinates in the first column and the corresponding *y*-coordinates in the second column.

x	*y*

Graph
Graph each ordered pair on a coordinate plane.

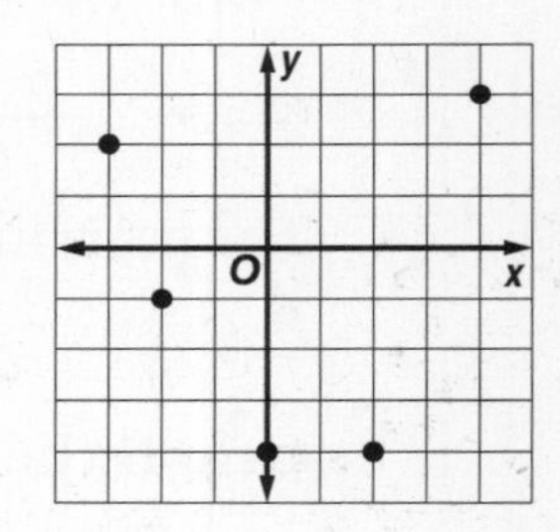

Mapping
List the *x* values in set *X* and the *y* values in set *Y*. Draw an arrow from each *x* value in *X* to the corresponding *y* value in *Y*.

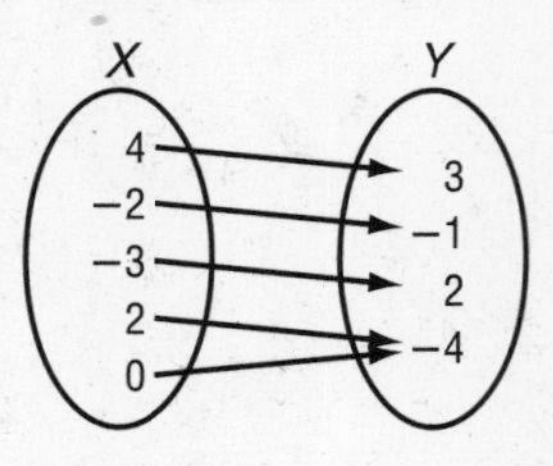

Your Turn Express the relation {(3, −2), (4, 6), (5, 2), (−1, 3)} as a table, a graph, and a mapping.

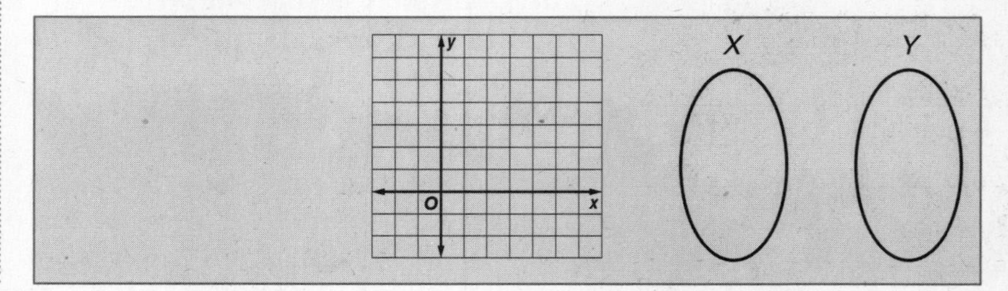

EXAMPLE Use a Relation

2 OPINION POLLS The table shows the percent of people satisfied with the way things were going in the U.S. at the time of the survey.

Year	1992	1995	1998	2001
Percent Satisfied	21	32	60	51

a. Determine the domain and range of the relation.

The domain is ______.

The range is ______.

b. Graph the data.

The values of the x-axis need to go from 1992 to 2001. Begin at 1992 and extend to 2001 to include all of the data. The units can be 1 unit per grid square.

The values on the y-axis need to go from 21 to 60. Begin at 0 and extend to 70. You can use units of 10.

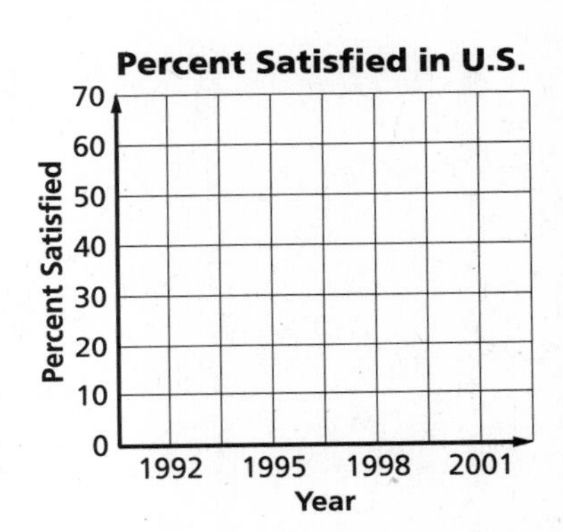

c. What conclusions might you make from the graph of the data?

Americans became more satisfied with the country from ______, but the percentage dropped from ______.

Your Turn **The table shows the approximate world population of the Indian Rhinoceros from 1982 to 1998.**

Indian Rhinoceros Population					
Year	1982	1986	1990	1994	1998
Population	1000	1700	1700	1900	2100

a. Determine the domain and range of the relation.

KEY CONCEPT

Inverse of a Relation Relation *Q* is the inverse of relation *S* if and only if for every ordered pair (*a*, *b*) in *S*, there is an ordered pair (*b*, *a*) in *Q*.

FOLDABLES Under the tab for Lesson 4–3. Write a relation with four ordered pairs. Then find the inverse of the relation.

b. Graph the data.

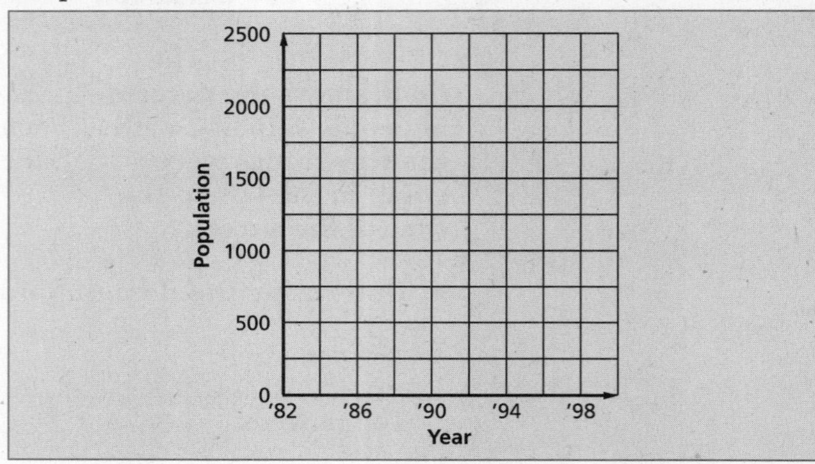

c. What conclusions might you make from the graph of the data?

EXAMPLE Inverse Rotation

3 **Express the relation shown in the mapping as a set of ordered pairs. Then write the inverse of the relation.**

Relation Notice that both 7 and 0 in the domain are paired with 2 in the range.

{(5, 1), (7, 2), ______}

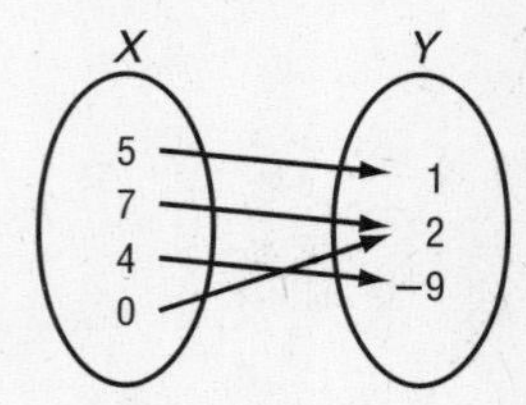

Inverse Exchange X and Y in each ordered pair to write the inverse relation.

{(1, 5), ______}

Your Turn Express the relation shown in the mapping as a set of ordered pairs. Then write the inverse of the relation.

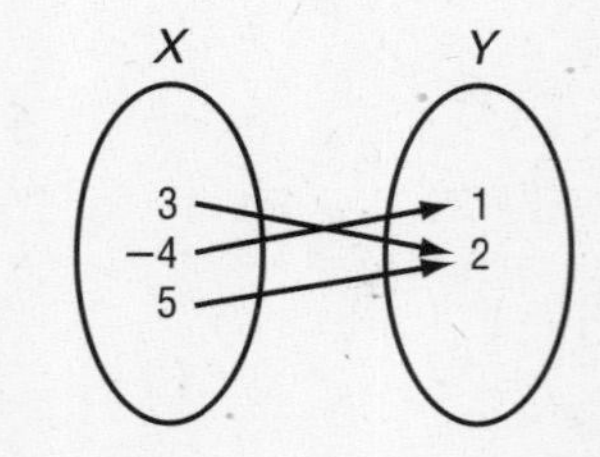

HOMEWORK ASSIGNMENT

Page(s):

Exercises:

4–4 Equations as Relations

WHAT YOU'LL LEARN

- Use an equation to determine the range for a given domain.
- Graph the solution set for a given domain.

EXAMPLE Solve Using a Replacement Set

1 **Find the solution set for $y = 7 + 3x$, given the replacement set {(−5, 0), (−3, −2), (2, 13), (4, 19)}.**

Make a table. Substitute each ordered pair into the equation.

x	y	$y = 7 + 3x$	True or False?
−5	0	$0 = 7 + 3(-5)$	
−3	−2	$-2 = 7 + 3(-3)$	
2	13	___ = ___	
		$19 = 7 + 3(4)$	

The solution set is ______.

EXAMPLE Solve Using a Given Domain

2 **Solve $d = 8 - c$ if the domain is {−2, 0, 3, 5, 8}.**

Make a table. The values of c come from the domain. Substitute each value of c into the equation to determine the values of d in the range.

c	$8 - c$	d	(c, d)
−2	$8 - (-2)$	10	(−2, 10)
0	$8 - 0$		
3	$8 - 3$	5	(3, 5)
5	$8 - 5$		
8	$8 - 8$	0	(8, 0)

The solution set is {(−2, 10), ______, (3, 5), ______, (8, 0)}.

Your Turn

a. Find the solution set for $y = 3x + 2$ given the replacement set $\{(3, 1), (6, 8), (1, 5), (-1, -1)\}$.

b. Solve $y = 2x - 4$ if the domain is $\{-1, 0, 2, 5\}$.

WRITE IT

What is the domain of a function?

EXAMPLE Solve and Graph the Solution Set

3 Solve $9x + 3y = 15$ if the domain is $\{0, 1, 2, 3\}$. Graph the solution set.

First solve the equation for y in terms of x. This makes creating a table of values easier.

$9x + 3y = 15$	Original equation
$9x + 3y - \square = 15 - \square$	Subtract $9x$ from each side.
$3y = 15 - 9x$	Simplify.
$\frac{3y}{y} = \frac{15 - 9x}{3}$	Divide each side by 3.
$y = \square$	Simplify.

Substitute each value of x from the domain to determine the corresponding values of y in the range.

x	$5 - 3x$	y	(x, y)
0	$5 - 3(0)$	5	
1		2	$(1, 2)$
2	$5 - 3(2)$	-1	$(2, -1)$
3	$5 - 3(3)$	-4	$(3, -4)$

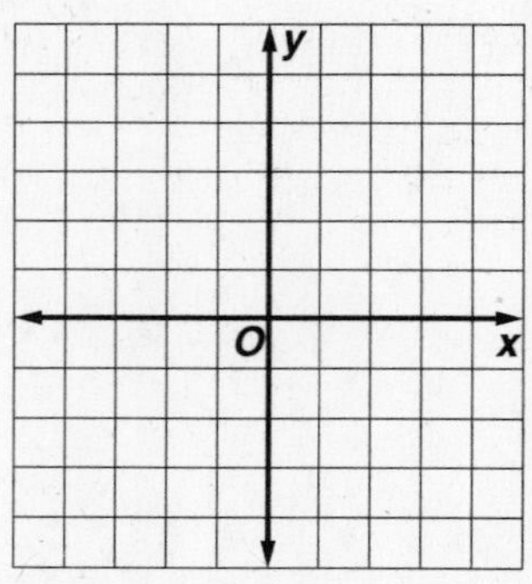

Graph the solution set $\{(0, 5)$, $\square$, $\square$, $\square\}$.

Your Turn Solve $6x + 2y = 8$ if the domain is $\{0, 1, 2, 3\}$. Graph the solution set.

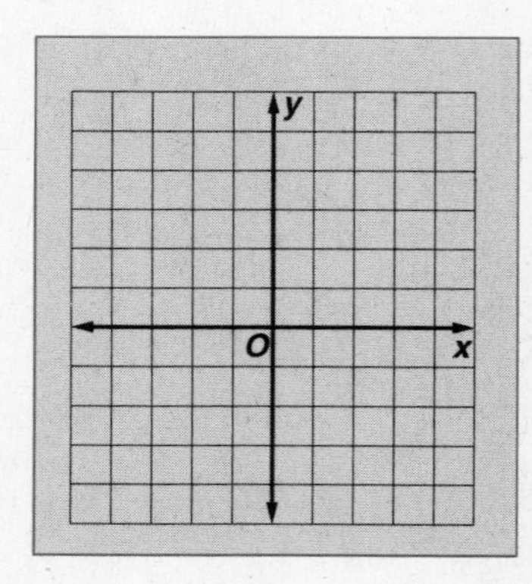

HOMEWORK ASSIGNMENT

Page(s):

Exercises:

4–5 Graphing Linear Equations

WHAT YOU'LL LEARN

- Determine whether an equation is linear.
- Graph linear equations.

BUILD YOUR VOCABULARY (pages 84–85)

A **linear equation** is the equation of a line. When an equation is written in the form $Ax + By = C$, it is said to be in **standard form**.

KEY CONCEPT

Standard Form of a Linear Equation The standard form of a linear equation is $Ax + By = C$, where $A \geq 0$, A and B are not both zero, and A, B, and C are integers whose greatest common factor is 1.

FOLDABLES™ On the Lesson 4–5 tab, write an example of a linear equation and one that is not linear. Draw a graph of the linear equation.

EXAMPLE Identifying Linear Equations

1 **Determine whether each equation is a linear equation. If so, write the equation in standard form.**

a. $5x + 3y = z + 2$

Rewrite the equation with the variables on one side.

$5x + 3y = z + 2$	Original equation
$5x + 3y - z = z + 2 - z$	Subtract.
$5x + 3y - z = 2$	Simplify.

Since there are ____ different variables on the left side of the equation, it ____ be written in the form $Ax + By = C$.

This is not a ____.

b. $\frac{3}{4}x = y + 8$

Rewrite the equation with the variables on one side.

$\frac{3}{4}x = y + 8$	Original equation
$\frac{3}{4}x - y = y + 8 - y$	Subtract y from each side.
$\frac{3}{4}x - y = 8$	Simplify.

Write the equation with integer coefficients.

$\frac{3}{4}x - y = 8$	
$4\left(\frac{3}{4}x\right) - 4(y) = 8(4)$	Multiply each side by 4.
$3x - 4y = 32$	Simplify.

The equation is now in standard form where $A =$ ____, $B =$ ____, and $C =$ ____. This is a ____ equation.

Your Turn **Determine whether each equation is a linear equation. If so, write the equation in standard form.**

a. $y = 4x - 5$

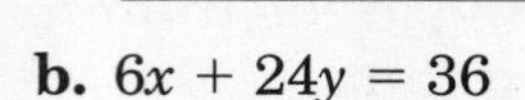

b. $6x + 24y = 36$

c. $\frac{2}{3}x = 9$

EXAMPLE Graph by Making a Table

2 **Graph $\frac{1}{2}y - x = 1$**

In order to find values for y more easily, solve the equation for y.

$\frac{1}{2}y - x = 1$	Original equation
$\frac{1}{2}y - x + \square = 1 + \square$	Add x to each side.
$\frac{1}{2}y = \square$	Simplify.
$2\left(\frac{1}{2}y\right) = 2(1 + x)$	Multiply each side by $\square$.
$y = \square$	Simplify.

Select five values for the domain and make a table. Then graph the ordered pairs.

x	$2 + 2x$	y	(x, y)
-3	$2 + 2(-3)$		$(-3, -4)$
-1		0	$(-1, 0)$
0	$2 + 2(0)$	2	
2	$2 + 2(2)$		
3			

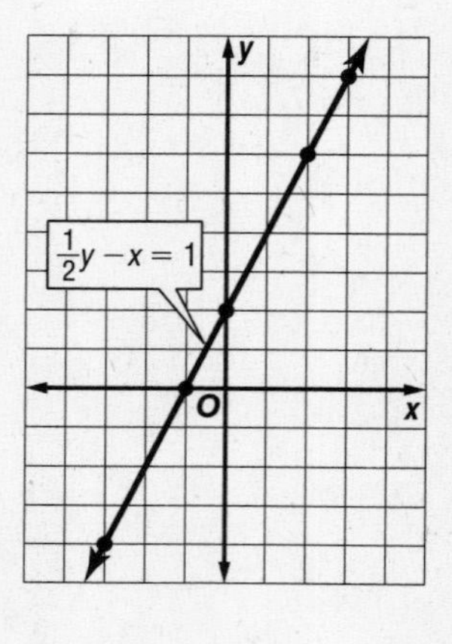

REVIEW IT
What do the arrowheads at the ends of a number line mean? *(Lesson 2–1)*

Your Turn Graph $3x + 2y = 6$.

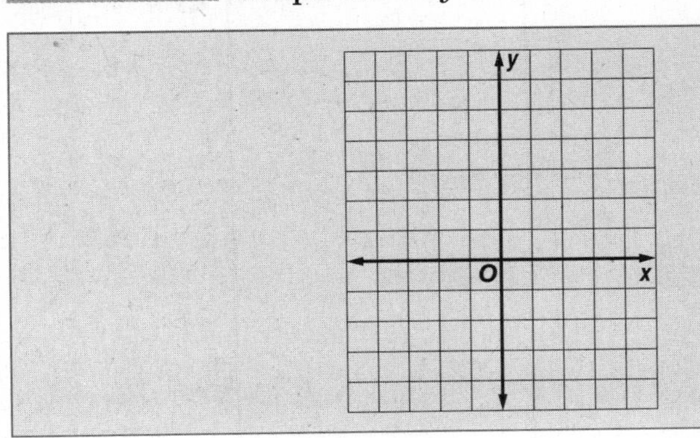

EXAMPLE Graph Using Intercepts

3 Determine the x-intercept and the y-intercept of $4x - y = 4$. Then graph the equation.

To find the x-intercept, let $y = 0$.

$4x - y = 4$

$4x - \square = 4$

$4x = 4$

$x = \square$

To find the y-intercept, let $x = 0$.

$4x - y = 4$

$4\square - y = 4$

$-y = 4$

$y = \square$

The x-intercept is 1, so the graph intersects the x-axis at $\square$.

The y-intercept is -4, so the graph

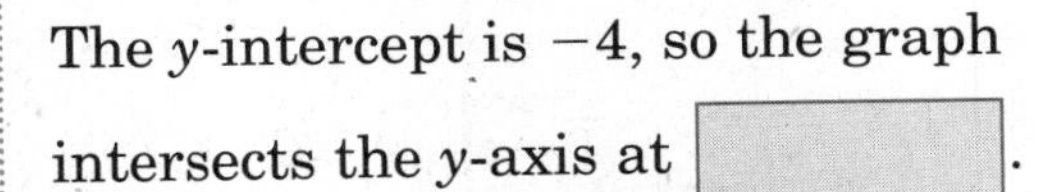

intersects the y-axis at $\square$.

Plot these points. Then draw a line that connects them.

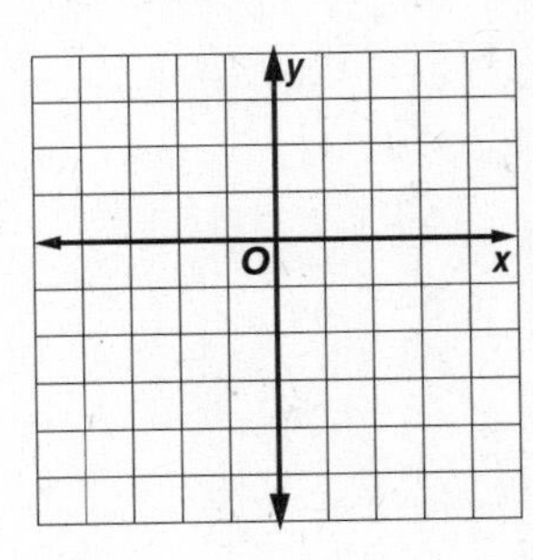

Your Turn Determine the x-intercept and the y-intercept of $2x + 5y = 10$. Then graph the equation.

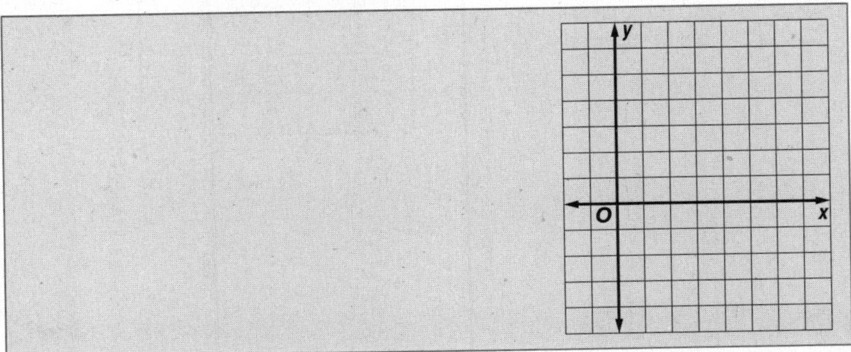

4–6 Functions

What You'll Learn

- Determine whether a relation is a function.
- Find function values.

Key Concept

A **function** is a relation in which each element of the domain is paired with exactly one element of the range.

Foldables

Use the tab for Lesson 4–6. Explain two ways to determine whether a relation is a function.

EXAMPLE Identify Functions

1 a. Determine whether each relation is a function. Explain.

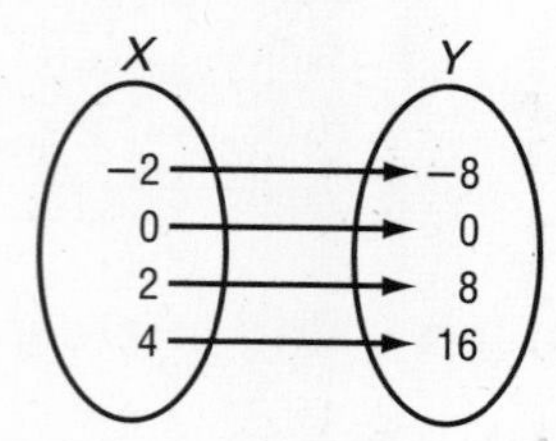

This is a function because the mapping shows each element of the ________ paired with exactly one member of the ________.

b.

x	y
−7	−12
−4	−9
2	−3
5	0

This table represents a function because the table shows each element of the domain paired with ________ element of the range.

c. {(−5, 2), (−2, 5), (0, 7), (0, 9)}

This relation is not a function because the element ____ in the ________ is paired with both ____ and 9 in the range.

Your Turn **Determine whether each relation is a function. Explain.**

a.

X → Y: 3 → −2, 4 → 6, 5 → 2, −1 → 3

b.

X	Y
3	2
1	−2
2	−4
3	−1

c. {(3, 0), (1, 2), (4, 0), (5, −1)}

EXAMPLE Function Values

2 **a. If $f(x) = 3x - 4$, find $f(4)$.**

$f\square = 3\square - 4$ — Replace x with 4.

$= 12 - 4$ — Multiply.

$= \square$ — Subtract.

b. If $f(x) = 3x - 4$, find $f(-5)$.

$f(\square) = 3(\square) - 4$ — Replace x with -5.

$= \square - 4$ — Multiply.

$= -19$ — Subtract.

Your Turn **If $f(x) = 2x + 5$, find each value.**

a. $f(-8)$

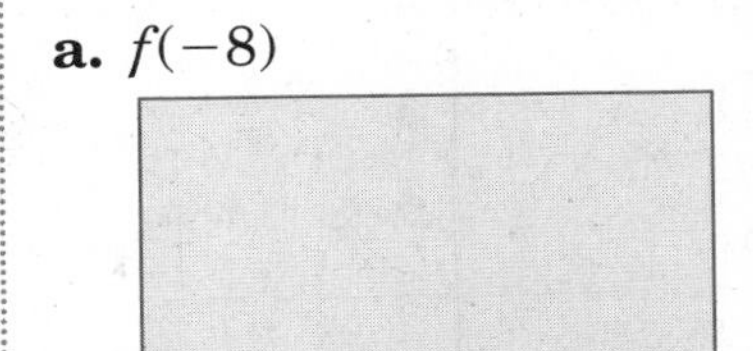

b. $f(x + 3)$

EXAMPLE Non-Linear Functions

3 **a. If $k(m) = m^2 - 4m + 5$, find $k(-3)$.**

$k(-3) = (-3)^2 - 4(-3) + 5$ — Replace m with -3.

$= 9 + 12 + 5$ — Multiply.

$= 26$ — Simplify.

b. If $k(m) = m^2 - 4m + 5$, find $k(6z)$.

$k(\square) = (6z)^2 - 4(6z) + 5$ — Replace m with $6z$.

$= \square - 24z + 5$ — Simplify.

Your Turn **If $h(x) = 3x^2 - 4$, find each value.**

a. $h(2)$

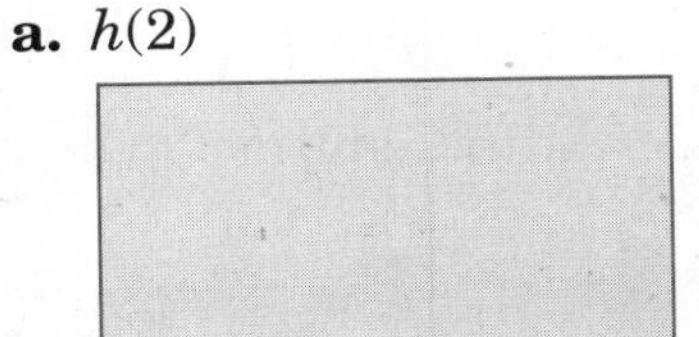

b. $h(3t)$

HOMEWORK ASSIGNMENT

Page(s):

Exercises:

4–7 Arithmetic Sequences

BUILD YOUR VOCABULARY (pages 84–85)

A **sequence** is a set of numbers in a specific order.

The numbers in a sequence are called **terms**.

If the differences between successive terms is constant, then it is called an **arithmetic sequence**.

The difference between the terms is called the **common difference.**

WHAT YOU'LL LEARN

- Recognize arithmetic sequences.
- Extend and write formulas for arithmetic sequences.

KEY CONCEPT

Arithmetic Sequence An arithmetic sequence is a numerical pattern that increases or decreases at a constant rate or value called the common difference.

EXAMPLE Identify an Arithmetic Sequence

1 **Determine whether the sequence −15, −13, −11, −9, ... is arithmetic. Justify your answer.**

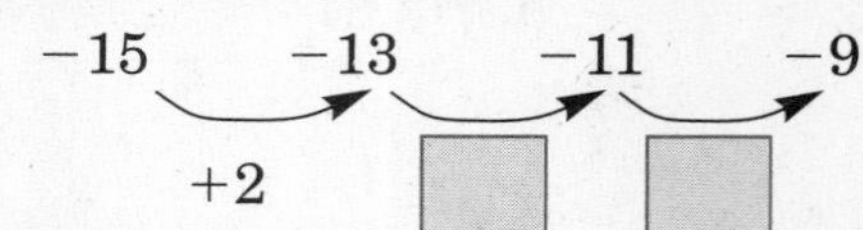

This is an ______ because the difference between terms is ______.

EXAMPLE Extend a Sequence

2 **Find the next three terms of the arithmetic sequence −8, −11, −14, −17, ...**

−8 −11 −14 −17

−3 −3 ______ The common difference is −3.

The next three terms are ______, ______, and ______.

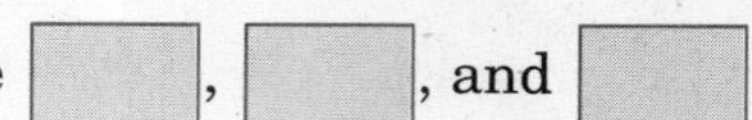

Your Turn

a. Determine whether the sequence 2, 4, 8, 10, 12, ... is arithmetic. Justify your answer.

b. Find the next three terms of the arithmetic sequence 5, 12, 19, 26, ...

KEY CONCEPT

nth Term of an Arithmetic Sequence The nth term a_n of an arithmetic sequence with first term a_1 and common difference d is given by $a_n = a_1 + (n - 1)d$, when n is a positive integer.

FOLDABLES

Use the tab for Lesson 4–7. Write the general form for an arithmetic sequence. Explain what each of the variables means.

KEY CONCEPT

Writing Arithmetic Sequences Each term of an arithmetic sequence after the first term can be found by adding the common difference to the preceding term.

HOMEWORK ASSIGNMENT

Page(s):

Exercises:

EXAMPLE Find a Specific Term

3 Find the 9th term of the arithmetic sequence 7, 11, 15, 19, … .

In this sequence, the first term, a_1, is ____. You want to find the 9th term, so $n = 9$. The common difference is ____.

Use the formula for the nth term of an arithmetic sequence.

$a_n = a_1 + (n - 1)d$	Formula for the nth term
$a_n = 7 + (9 - 1)(4)$	$a_1 = 7$, $n =$ ____, $d =$ ____
$a_n = 7 + (\ \_\_\_\_\ )(4)$ or ____	Simplify.

EXAMPLE Write an Equation for a Sequence

4 Consider the arithmetic sequence −8, 1, 10, 19, … Write an equation for the nth term of the sequence.

In this sequence, the first term, a_1, is ____. The common difference is ____.

Use the formula for the nth term to write an equation.

$a_n = a_1 + (n - 1)d$	Formula for the nth term
$a_n = -8 + (n - 1)9$	$a_1 = -8$, $d =$ ____
$a_n = -8 +$ ____	Distributive Property
$a_n =$ ____	Simplify.

An equation for the nth term in this sequence is

$a_n =$ ____.

Your Turn

a. Find the 12th term in the arithmetic sequence 12, 17, 22, 27, … .

b. Consider the arithmetic sequence −3, 0, 3, 6, … Write an equation for the nth term of the sequence.

4–8 Writing Equations from Patterns

BUILD YOUR VOCABULARY (page 84)

When you make a conclusion based on a ______ of examples, you are using **inductive reasoning**.

WHAT YOU'LL LEARN

- Look for a pattern.
- Write an equation given some of the solutions.

FOLDABLES

ORGANIZE IT

On the tab for Lesson 4–8, write 2 ways you can decide what a pattern is in a sequence.

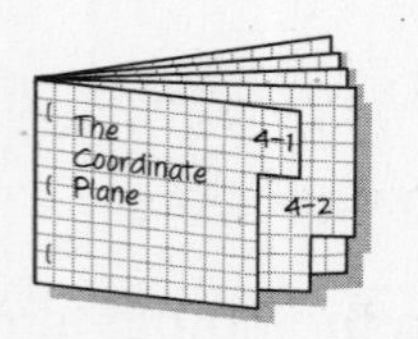

EXAMPLE Extend a Pattern

1 Study the pattern.

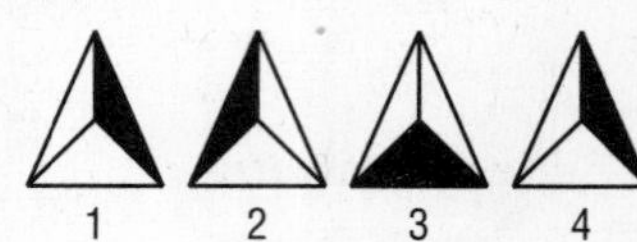

a. Draw the next three figures in the pattern.

5 6

The pattern consists of triangles with one-third shaded. The section that is shaded is rotated in a counterclockwise direction.

b. Draw the 17th triangle in the pattern.

The pattern repeats every ______ design.

So, designs 3, 6, 9, 12, 15, and so on will all be the ______.

Since 15 is the greatest number less than 17 that is a multiple of ______, the 17th triangle in the pattern will be the same as the second triangle.

Your Turn **Study the pattern below.**

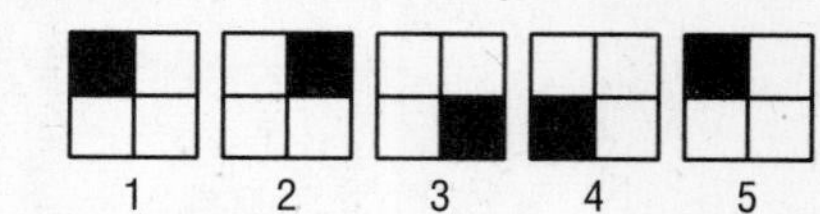

a. Draw the next three figures in the pattern.

b. Draw the 19th square in the pattern.

EXAMPLE Write an Equation from Data

2 **The table shows the number of miles driven for each hour of driving.**

Hours	1	2	3	4
Miles	50	100	150	200

a. Graph the data. What conclusion can you make about the relationship between the number of hours driving, *h* and the number of miles driven, *m*?

This graph shows a linear relationship between the number of hours driving and the number of miles driven.

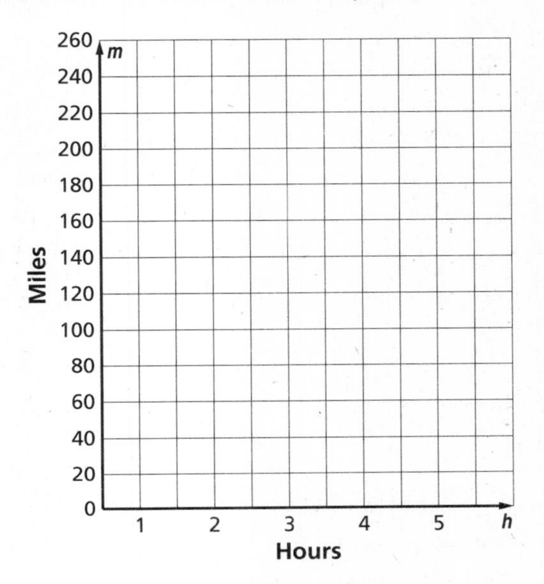

b. Write an equation to describe this relationship.

Look at the relationship between the domain and the range to find a pattern that can be described as an equation.

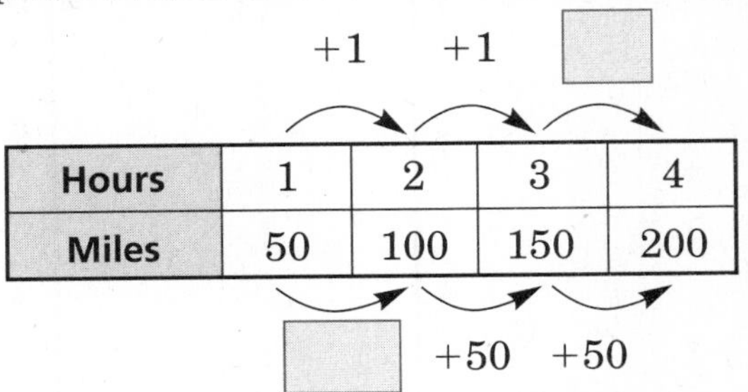

Hours	1	2	3	4
Miles	50	100	150	200

The difference of the values for h is ______ and the difference of the values for m is ______. This suggests that $m =$ ______.

Since the relation is also a ______, we can write the equation as $f(h) =$ ______, where $f(h)$ represents the number of ______.

Your Turn **The table below shows the number of miles walked for each hour of walking.**

Hours	1	2	3	4	5
Miles	1.5	3	4.5	6	7.5

a. Graph the data. What conclusion can you make about the relationship between the number of miles and the time spent walking?

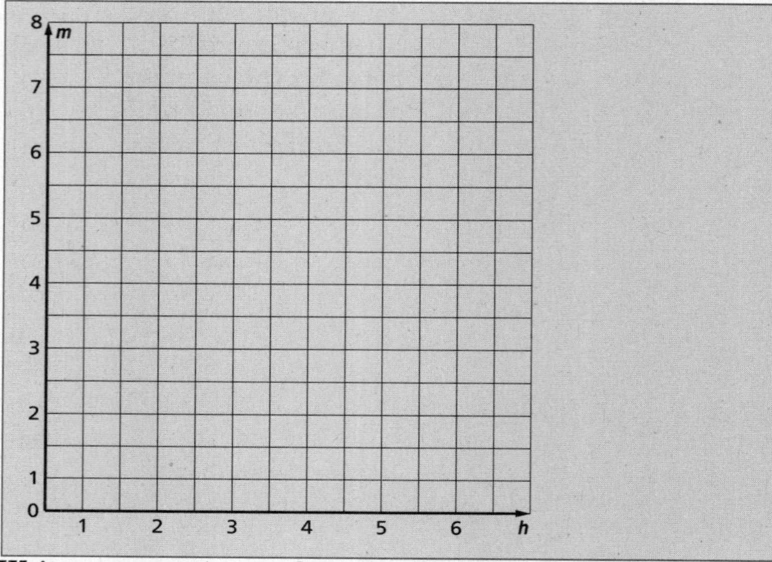

b. Write an equation to describe the relationship.

CHAPTER 4

BRINGING IT ALL TOGETHER

STUDY GUIDE

FOLDABLES™	VOCABULARY PUZZLEMAKER	BUILD YOUR VOCABULARY
Use your **Chapter 4 Foldable** to help you study for your chapter test.	To make a crossword puzzle, word search, or jumble puzzle of the vocabulary words in Chapter 4, go to www.glencoe.com/sec/math/t_resources/free/index.php	You can use your completed **Vocabulary Builder** (pages 84–85) to help you solve the puzzle.

4–1

The Coordinate Plane

Use the ordered pair (−2, 3).

1. Explain how to identify the x- and y-coordinates.

2. Name the x- and y- coordinates.

3. Describe the steps you would use to locate the point for (−2, 3) on the coordinate plane.

4–2

Transformations on the Coordinate Plane

Write the letter of the term and the Roman numeral of the figure that best matches each statement.

4. A figure is flipped over a line.

5. A figure is turned around a point.

6. A figure is enlarged or reduced.

7. A figure is slid horizontally, vertically, or both.

A. dilation

B. translation

C. reflection

D. rotation

I.

II.

III.

IV.

4–3

Relations

8. Write the relation shown in the table.

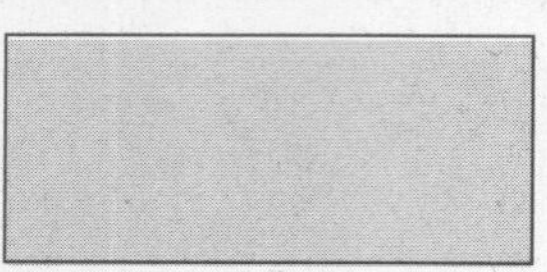

x	y
0	−2
1	4
−3	5
−2	0

9. Write the inverse of the relation {(1, 2), (2, 4), (3, 6), (4, 8)}.

4–4

Equations as Relations

The solution set of the equation $y = 2x$ for a given domain is {(−2, −4), (0, 0), (2, 4), (7, 14)}. Tell whether each sentence is *true* or *false*. If false, replace the underlined word(s) to make a true sentence.

10. The domain contains the values represented by the independent variable.

11. The domain is {−4, 0, 4, and 14}.

12. What is meant by "solving an equation for y in terms of x"?

4–5

Graphing Linear Equations

Determine whether each equation is a linear equation. Explain.

	Equation	Linear or non-linear?	Explanation
13.	$4xy + 2y = 7$		
14.	$\frac{x}{5} - \frac{4y}{3} = 2$		

4–6 Functions

15. Describe how the mapping shows that the relation represented is a function.

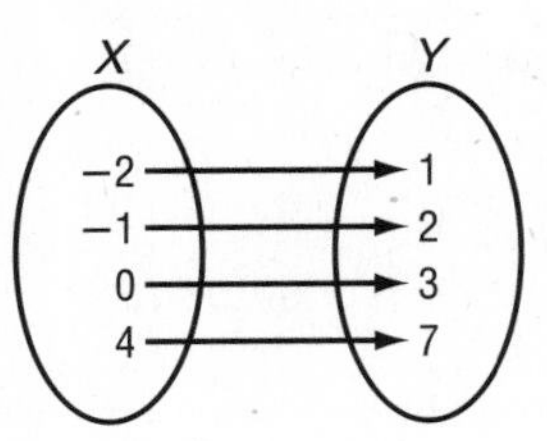

4–7 Arithmetic Sequences

Complete the table.

	Pattern	Is the sequence increasing or decreasing?	Is there common difference? If so, what is it?
16.	55, 50, 45, 40, …		
17.	1, 2, 4, 9, 16 …		
18.	$\frac{1}{2}$, 0, $-\frac{1}{2}$, −1, …		

4–8 Writing Equations from Patterns

19. For the figures below, explain why Figure 5 does not follow the pattern.

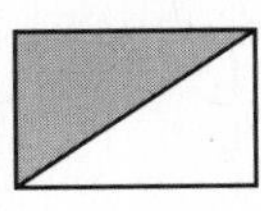
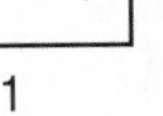
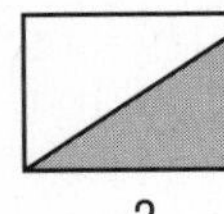
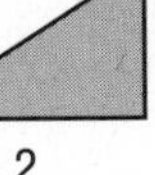
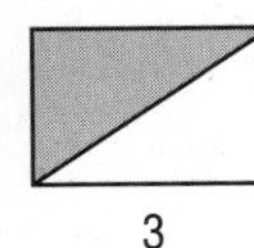
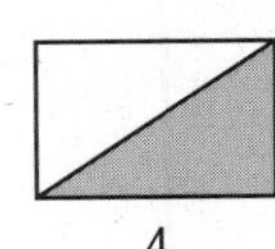
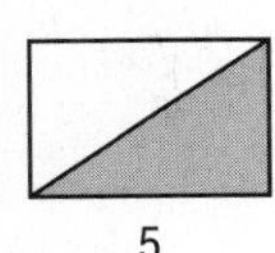

1 2 3 4 5

20. Write the next 3 terms of the sequence 1, 5, 25, 125, … .

ARE YOU READY FOR THE CHAPTER TEST?

Math Online

Visit **www.algebra1.com** to access your textbook, more examples, self-check quizzes, and practice tests to help you study the concepts in Chapter 4.

Check the one that applies. Suggestions to help you study are given with each item.

☐ **I completed the review of all or most lessons without using my notes or asking for help.**

- You are probably ready for the Chapter Test.
- You may want to take the Chapter 4 Practice Test on page 251 of your textbook as a final check.

☐ **I used my Foldable or Study Notebook to complete the review of all or most lessons.**

- You should complete the Chapter 4 Study Guide and Review on pages 246–250 of your textbook.
- If you are unsure of any concepts or skills, refer back to the specific lesson(s).
- You may also want to take the Chapter 4 Practice Test on page 251.

☐ **I asked for help from someone else to complete the review of all or most lessons.**

- You should review the examples and concepts in your Study Notebook and Chapter 4 Foldable.
- Then complete the Chapter 4 Study Guide and Review on pages 246–250 of your textbook.
- If you are unsure of any concepts or skills, refer back to the specific lesson(s).
- You may also want to take the Chapter 4 Practice Test on page 251.

Student Signature

Parent/Guardian Signature

Teacher Signature

Analyzing Linear Equations

Use the instructions below to make a Foldable to help you organize your notes as you study the chapter. You will see Foldable reminders in the margin of this Interactive Study Notebook to help you in taking notes.

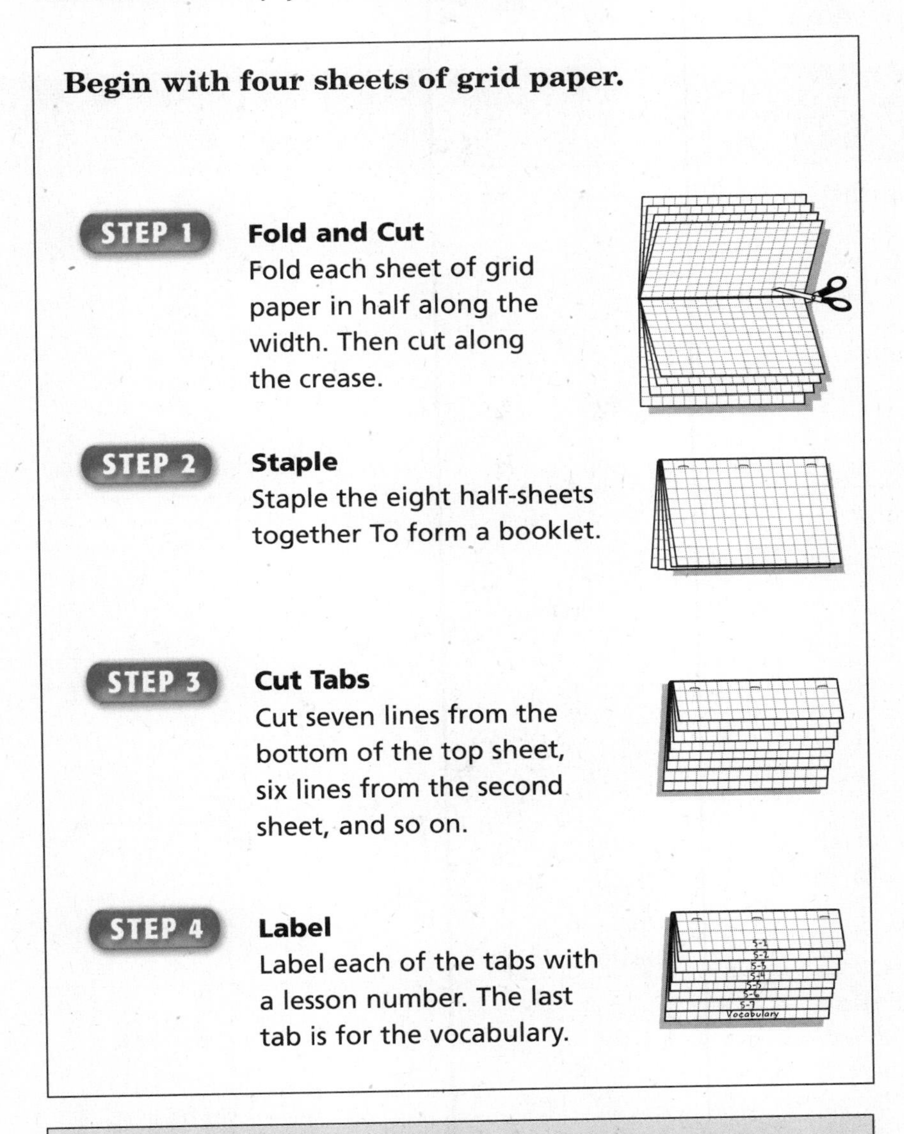

Begin with four sheets of grid paper.

STEP 1 **Fold and Cut**
Fold each sheet of grid paper in half along the width. Then cut along the crease.

STEP 2 **Staple**
Staple the eight half-sheets together To form a booklet.

STEP 3 **Cut Tabs**
Cut seven lines from the bottom of the top sheet, six lines from the second sheet, and so on.

STEP 4 **Label**
Label each of the tabs with a lesson number. The last tab is for the vocabulary.

NOTE-TAKING TIP: When you take notes, circle, underline, or star anything the teacher emphasizes. When your teacher emphasizes a concept, it will usually appear on a test, so make an effort to include it in your notes.

BUILD YOUR VOCABULARY

This is an alphabetical list of new vocabulary terms you will learn in Chapter 5. As you complete the study notes for the chapter, you will see Build Your Vocabulary reminders to complete each term's definition or description on these pages. Remember to add the textbook page number in the second column for reference when you study.

Vocabulary Term	Found on Page	Definition	Description or Example
constant of variation			
direct variation			
family of graphs			
line of fit			
linear extrapolation [ihk-STRA-puh-LAY-shun]			
linear interpolation [ihk-TUHR-puh-LAY-shun]			
negative correlation [KAWR-uh-LAY-shun]			

Vocabulary Term	Found on Page	Definition	Description or Example
parallel lines			
perpendicular lines [PUHR-puhn-DIH-kyuh-luhr]			
point-slope form			
positive correlation			
rate of change			
scatter plot			
slope			
slope-intercept form [IHN-tuhr-SEHPT]			

5-1 Slope

What You'll Learn

- Find the slope of a line.
- Use the rate of change to solve problems.

Build Your Vocabulary (page 112–113)

The **slope** of a line is a number determined by any two points on the line.

The **rate of change** tells, on average, how a quantity is changing over time.

Key Concept

Slope of a Line The slope of a line is the ratio of the rise to the run.

Foldables

Write the formula for finding the slope of a line under the tab for Lesson 5-1.

Example Positive Slope

1 Find the slope of the line that passes through (−3, 2) and (5, 5).

$m = \frac{y_2 - y_1}{x_2 - x_1}$ — $\frac{\text{rise}}{\text{run}}$

$= \frac{\square - \square}{\square - \square}$ — $(-3, 2) = (x_1, y_1)$ and $(5, 5) = (x_2, y_2)$

$= \square$ — Simplify.

Example Negative Slope

2 Find the slope of the line that passes through (−3, −4) and (−2, −8).

$m = \frac{y_2 - y_1}{x_2 - x_1}$ — $\frac{\text{rise}}{\text{run}}$

$= \frac{\square - \square}{-2 - (-3)}$ — $(-3, -4) = (x_1, y_1)$ and $(-2, -8) = (x_2, y_2)$

$= \frac{-4}{1}$ or -4 — Simplify.

Example Zero Slope

3 Find the slope of the lines that passes through (−3, 4) and (4, 4).

$m = \frac{y_2 - y_1}{x_2 - x_1}$ — $\frac{\text{rise}}{\text{run}}$

$= \frac{4 - 4}{\square - \square}$ — $(-3, 4) = (x_1, y_1)$ and $(4, 4) = (x_2, y_2)$

$= \square$ or $\square$ — Simplify.

EXAMPLE Undefined Slope

4 **Find the slope of the line that passes through (−2, −4) and (−2, 3).**

$m = \frac{y_2 - y_1}{x_2 - x_1}$ — rise/run

$= \frac{3 - (-4)}{-2 - (-2)}$ or ☐ — $(-2, -4) = (x_1, y_1)$ and $(-2, 3) = (x_2, y_2)$

The ☐ is ☐.

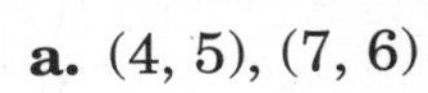

Your Turn **Find the slope of the line that passes through each pair of points.**

a. (4, 5), (7, 6)

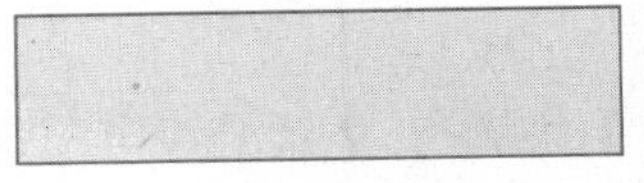

b. (−3, −5), (−2, −7)

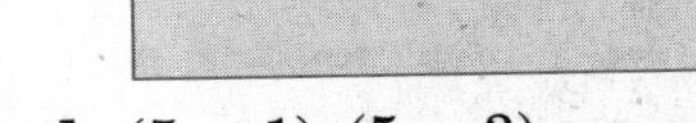

c. (−3, −1), (5, −1)

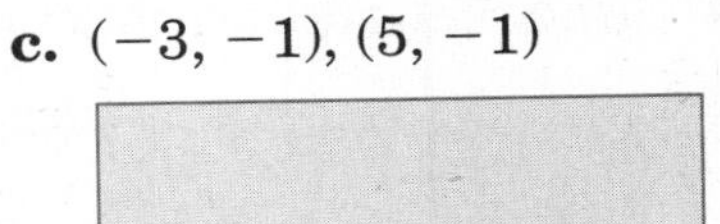

d. (5, −1), (5, −3)

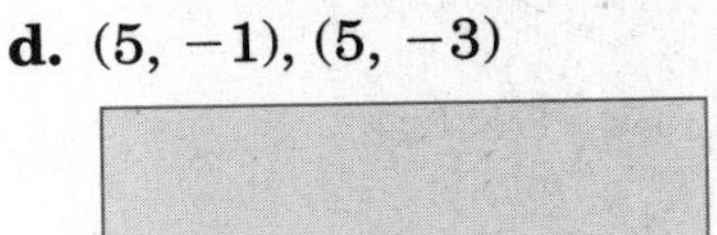

EXAMPLE Find Coordinates Given Slope

5 **Find the value of *r* so that the line through (6, 3) and (*r*, 2) has a slope of $\frac{1}{2}$.**

Let $(r, 2) = (x_1, y_1)$ and $(6, 3) = (x_2, y_2)$.

$m = \frac{y_2 - y_1}{x_2 - x_1}$	Slope Formula.
$\frac{1}{2} = \frac{2 - 3}{r - 6}$	Substitute.
$\frac{1}{2} = \frac{☐}{r - 6}$	Subtract.
☐$(-1) =$ ☐$(r - 6)$	Find the ☐.
☐ = ☐	Simplify.
$-2 +$ ☐ $= r - 6 +$ ☐	Add ☐ to each side.
☐ $= r$	Simplify.

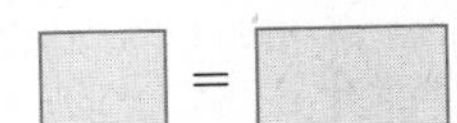

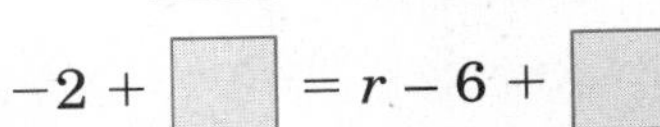

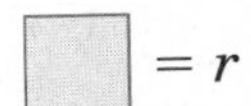

Your Turn Find the value of p so that the line through $(p, 4)$ and $(3, -1)$ has a slope of $-\frac{5}{8}$.

REVIEW IT

Describe how you find cross products.
(Lesson 3-6)

HOMEWORK ASSIGNMENT

Page(s):
Exercises:

5–2 Slope and Direct Variation

WHAT YOU'LL LEARN

- Write and graph direct variation equations.
- Solve problems involving direct variation.

BUILD YOUR VOCABULARY (pages 112–113)

A **direct variation** is described by an equation of the form ______, where $k \neq 0$.

In the equation $y = kx$, ______ is the **constant of variation.**

A **family of graphs** includes graphs and equations of graphs that have at least ______ characteristic in common.

EXAMPLE Direct Variation with $k > 0$

1 Graph $y = x$.

Recall that the slope of the graph of $y = kx$ is k.

Step 1 Write the slope as a ratio.

$1 = $ ______ $\frac{\text{rise}}{\text{run}}$

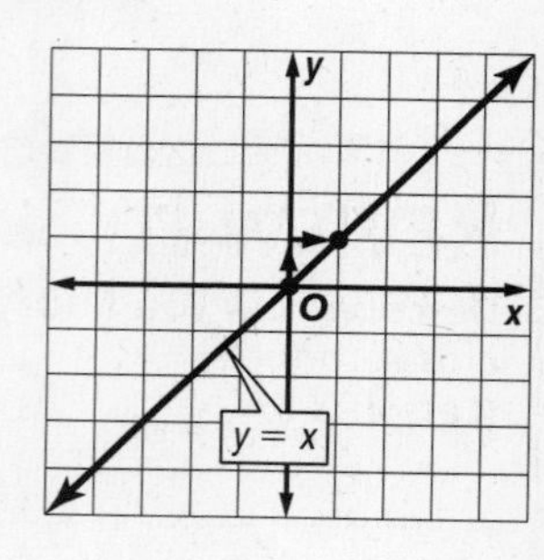

Step 2 Graph (0, 0).

Step 3 From the point (0, 0), move up

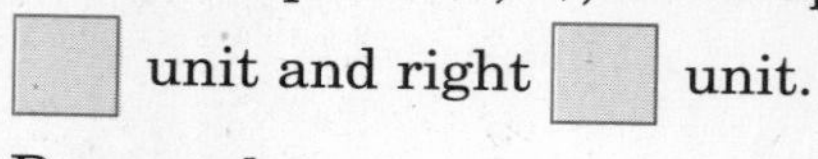

______ unit and right ______ unit.

Draw a dot.

Step 4 Draw a line containing the points.

EXAMPLE Direct Variation with $k < 0$

2 Graph $y = -\frac{3}{2}x$.

Step 1 Write the slope as a ratio.

$-\frac{3}{2} = $ ______ $\frac{\text{rise}}{\text{run}}$

Step 2 Graph (0, 0).

Step 3 From the point (0, 0), move

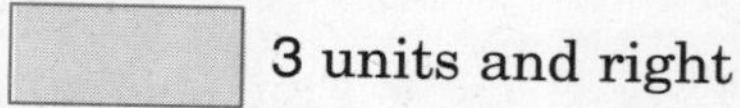

______ 3 units and right

______ units. Draw a dot.

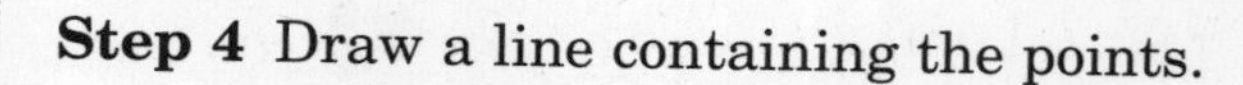

Step 4 Draw a line containing the points.

FOLDABLES™

ORGANIZE IT

Under the tab for Lesson 5-2, give an example of a direct variation equation and its graph.

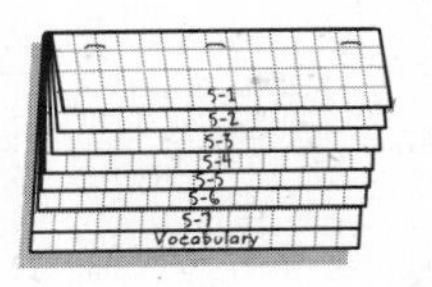

Your Turn **Graph each equation.**

a. $y = 2x$

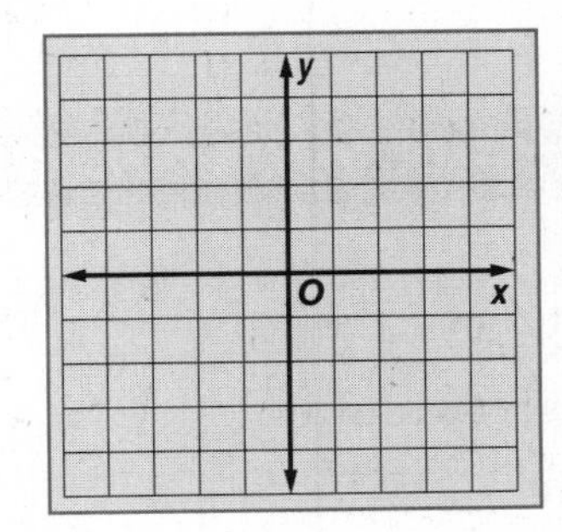

b. $y = -\frac{2}{3}x$

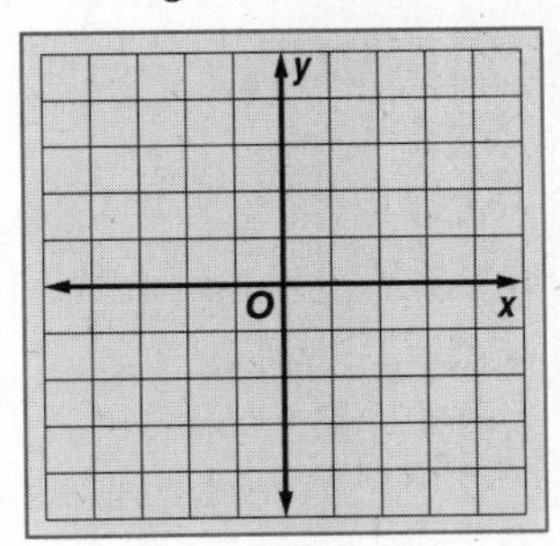

EXAMPLE **Write and Solve a Direct Variation Equation**

3 **Suppose y varies directly as x, and $y = 9$ when $x = -3$.**

a. Write a direct variation equation that relates x and y.

$y = kx$	Direct variation formula
______ $= k$ ______	Replace y with ______ and x with ______.
$\frac{9}{-3} = \frac{k(-3)}{-3}$	Divide each side by  ______.
______ $= k$	Simplify.

Therefore $y =$ ______.

b. Use the direct variation equation to find x when $y = 15$.

$y =$ ______	Direct variation formula
______ $=$ ______	Replace y with ______.
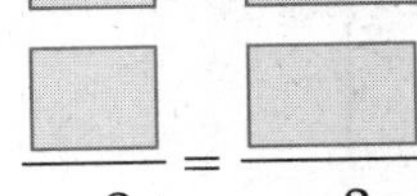 $\frac{\text{______}}{-3} = \frac{\text{______}}{-3}$	Divide each side by -3.
______ $= x$	Simplify.

Therefore, $x =$ ______ when $y =$ ______.

Your Turn **Suppose y varies directly as x, and $y = 15$ when $x = 5$.**

a. Write a direct variation equation that relates x and y.

______

b. Use the direct variation equation to find x when $y = -45$.

______

HOMEWORK ASSIGNMENT

Page(s):

Exercises:

5–3 Slope-Intercept Form

BUILD YOUR VOCABULARY (pages 112–113)

An equation of the form ______ is in the **slope-intercept form**.

WHAT YOU'LL LEARN

- Write and graph linear equations in slope-intercept form.
- Model real-world data with an equation in slope-intercept form.

EXAMPLE Write an Equation Given Slope and *y*-intercept

1 **Write an equation of the line whose slope is $\frac{1}{4}$ and whose *y*-intercept is −6.**

$y = mx + b$ Slope-intercept form

$y = \frac{1}{4}x - 6$ Replace *m* with ______ and *b* with ______.

Your Turn Write an equation of the line whose slope is 4 and whose *y*-intercept is 3.

______

EXAMPLE Write an Equation Given Two Points

2 **Write an equation of the line shown in the graph.**

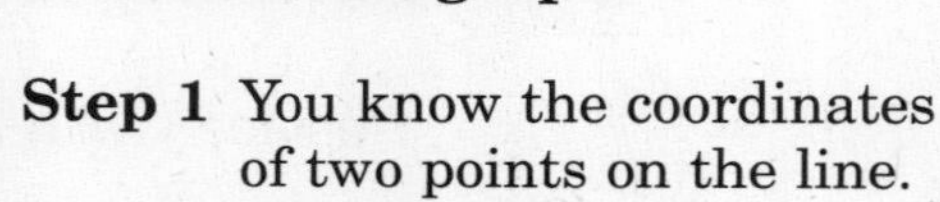

Step 1 You know the coordinates of two points on the line. Find the slope. Let $(x_1, y_1) = (0, -3)$ and $(x_2, y_2) = (2, 1)$.

KEY CONCEPT

Slope-Intercept Form The linear equation $y = mx + b$ is written in slope-intercept form, where *m* is the slope and *b* is the *y*-intercept.

$m = \frac{y_2 - y_1}{x_2 - x_1}$ $\frac{\text{rise}}{\text{run}}$

$m = \frac{\square - \square}{\square - \square}$ $x_1 = 0, x_2 = 2$; $y_1 = -3, y_2 = 1$

$m = \square$ or $\square$ Simplify.

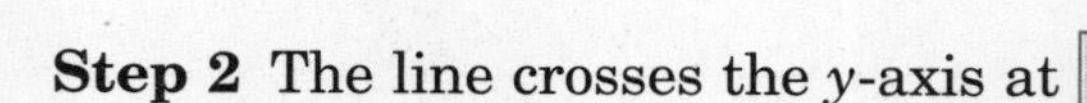

Step 2 The line crosses the *y*-axis at ______.

So, the *y*-intercept is ______.

Step 3 Finally, write the equation.

$y = mx + b$ Slope-intercept form

$y = 2x - 3$ Replace *m* with 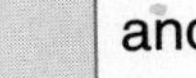 and *b* with .

The equation of the line is $y =$ ______.

Your Turn Write an equation of the line shown in the graph.

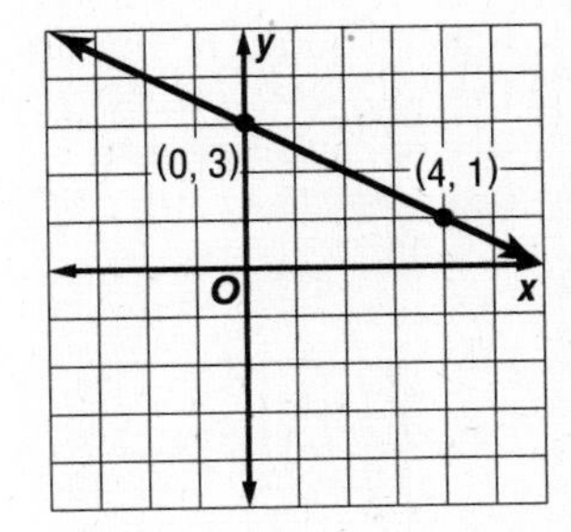

EXAMPLE Graph an Equation in Slope-Intercept Form

3 **Graph $y = 0.5x - 7$.**

Step 1 The *y*-intercept is ______.

So graph .

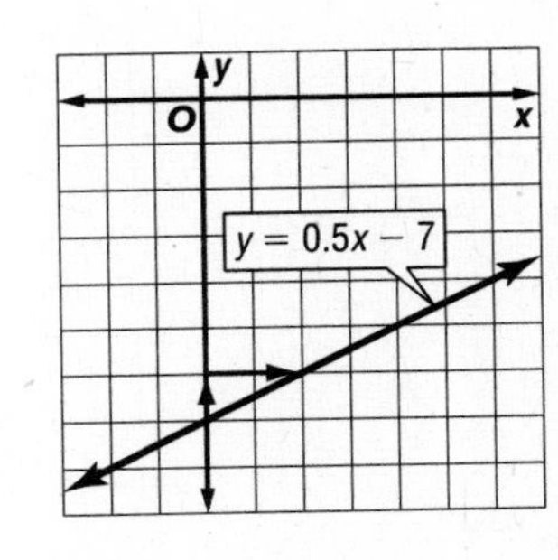

Step 2 The slope is 0.5 or ______.

From (0, −7), move up

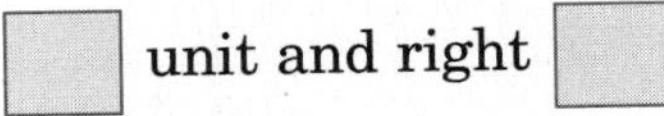

______ unit and right ______ units.

Draw a dot.

Step 3 Draw a line connecting the points.

WRITE IT

Which type of lines, vertical or horizontal, can be written in slope-intercept form?

______

______

______

______

Your Turn Graph $y = 2x - 4$.

EXAMPLE Graph an Equation in Standard Form

4 **Graph $5x + 4y = 8$.**

Step 1 Solve for y to find the slope-intercept form.

$5x + 4y = 8$	Original Equation
$5x + 4y - \square = 8 - \square$	Subtract $\square$ from each side.
$4y = 8 - 5x$	Simplify.
$4y = -5x + 8$	$8 - 5x = 8 + (-5x)$ or $-5x + 8$
$\frac{4y}{\square} = \frac{-5x + 8}{\square}$	Divide each side by $\square$.
$\frac{4y}{4} = \frac{-5x}{4} + \frac{8}{4}$	Divide each term in the numerator by 4.
$y = \square$	

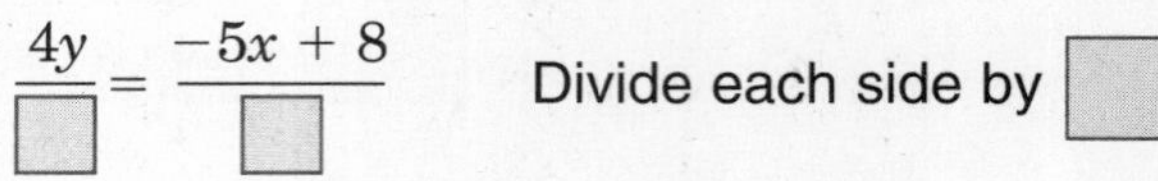

Step 2 The y-intercept of $y = -\frac{5}{4}x + 2$ is $\square$.

So graph $\square$.

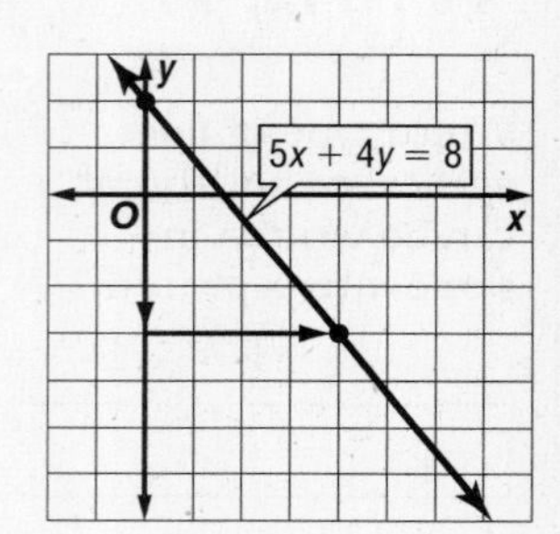

Step 3 The slope is $\square$. From (0, 2), move $\square$ 5 units and $\square$ 4 units. Draw a dot.

Step 4 Draw a line connecting the points.

Your Turn Graph $3x + 2y = 6$.

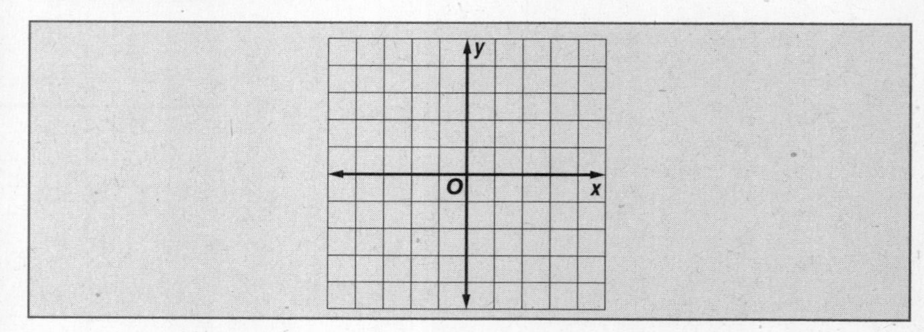

FOLDABLES™

ORGANIZE IT

On the tab for Lesson 5–3, write the slope-intercept form of a linear equation. Under the tab, describe how to use the slope and intercept to graph of $y = 4x + 3$.

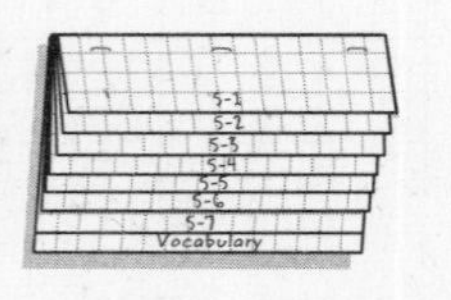

HOMEWORK ASSIGNMENT

Page(s):

Exercises:

5–4 Writing Equations in Slope-Intercept Form

What You'll Learn

- Write an equation of a line given the slope and one point on a line.
- Write and equation of a line given two points on the line.

BUILD YOUR VOCABULARY (pages 112–113)

When you use a linear equation to ______ values that are beyond the range of the data, you are using **linear extrapolation**.

FOLDABLES

ORGANIZE IT

Under the tab for Lesson 5-4, explain how to write an equation given slope and one point and given two points. Include examples.

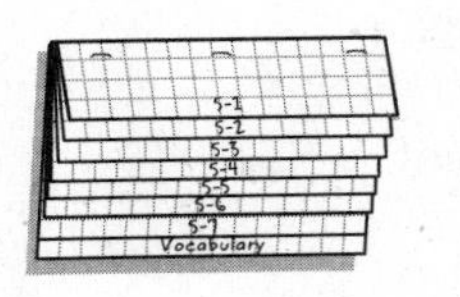

EXAMPLE Write an Equation Given Slope and One Point

1 Write an equation of a line that passes through (2, −3) with slope $\frac{1}{2}$.

Step 1 The line has slope $\frac{1}{2}$. To find the y-intercept, replace m with $\frac{1}{2}$ and (x, y) with $(2, -3)$ in the slope-intercept form. Then, solve for b.

$y = mx + b$	Slope-intercept form
$-3 = \frac{1}{2}(2) + b$	Replace m with $\frac{1}{2}$, y with −3, and x with 2.
$-3 = \square + b$	______.
$-3 - \square = 1 + b - \square$	Subtract ______ from each side.
$\square = b$	Simplify.

Step 2 Write the slope-intercept form using $m =$ ______ and b with ______.

$y = mx + b$	Slope-intercept form
$y = \square$	Replace m with ______ and b with ______.

The equation is $y =$ ______.

Your Turn Write an equation of a line that passes through (1, 4) and has a slope of −3.

______

EXAMPLE Write an Equation Given Two Points

2 The table of ordered pairs shows the coordinates of two points on the graph of a function. Write an equation that describes that function.

x	y
−3	−4
−2	−8

The table represents the ordered pairs

_____ and _____.

Step 1 Find the slope of the line containing the points. Let $(x_1, y_1) = (-3, -4)$ and $(x_2, y_2) = (-2, -8)$.

$m = \dfrac{y_2 - y_1}{x_2 - x_1}$ 

$m = \dfrac{\square - \square}{\square - \square}$ $x_1 = -3$, $x_2 = -2$, $y_1 = -4$, $y_2 = -8$

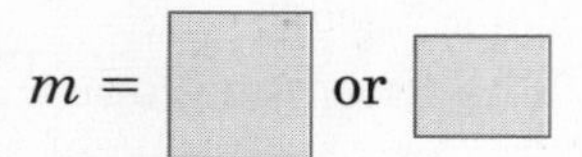

$m = \square$ or $\square$ Simplify.

Step 2 You know the slope and two points. Choose one point and find the y-intercept. In this case, we chose $(-3, -4)$.

$y = mx + b$ Slope-intercept form

$-4 = -4(-3) + b$ Replace m with 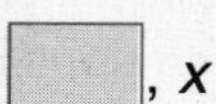, x with _____ and y with _____.

$-4 = 12 + b$ Multiply.

$-4 - \square = 12 + b - \square$ Subtract.

$\square = b$ Simplify.

Step 3 Write the slope-intercept form using

$m = \square$ and $b = \square$.

$y = mx + b$ Slope-intercept form

$y = \square$ Replace m with −4 and b with −16.

The equation is $y =$ _____.

Your Turn The table of ordered pairs shows the coordinates of two points on the graph of a function. Write an equation that describes the function.

x	y
−1	3
2	6

_____

HOMEWORK ASSIGNMENT

Page(s):

Exercises:

5–5 Writing Equations in Point-Slope Form

What You'll Learn

- Write the equation of a line in point-slope form.
- Write linear equations in different forms.

EXAMPLE Write an Equation Given Slope and a Point

1 **Write the point-slope form of an equation for a line that passes through (−2, 0) with slope $-\frac{3}{2}$.**

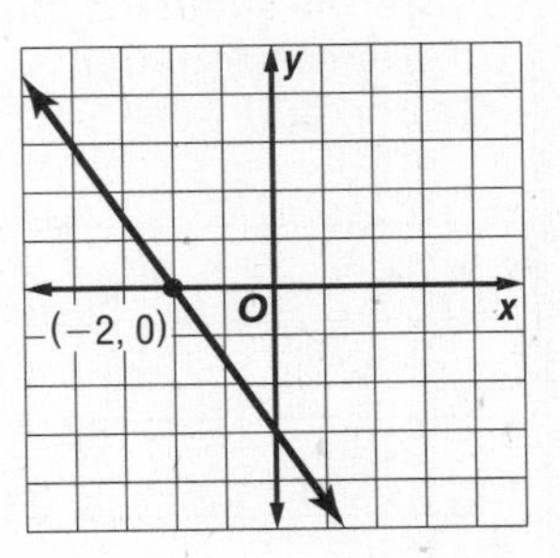

$y - y_1 = m(x - x_1)$ Point-slope form

$y - 0 = -\frac{3}{2}[x - (-2)]$ $(x_1, y_1) = (-2, 0)$

$y =$ ______ Simplify.

The equation is $y =$ ______.

Key Concepts

Point-Slope Form The linear equation $y - y_1 = m(x - x_1)$ is written in point-slope form, where (x_1, y_1) is a given point on a nonvertical line and m is the slope of the line.

Foldables

Under the tab for Lesson 5-5, draw a graph that goes through (3, −3) and has slope of −5. Explain how to find the equation of this line.

EXAMPLE Write an Equation of a Horizontal Line

2 **Write the point-slope form of an equation for a horizontal line that passes through (0, 5).**

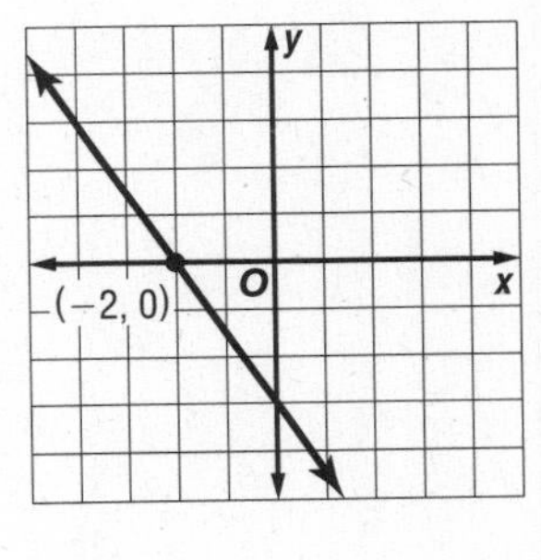

$y - y_1 = m(x - x_1)$ Point-slope form

______ = ______ $(x_1, y_1) = (0, 5)$

______ $= 0$ Simplify.

The equation is ______.

Your Turn

a. Write the point-slope form of an equation for a line that passes through (4, −3) with slope −2.

b. Write the point-slope form of an equation for a horizontal line that passes through (−3, −4).

EXAMPLE Write an Equation in Standard Form

3 **Write $y = \frac{3}{4}x - 5$ in standard form.**

In standard form, the variables are on the left side of the equation. A, B, and C are all integers.

$y = \frac{3}{4}x - 5$	Original equation
$\square(y) = \square\left(\frac{3}{4}x - 5\right)$	Multiply each side by $\square$ to eliminate the fraction.
$\square = \square - \square$	Distributive Property
$4y - \square = 3x - 20 - \square$	Subtract $\square$ from each side.
$3x - 4y = 20$	Simplify.

Your Turn Write $y - 3 = 2(x + 4)$ in standard form.

REVIEW IT

Write the standard form of a linear equation. *(Lesson 4-5)*

______

______

______

______

EXAMPLE Write an Equation in Slope-Intercept Form

4 **Write $y - 5 = \frac{4}{3}(x - 3)$ in slope-intercept form.**

In slope-intercept form, y is on the left side of the equation. The constant and x are on the right side.

$y - 5 = \frac{4}{3}(x - 3)$	Original equation
$y - 5 = \square$	Distributive Property
$y - 5 + \square = \frac{4}{3}x - 4 - \square$	Add $\square$ to each side.
$y = \square$	Simplify.

Your Turn Write $3x + 2y = 6$ in slope-intercept form.

HOMEWORK ASSIGNMENT

Page(s):

Exercises:

5–6 Geometry: Parallel and Perpendicular Lines

BUILD YOUR VOCABULARY (pages 112–113)

Lines in the same plane that do not ________ are called **parallel lines**.

Lines that intersect at ________ are called **perpendicular lines**.

WHAT YOU'LL LEARN

- Write an equation of the line that passes through a given point, parallel to a given line.
- Write an equation of the line that passes through a given point, perpendicular to a given line.

KEY CONCEPT

Parallel Lines in a Coordinate Plane Two nonvertical lines are parallel if they have the same slope. All vertical lines are parallel.

EXAMPLE Parallel Line Through a Given Point

1 **Write the slope-intercept form of an equation for the line that passes through (4, −2) and is parallel to the graph of $y = \frac{1}{2}x - 7$.**

The line parallel to $y = x - 7$ has the same slope, $\frac{1}{2}$. Replace m with $\frac{1}{2}$ and (x, y) with $(4, -2)$ in the point-slope form.

$y - y_1 = m(x - x_1)$	Point-slope form
$y - (-2) = \frac{1}{2}(x - 4)$	Replace m with $\frac{1}{2}$, y with −2, and x with 4.
______ $= \frac{1}{2}(x - 4)$	Simplify.
$y + 2 =$ ______	Distributive Property
$y + 2 -$ ____ $= \frac{1}{2}x - 2 -$ ____	Subtract ____ from each side.
$y =$ ______	Write the equation in slope-intercept form.

Your Turn Write the slope-intercept form of an equation for the line that passes through (2, 3) and is parallel to the graph of $y = \frac{1}{2}x - 1$.

EXAMPLE Determine Whether Lines are Perpendicular

KEY CONCEPT

Perpendicular Lines in a Coordinate Plane Two nonvertical lines are perpendicular if the product of their slopes is −1. That is, the slopes are *opposite reciprocals* of each other. Vertical lines and horizontal lines are also perpendicular.

2 GEOMETRY **The height of a trapezoid is measured on a segment that is perpendicular to a base. In a trapezoid $ARTP$, $\overline{RT}$ and $\overline{AP}$ are bases. Can $\overline{EZ}$ be used to measure the height of the trapezoid? Explain.**

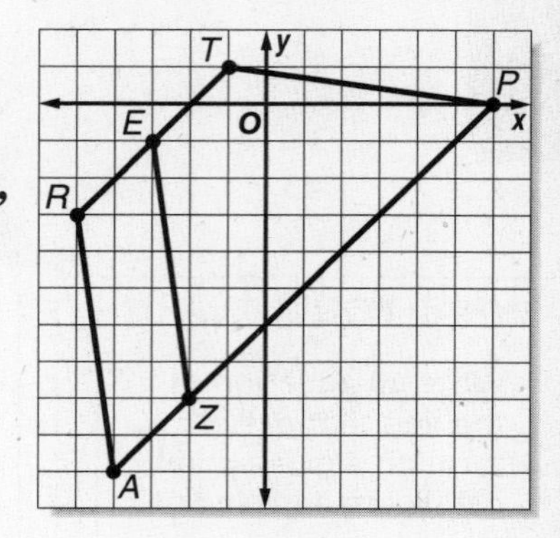

Find the slope of each segment.

Slope of $\overline{RT}$: $m = \frac{-3-1}{-5-(-1)}$ or ______

Slope of $\overline{AP}$: $m =$ ______ or ______

Slope of $\overline{EZ}$: $m =$ ______ or ______

The slope of $\overline{RT}$ and $\overline{AP}$ is ______ and the slope of $\overline{EZ}$ is ______.

$-7 \cdot 1 \neq$ ______. $\overline{EZ}$ is not ______ to $\overline{RT}$ and $\overline{AP}$, so it cannot be used to measure height.

Your Turn The graph shows the diagonals of a rectangle. Determine whether $\overline{JL}$ is perpendicular to $\overline{KM}$.

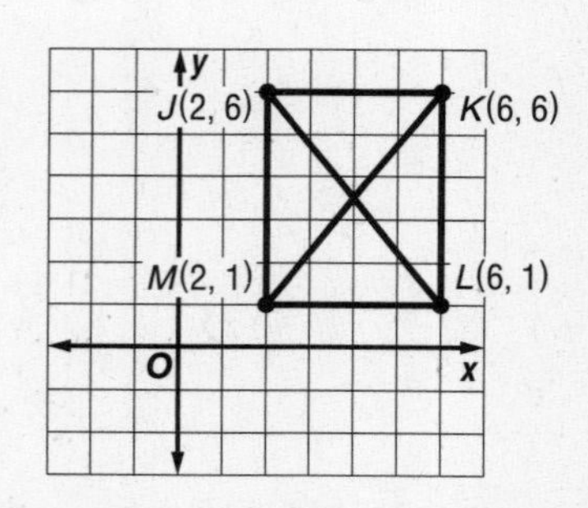

EXAMPLE Perpendicular Line Through a Given Point

3 Write the slope-intercept form for an equation of a line that passes through (4, −1) and is perpendicular to the graph of $7x - 2y = 3$.

Step 1 Find the slope of the given line.

$7x - 2y = 3$	Original equation
$7x - 2y - 7x = 3 - 7x$	Subtract ____ from each side.
____ $= -7 + 3$	Simplify.
$\frac{-2y}{-2} = \frac{-7x + 3}{-2}$	Divide each side by .
$y =$ ____	Simplify.

Step 2 The slope of the given line is ____. So, the slope of the line perpendicular to this line is the opposite reciprocal of $\frac{7}{2}$, or $-\frac{2}{7}$.

Step 3 Use the point-slope form to find the equation.

$y - y_1 = m(x - x_1)$	Point-slope form
$y - (-1) = -\frac{2}{7}(x - 4)$	$(x_1, y_1) = (4, -1)$, $m =$ ____
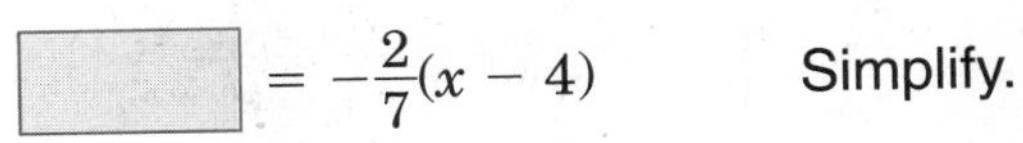 ____ $= -\frac{2}{7}(x - 4)$	Simplify.
$y + 1 = -\frac{2}{7}x + \frac{8}{7}$	Distributive Property
$y + 1$ ____ $= -\frac{2}{7}x + \frac{8}{7}$ ____	Subtract.
$y =$ ____	Simplify.

Your Turn Write the slope-intercept form for an equation of a line that passes through (−3, 6) and is perpendicular to the graph of $3x + 2y = 6$.

HOMEWORK ASSIGNMENT

Page(s): ____

Exercises: ____

5–7 Statistics: Scatter Plots and Lines of Fit

BUILD YOUR VOCABULARY (pages 112–113)

A **scatter plot** is a graph in which two sets of data are plotted as ordered pairs in a coordinate plane.

A ________ correlation exists when as x increases, y increases.

A ________ correlation exists when as x increases, y decreases.

A **line of fit** describes the trend of the data.

When you use a linear equation to predict the values that are inside the range of the data this is called **linear interpolation**.

WHAT YOU'LL LEARN

- Interpret points on a scatter plot.
- Write equations for lines of fit.

EXAMPLE Analyze Scatter Plots

1 **Determine whether the graph shows a *positive correlation*, a *negative correlation*, or *no correlation*. If there is a positive or negative correlation describe it.**

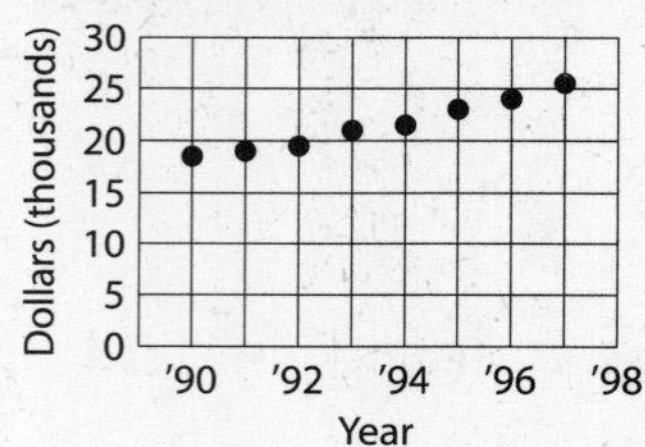

The graph shows average personal income for U.S. citizens.

The graph shows a ________ correlation. With each year, the average personal income rose.

Your Turn The graph shows the percentage of voter participation in Presidential Elections. Determine whether the graph shows a *positive correlation*, a *negative correlation*, or *no correlation*. If there is a positive or negative correlation describe it.

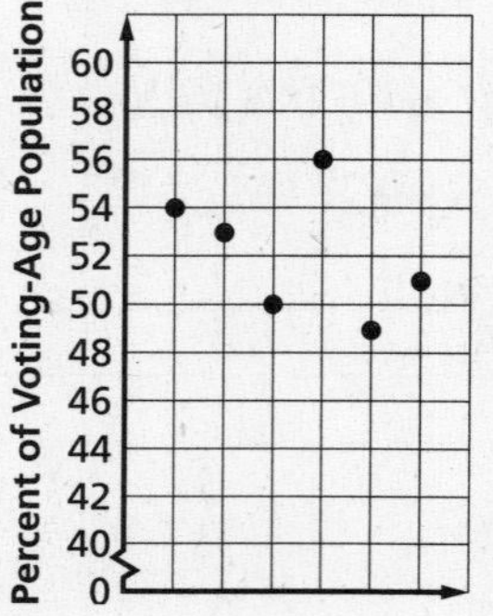

____

FOLDABLES ORGANIZE IT

Write the definitions of the vocabulary builder words under the vocabulary tab.

REMEMBER IT

When graphing, the line of fit is only an approximation.

EXAMPLE Find a Line of Fit

2 The table shows the world's population growing at a rapid rate.

Year	Population (millions)
1650	500
1850	1000
1930	2000
1975	4000
1998	5900

a. Draw a scatter plot and determine what relationship exists, if any, in the data.

Let the independent variable x be the year and let the dependent variable y be the population (in millions).

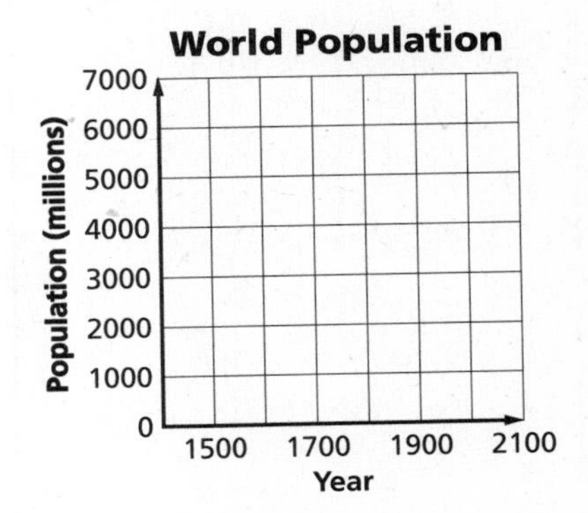

The scatter plot seems to indicate that as the year ______, the population ______. There is a ______ correlation between the two variables.

b. Draw a line of fit for the scatter plot.

No one line will pass through all of the data points. Draw a ______ that passes ______ to the points. A line is shown in the scatter plot.

c. Write the slope intercept form of an equation for the equation for the line of fit.

The line of fit shown passes through the data points (1850, 1000) and (1998, 5900).

Step 1 Find the slope.

$m = \frac{y_2 - y_1}{x_2 - x_1}$ Slope formula

$m =$ ______ Let $(x_1, y_1) = (1850, 1000)$ and $(x_2, y_2) = (1998, 5900)$.

$m =$ ______ or ≈ 33.1 Simplify.

Step 2 Use $m = 33.1$ and either the point-slope form or the slope-intercept form to write the equation. You can use either data point. We chose (1850, 1000).

Point-slope form	**Slope-intercept form**
$y - y_1 = m(x - x_1)$	$y = mx + b$
$y - 1000 \approx 33.1\,(x - 1850)$	$1000 \approx 33.1(1850) + b$
$y - 1000 \approx 33.1x - 61{,}235$	$1000 \approx 61{,}235 + b$
$y \approx$ ______	$-60{,}235 \approx b$
	$y \approx$ ______

The equation of the line is $y \approx$ ______.

Your Turn **The table shows the number of bachelor's degrees received since 1988.**

Years since 1988	2	4	6	8	10
Bachelor's Degrees Received (thousands)	1051	1136	1169	1165	1184

Source: National Center for Education Statistics

a. Draw a scatter plot and determine what relationship exists, if any, in the data.

b. Draw a line of best fit for the scatter plot

c. Write the slope-intercept form of an equation for the line of fit.

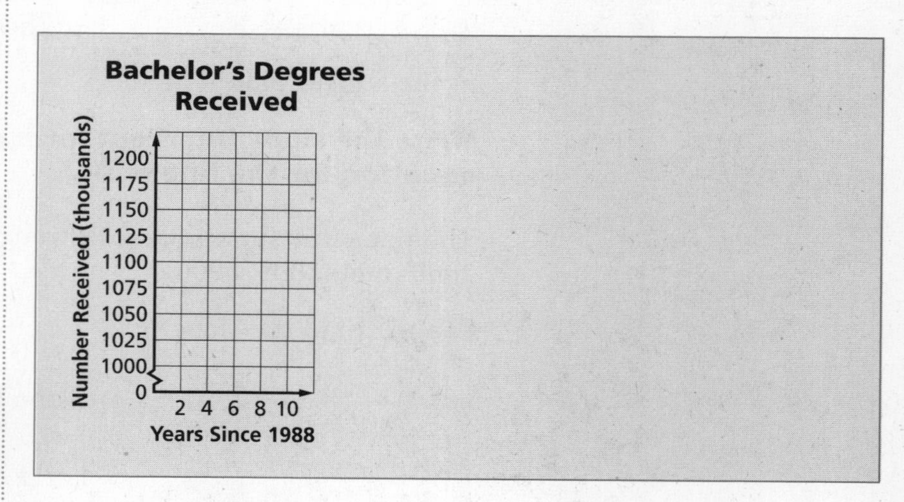

HOMEWORK ASSIGNMENT

Page(s): ______

Exercises: ______

CHAPTER 5

BRINGING IT ALL TOGETHER

STUDY GUIDE

FOLDABLES	VOCABULARY PUZZLEMAKER	BUILD YOUR VOCABULARY
Use your **Chapter 5 Foldable** to help you study for your chapter test.	To make a crossword puzzle, word search, or jumble puzzle of the vocabulary words in Chapter 5, go to: www.glencoe.com/sec/math/t_resources/free/index.php	You can use your completed **Vocabulary Builder** (pages 112–113) to help you solve the puzzle.

5-1 Slope

Describe each type of slope.

	Type of Slope	Description of Graph
1.	positive	
2.	negative	
3.	zero	

5-2 Slope and Direct Variation

For each situation, write an equation with the proper constant of variation.

4. The distance d varies directly as time t, and a cheetah can travel 88 feet in 1 second.

5. The perimeter p of a pentagon with all sides of equal length varies directly as the length s of a side of the pentagon. A pentagon has 5 sides.

5-3 Slope-Intercept Form

6. Fill in the boxes with the correct words to describe what m and b represent.

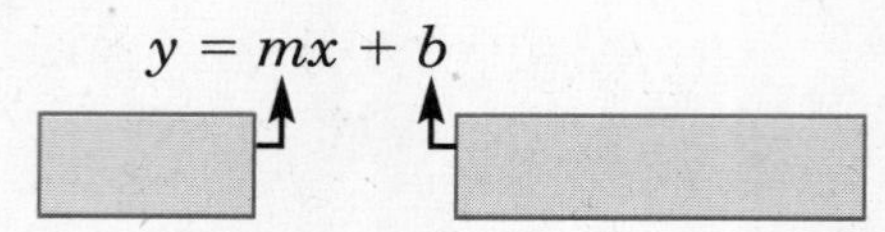

7. What are the slope and y-intercept of a vertical line?

8. What are the slope and y-intercept of a horizontal line?

5-4 Writing Equations in Slope-Intercept Form

9. Suppose you are given that a line goes through (2, 5) and has a slope of −2. Use this information to complete the following equation.

10. What must you first do if you are not given the slope in the problem?

Write an equation of the line that passes through each pair of points.

11. $(-5, 4)$, $m = -3$

12. $(-2, -3)$, $(4, 5)$

5-5 Writing Equations in Point-Slope Form

13. In the formula $y - y_1 = m(x - x_1)$, what do x_1 and y_1 represent?

Complete the chart.

	Form of Equation	Formula	Example
14.	slope-intercept		$y = 3x + 2$
15.	point-slope		$y - 2 = 4(x + 3)$
16.	standard		$3x - 5y = 15$

5-6 Geometry: Parallel and Perpendicular Lines

Write the slope-intercept form for an equation of the line that passes through the given point and is either parallel or perpendicular to the graph of the equation.

17. $(-2, 2)$, $y = 4x - 2$ (parallel)

18. $(4, 2)$, $y = \frac{1}{2}x + 1$ (perpendicular)

5-7 Statistics: Scatter Plots and Lines of Fit

19. What is a *line of fit*? How many data points fall on the line of fit?

ARE YOU READY FOR THE CHAPTER TEST?

Visit **www.algebra1.com** to access your textbook, more examples, self-check quizzes, and practice tests to help you study the concepts in Chapter 5.

Check the one that applies. Suggestions to help you study are given with each item.

☐ **I completed the review of all or most lessons without using my notes or asking for help.**

- You are probably ready for the Chapter Test.
- You may want take the Chapter 5 Practice Test on page 313 of your textbook as a final check.

☐ **I used my Foldable or Study Notebook to complete the review of all or most lessons.**

- You should complete the Chapter 5 Study Guide and Review on pages 308–312 of your textbook.
- If you are unsure of any concepts or skills, refer to the specific lesson(s).
- You may also want to take the Chapter 5 Practice Test on page 313.

☐ **I asked for help from someone else to complete the review of all or most lessons.**

- You should review the examples and concepts in your Study Notebook and Chapter 5 Foldable.
- Then complete the Chapter 5 Study Guide and Review on pages 308–312 of your textbook.
- If you are unsure of any concepts or skills, refer to the specific lesson(s).
- You may also want to take the Chapter 5 Practice Test on page 313.

Student Signature

Parent/Guardian Signature

Teacher Signature

Solving Linear Inequalities

FOLDABLES™ Use the instructions below to make a Foldable to help you organize your notes as you study the chapter. You will see Foldable reminders in the margin of this Interactive Study Notebook to help you in taking notes.

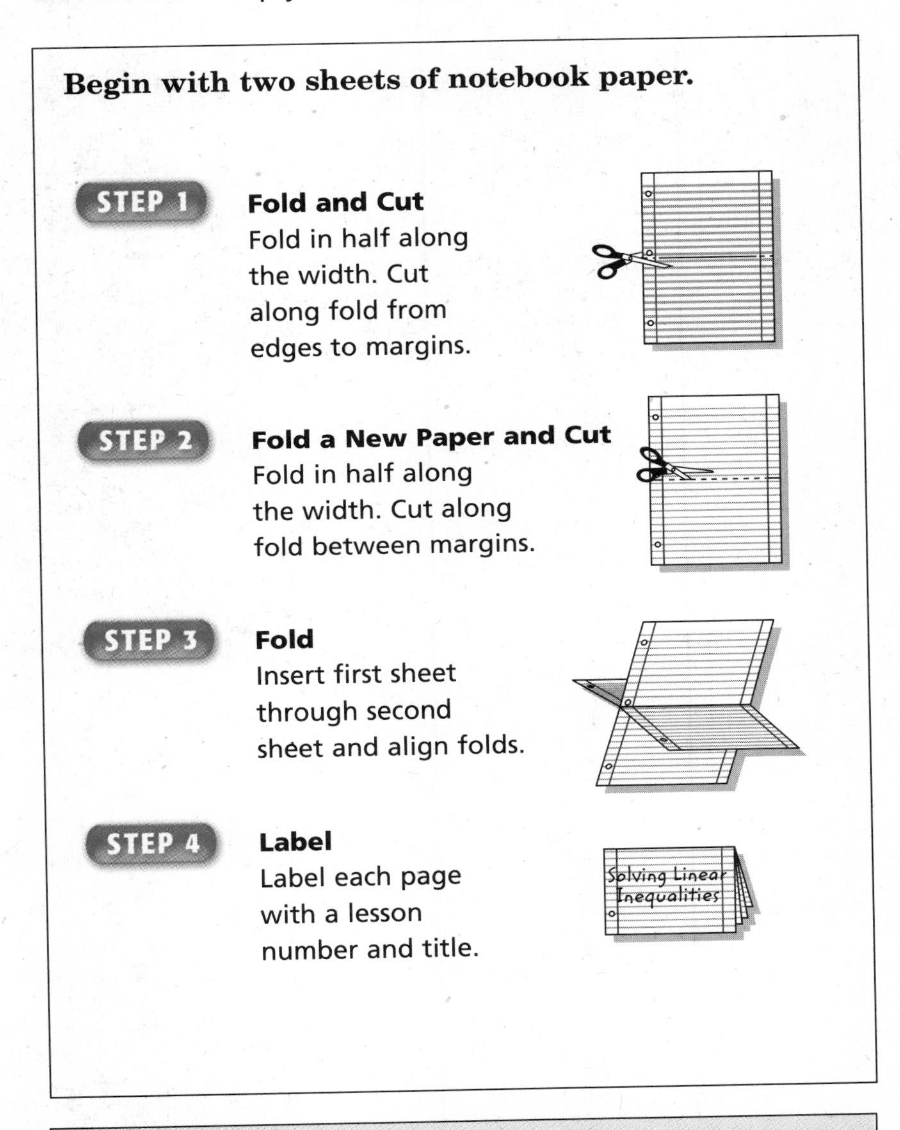

Begin with two sheets of notebook paper.

STEP 1 **Fold and Cut**
Fold in half along the width. Cut along fold from edges to margins.

STEP 2 **Fold a New Paper and Cut**
Fold in half along the width. Cut along fold between margins.

STEP 3 **Fold**
Insert first sheet through second sheet and align folds.

STEP 4 **Label**
Label each page with a lesson number and title.

NOTE-TAKING TIP: When you take notes, write down the math problem and each step in the solution using math symbols. Next to each step, write down, in your own words, exactly what you are doing.

CHAPTER 6

BUILD YOUR VOCABULARY

This is an alphabetical list of new vocabulary terms you will learn in Chapter 6. As you complete the study notes for the chapter, you will see Build Your Vocabulary reminders to complete each term's definition or description on these pages. Remember to add the textbook page number in the second column for reference when you study.

Vocabulary Term	Found on Page	Definition	Description or Example
Addition Property of Inequalities			
boundary			
compound inequality			
Division Property of Inequalities			
half-plane			

Vocabulary Term	Found on Page	Definition	Description or Example
intersection			
Multiplication Property of Inequalities			
set-builder notation			
Subtraction Property of Inequalities			
union			

6–1 Solving Inequalities by Addition and Subtraction

WHAT YOU'LL LEARN

- Solve linear inequalities by using addition.
- Solve linear inequalities by using subtraction.

KEY CONCEPT

Addition Property of Inequalities If any number is added to each side of a true inequality, the resulting inequality is also true.

EXAMPLE Solve by Adding

1 Solve $s - 12 > 65$. Then check your solution.

$s - 12 > 65$	Original inequality
$s - 12 + \square > 65 + \square$	Add 12 to each side.
$s > \square$	This means all numbers greater than $\square$.

Check Substitute 77, a number less than 77, and a number greater than 77.

Let $s = 77$.	Let $s = 64$.	Let $s = 80$.
$77 - 12 \overset{?}{>} 65$	$64 - 12 \overset{?}{>} 65$	$80 - 12 \overset{?}{>} 65$
$65 \not> 65$	$52 \not> 65$	$68 > 65$ ✔

The solution is the set $\square$.

Your Turn Solve $k - 4 < 10$. Then check your solution.

EXAMPLE Graph the Solution

2 Solve $12 \geq y - 9$. Then graph it on a number line.

$12 \geq y - 9$	Original inequality
$12 + \square \geq y - 9 + \square$	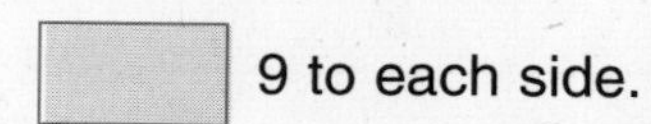 9 to each side.
$\square \geq y$	Simplify.

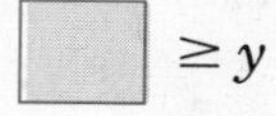

Since $\square \geq y$ is the same as $y \leq \square$, the solution set is $\{y \mid y \leq \square\}$.

EXAMPLE Solve by Subtracting

3 **Solve $q + 23 < 14$. Then graph the solution.**

$q + 23 < 14$ — Original inequality

$q + 23 - \square < 14 - \square$ — Subtract 23 from each side.

$q <$ 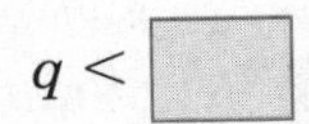 — Simplify.

The solution set is ______.

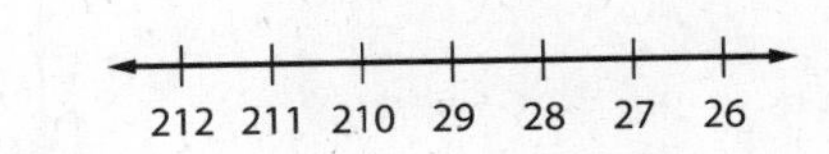

KEY CONCEPT

Subtraction Property of Inequalities If any number is subtracted from each side of a true inequality, the resulting inequality is also true.

FOLDABLES Include the Addition and Subtraction Properties of Inequalities in your Foldable. Be sure to show examples.

Your Turn **Solve each inequality. Then graph the solution.**

a. $8 > x - 2$

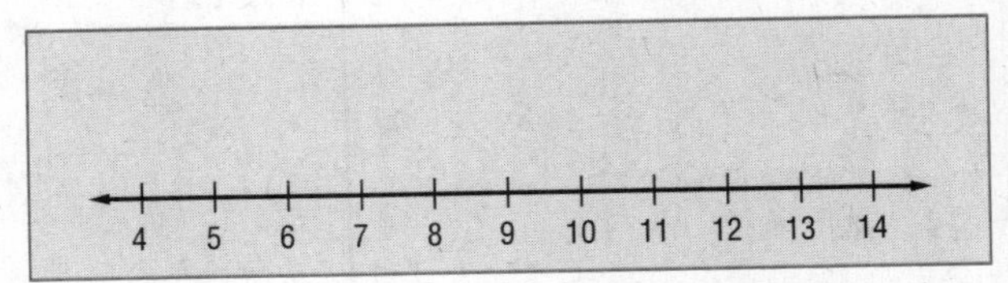

b. $m + 15 > 13$

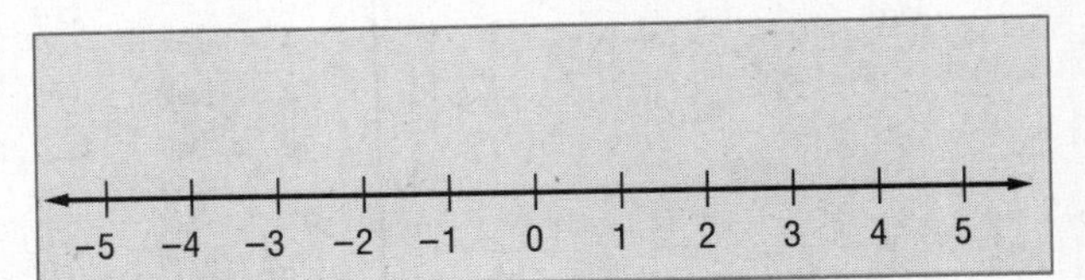

EXAMPLE Variables on Both Sides

4 **Solve $12n - 4 \leq 13n$. Then graph the solution.**

$12n - 4 \leq 13n$ — Original inequality

$12n - 4 - \square \leq 13n - \square$ — Subtract.

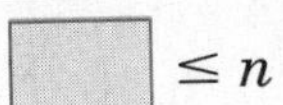 $\leq n$ — Simplify.

Since $-4 \leq n$ is the same as $n \geq -4$, the solution set is ______.

−6 −5 −4 −3 −2 −1 0

Your Turn Solve $3p - 6 \geq 4p$. Then graph the solution.

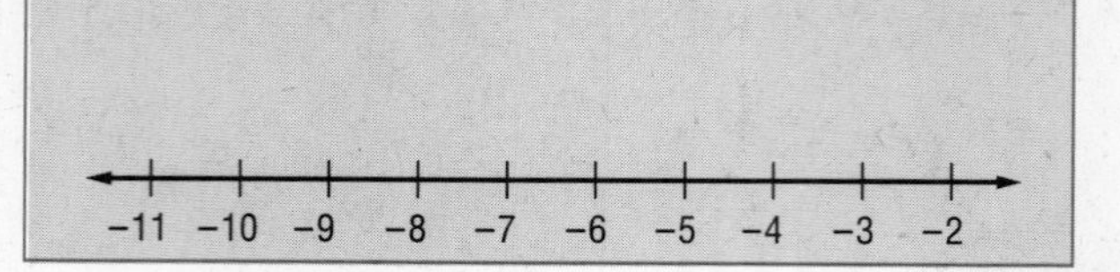

EXAMPLE Write and Solve an Inequality

5 **Write an inequality for the sentence below. Then solve the inequality.**

Seven times a number is greater than 6 times that number minus two.

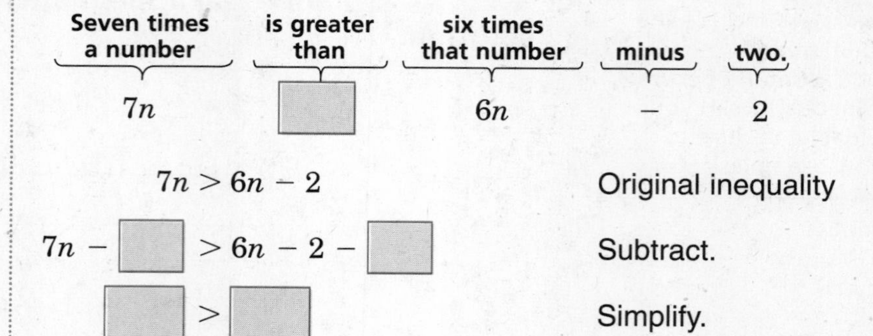

$7n > 6n - 2$ Original inequality

$7n - \square > 6n - 2 - \square$ Subtract.

$\square > \square$ Simplify.

The solution set is $\square$.

Your Turn Write an inequality for the sentence below. Then solve the inequality.

Three times a number is less than two times that number plus 5.

HOMEWORK ASSIGNMENT

Page(s):

Exercises:

6–2 Solving Inequalities by Multiplication and Division

WHAT YOU'LL LEARN

- Solve linear inequalities by using multiplication.
- Solve linear inequalities by using division.

KEY CONCEPTS

Multiplying by a Positive Number If each side of a true inequality is multiplied by the same positive number, the resulting inequality is also true

Multiplying by a Negative Number If each side of a true inequality is multiplied by the same negative number, the direction of the inequality symbol must be reversed so that the resulting inequality is also true.

EXAMPLE Multiply by a Positive Number

1 **Solve $\frac{g}{3} < 12$.**

$\frac{g}{3} < 12$	Original inequality.
$\square \frac{g}{3} < 12 \square$	Multiply each side by ____.
$g < \square$	Simplify.

The solution set is ____.

EXAMPLE Multiply by a Negative Number

2 **Solve $-\frac{3}{4}d \geq 6$.**

$-\frac{3}{4}d \geq 6$	Original inequality.
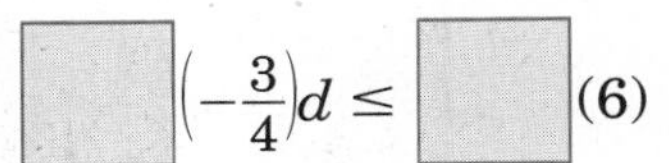 $\square\left(-\frac{3}{4}\right)d \leq \square(6)$	Multiply each side by ____ and change ____ to ____.
$d \leq \square$	Simplify.

The solution set is ____.

EXAMPLE Write and Solve an Inequality

3 **Write an inequality for the sentence *Four fifths of a number is at most twenty.* Then solve the inequality.**

Four-fifths	of	a number	is at most	twenty.
____	____	r	____	____

$\frac{4}{5}r \leq 20$	Original inequality
$\square \frac{4}{5}r \leq \square 20$	Multiply each side by ____.
$r \leq \square$	Simplify.

The solution set is ____.

Your Turn **Solve each inequality.**

a. $\frac{n}{6} \leq 2$

b. $-\frac{2}{3}m \geq 6$

c. Write an inequality for the sentence *Two-thirds of a number is less than 12*. Then solve the inequality.

EXAMPLE **Divide by a Positive Number**

4 **Solve $12s \geq 60$.**

$12s \geq 60$	Original inequality
$\frac{12s}{\square} \geq \frac{60}{\square}$	Divide each side by ☐ and do not change the direction of the inequality sign.
$s \geq \square$	Simplify.

The solution set is ☐.

EXAMPLE **Divide by a Negative Number**

5 **Solve $-8q < 136$.**

Method 1 Divide.

$-8q < 136$	Original inequality
$\frac{-8q}{\square} > \frac{136}{\square}$	Divide each side by ☐ and change $<$ to $>$.
$q > \square$	Simplify.

The solution set is ☐.

Your Turn **Solve each inequality.**

a. $15p < 60$

b. $-4z > 64$

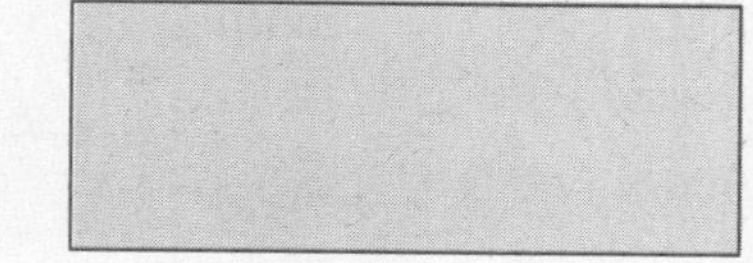

Key Concepts

Dividing by a Positive Number If each side of a true inequality is divided by the same positive number, the resulting inequality is also true.

Dividing by a Negative Number If each side of a true inequality is divided by the same negative number, the direction of the inequality symbol must be *reversed* so that the resulting inequality is also true.

FOLDABLES™ Be sure to write the Multiplication and Division Properties of Inequalities in your Foldable.

Homework Assignment

Page(s):

Exercises:

6–3 Solving Multi-Step Inequalities

WHAT YOU'LL LEARN

- Solve linear inequalities involving more than one operation.
- Solve linear inequalities involving the Distributive Property.

EXAMPLE Solve a Real-World Problem

1 SCIENCE **The inequality $F > 212$ represents the temperature in degrees Fahrenheit for which water is a gas (steam). Similarly, the inequality $\frac{9}{5}C + 32 > 212$ represents the temperature in degrees Celsius for which water is a gas. Find the temperature in degrees Celsius for which water is a gas.**

$\frac{9}{5}C + 32 > 212$	Original inequality
$\frac{9}{5}C + 32 - \square > 212 - \square$	Subtract $\square$ from each side.
$\frac{9}{5}C > \square$	Simplify.
$\square\,\frac{9}{5}C > \square\,180$	Multiply each side by $\square$.
$C > \square$	Simplify.

Water will be a gas for all temperatures greater than 100°C.

EXAMPLE Inequality Involving a Negative Coefficient

2 **Solve $13 - 11d \geq 79$.**

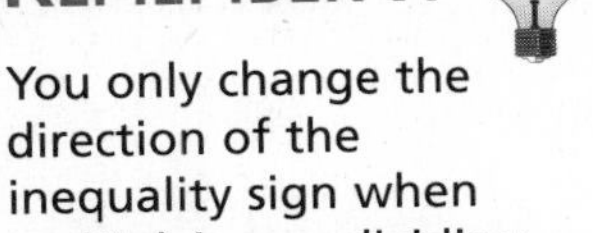

REMEMBER IT

You only change the direction of the inequality sign when multiplying or dividing both sides by a negative number.

$13 - 11d \geq 79$	Original inequality
$13 - 11d - \square \geq 79 - \square$	Subtract $\square$ from each side.
$\square \geq \square$	Simplify.
$\square \leq \square$	Divide each side by $\square$ and change $\geq$ to $\leq$.
$d \leq \square$	Simplify.

The solution set is $\square$.

Your Turn

a. The boiling point of helium is $-452°F$. Solve $\frac{9}{5}C + 32 > -452$ to find the temperatures in degrees Celsius for which helium is a gas.

b. Solve $-8y + 3 > -5$.

FOLDABLES™

ORGANIZE IT

In Lesson 6-3 of your journal, explain how solving an inequality is different from solving an equation.

EXAMPLE Distributive Property

3 **Solve $8 - (c + 3) \leq 6c + 3(2 - c)$.**

$8 - (c + 3) \leq 6c + 3(2 - c)$	Original inequality
$8 - c - 3 \leq 6c + \square$	Distributive Property
$\square \leq \square$	Combine like terms.
$5 - c + \square \leq 3c + 6 + \square$	Add $\square$ on each side.
$\square \leq 4c + 6$	Simplify.
$5 - \square \leq 4c + 6 - \square$	Subtract $\square$ to each side.
$-1 \leq 4c$	Simplify.
$\square \leq c$	Divide each side by $\square$.

Since $-\frac{1}{4} \leq c$ is the same as $c \; \square \; -\frac{1}{4}$, the solution set is $\square$.

REMEMBER IT

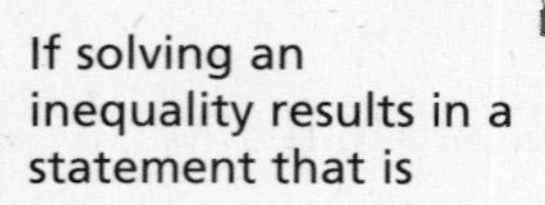

If solving an inequality results in a statement that is

- true, the solution is all real numbers.
- false, the solution is the empty set, Ø.

Your Turn $3p - 2(p - 4) < p - (2 - 3p)$

HOMEWORK ASSIGNMENT

Page(s):

Exercises:

6–4 Solving Compound Inequalities

What You'll Learn

- Solve compound inequalities containing the word *and* and graph their solution sets.
- Solve compound inequalities containing the word *or* and graph their solution sets.

Build Your Vocabulary (pages 136–137)

Two or more inequalities that are connected by the words ______ or ______ is a **compound inequality.**

The graph of a compound inequality containing ______ is the **intersection** of the graphs of the two inequalities.

The graph of a compound inequality containing ______ is the **union** of the graphs of the two inequalities.

Example Graph an Intersection

1 Graph the solution set of $y \geq 5$ and $y < 12$.

Graph $y \geq 5$.

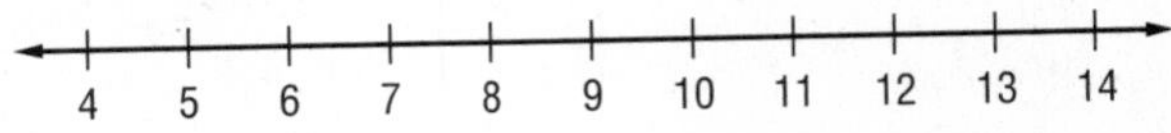

Graph $y < 12$.

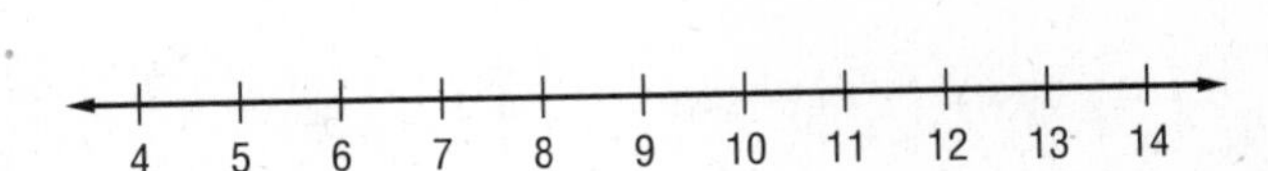

Find the ______.

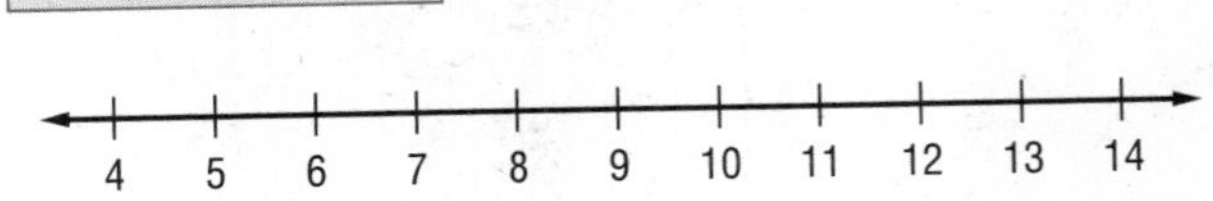

The solution set is ______. Note that the graph of $y \geq 5$ includes the point 5. The graph of $y < 12$ does *not* include ______.

Your Turn Graph the solution set of $y > 6$ and $y \leq 10$.

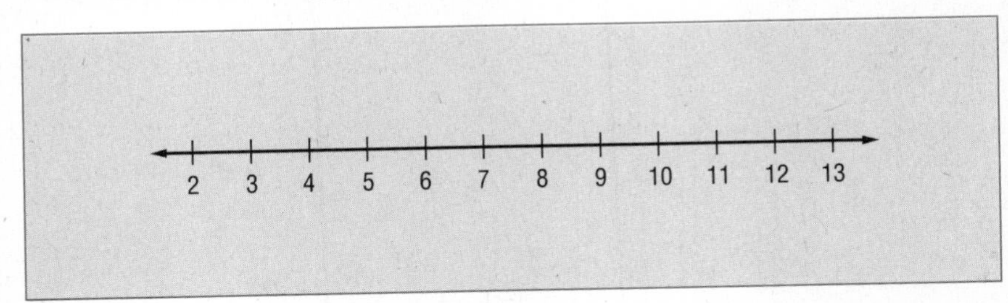

Write It

Use the words to describe the compound inequality $25 < x < 30$.

______

______

______

______

FOLDABLES™

ORGANIZE IT

In Lesson 6-4 of your journal, explain how the solution of an intersection is different from the solution of a union.

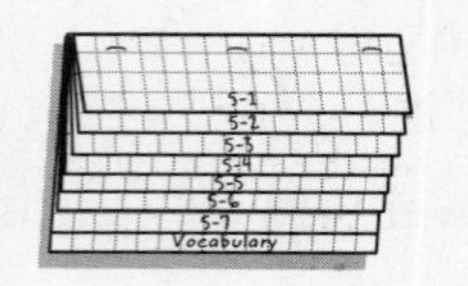

EXAMPLE Solve and Graph an Intersection

2 Solve $7 < z + 2 \leq 11$. Then graph the solution set.

First express $7 < z + 2 \leq 11$ using *and*. Then solve each inequality.

$7 < z + 2$	and	$z + 2 \leq 11$
7 ______ $< z + 2$ ______		$z + 2$ ______ ≤ 11 ______
______ $< z$		$z \leq$ ______

The solution set is the ______ of the two graphs.

Graph $5 < z$ or $z > 5$

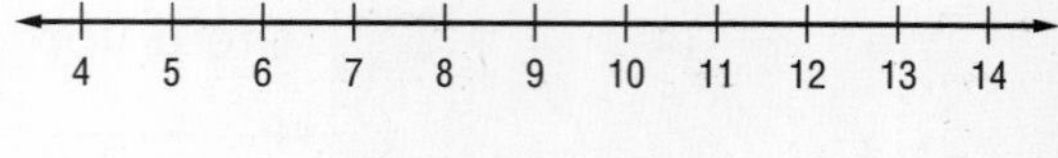

Graph $z \leq 9$

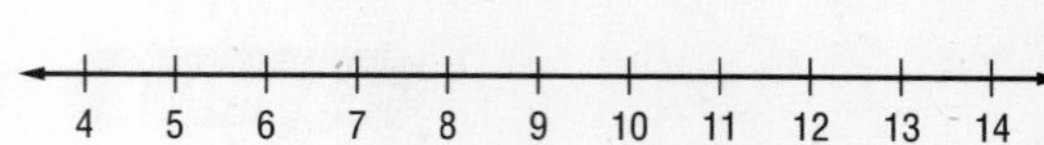

Find the ______.

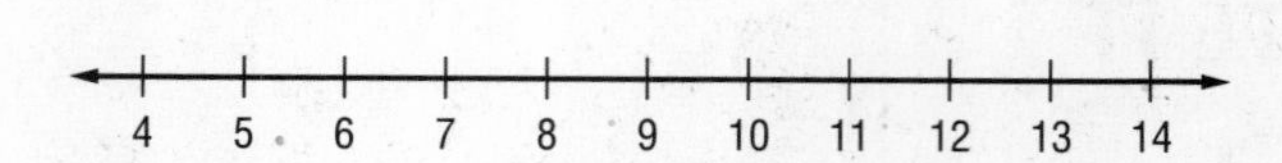

The solution set is ______.

Your Turn Solve $-3 < x - 2 < 5$. Then graph the solution set.

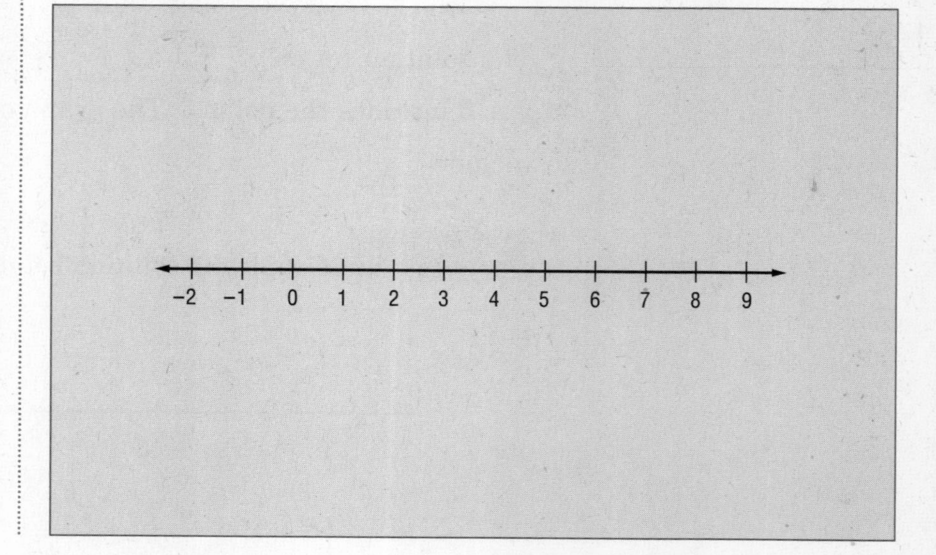

EXAMPLE Solve and Graph a Union

3 **Solve $4k - 7 \leq 25$ or $12 - 9k \geq 30$. Then graph the solution set.**

$4k - 7 \leq 25$ or $12 - 9k \geq 30$

$4k - 7$ ☐ ≤ 25 ☐ $12 - 9k$ ☐ ≥ 30 ☐

$4k \leq 32$ $-9k \geq 18$

$\frac{4k}{4} \leq \frac{32}{4}$ $\frac{-9k}{-9} \leq \frac{18}{-9}$

$k \leq$ ☐ $k \leq$ ☐

Graph $k \leq 8$.

–4 –3 –2 –1 0 1 2 3 4 5 6 7 8 9 10

Graph $k \leq -2$.

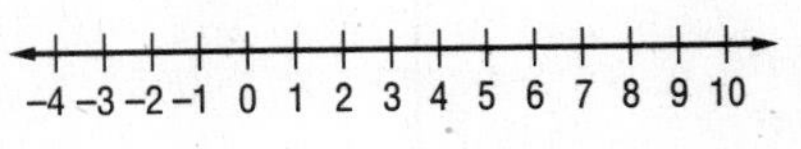

–4 –3 –2 –1 0 1 2 3 4 5 6 7 8 9 10

Notice that the graph of $k \leq 8$ contains ☐ point in the

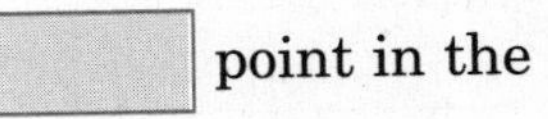

graph of $k \leq -2$. So, the ☐ is the graph of $k \leq 8$. The

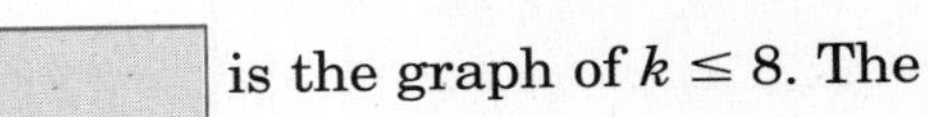

solution set is ☐.

Your Turn Solve $-2x + 5 < 15$ or $5x + 15 > 20$. Then graph the solution set.

–8 –7 –6 –5 –4 –3 –2 –1 0 1 2 3

HOMEWORK ASSIGNMENT

Page(s):

Exercises:

6–5 Solve an Absolute Value Equation

WHAT YOU'LL LEARN

- Solve absolute value equations.
- Solve absolute value inequalities.

REVIEW IT

Why is the absolute value of a number always greater than or equal to zero? *(Lesson 2-1)*.

____

____

____

____

____

EXAMPLE Solve an Absolute Value Equation

1 Solve $|b + 6| = 5$.

Method 1 Graphing

$|b + 6| = 5$ means that the distance between b and ____ is ____ units. To find b on the number line, start at ____ and move ____ units in ____ direction.

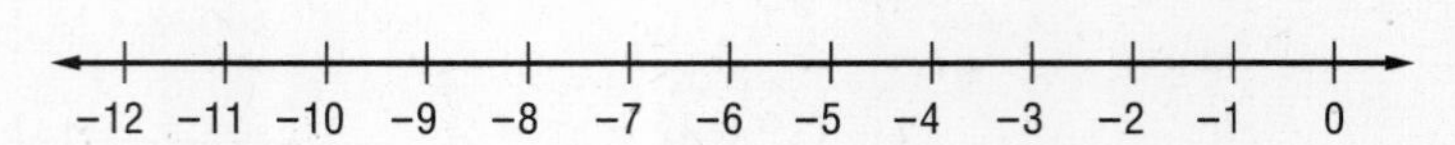

The distance from -6 to -11 is ____ units. The distance from -6 to -1 is ____ units. The solution set is ____.

Method 2 Compound Sentence

Write $|b + 6| = 5$ as $b + 6 = 5$ or $b + 6 = -5$.

Case 1		Case 2
$b + 6 = 5$	Original inequality	$b + 6 = -5$
$b + 6 - 6 = 5 - 6$	Subtract 6 from each side.	$b + 6 - 6 = -5 - 6$
$b =$ ____	Simplify.	$b =$ ____

The solution set is ____.

Your Turn Solve $|b - 5| = 7$.

EXAMPLE Solve an Absolute Value Inequality (<)

2 Solve $|s - 3| \leq 12$. Then graph the solution set.

Write $|s - 3| \leq 12$ as $s - 3 \leq 12$ and $s - 3 \geq -12$.

Case 1		Case 2
$s - 3 \leq 12$	Original inequality	$s - 3 \geq -12$
$s - 3 + 3 \leq 12 + 3$	Add 3 to each side.	$s - 3 + 3 \geq -12 + 3$
$s \leq$ ☐	Simplify.	$s \geq$ ☐

The solution set is ☐.

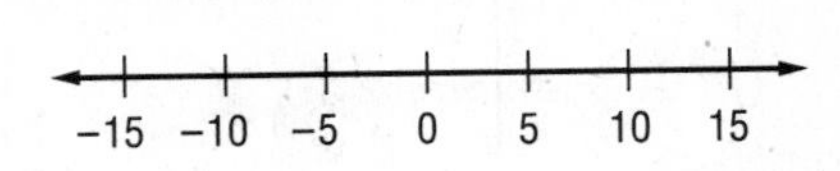

EXAMPLE Solve an Absolute Value Inequality (>)

3 Solve $|3y - 3| > 9$.

Write $|3y - 3| > 9$ as $3y - 3 > 9$ or $3y - 3 < -9$.

Case 1		Case 2
$3y - 3 > 9$	Original inequality	$3y - 3 < -9$
$3y - 3 + 3 > 9 + 3$	Add 3 to each side.	$3y - 3 + 3 < -9 + 3$
☐ > ☐	Simplify.	☐ < ☐
$\frac{3y}{3} > \frac{12}{3}$	Divide each side by 3.	$\frac{3y}{3} < \frac{-6}{3}$
$y >$ ☐	Simplify.	$y <$ ☐

The solution set is ☐.

Your Turn **Solve each open sentence. Then graph the solution set.**

a. $|p + 4| < 6$

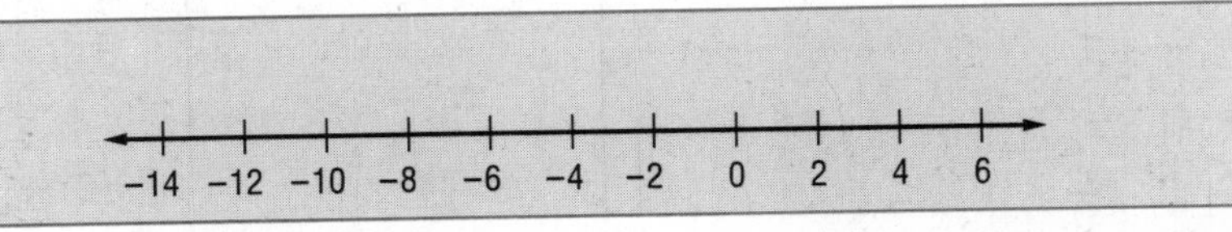

b. $|2m - 2| > 6$

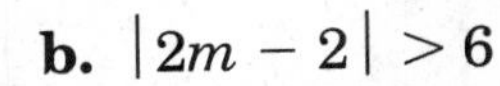

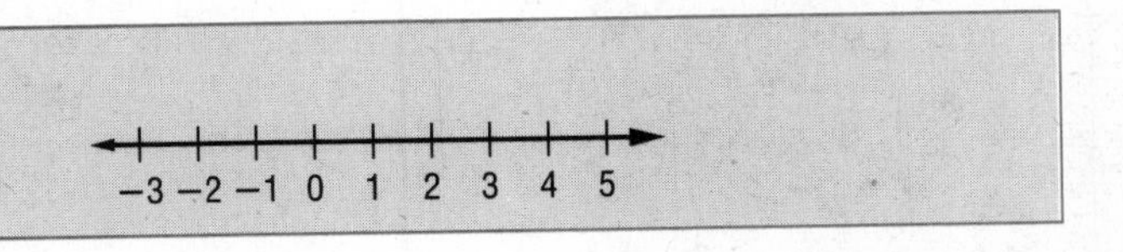

WRITE IT

Will the solution to $|x + 7| < 11$ require finding the intersection or union of the two cases? Explain.

FOLDABLES ORGANIZE IT

In Lesson 6-5 of your journal, write your own absolute value inequality. Then solve and graph it. Explain the steps you use.

Solving Linear Inequalities

HOMEWORK ASSIGNMENT

Page(s):

Exercises:

6–6 Graphing Inequalities in Two Variables

WHAT YOU'LL LEARN

- Graph inequalities on the coordinate plane.
- Solve real-world problems involving linear inequalities.

BUILD YOUR VOCABULARY (pages 136–137)

The region of the graph of an inequality on one side of the ______ is called a **half-plane**.

An ______ defines the **boundary** or edge for each half-plane.

KEY CONCEPT

Half-Planes and Boundaries Any line in the plane divides the plane into two regions called half-planes. The line is called the boundary of each of the two half-planes.

REMEMBER IT

A dashed line indicates that the boundary is *not* part of the solution set. A solid line indicates that the boundary line *is* part of the solution set.

EXAMPLE Graph an Inequality

1 **Graph $2y - 4x > 6$.**

Step 1 Solve for y in terms of x.

$2y - 4x > 6$	Original Inequality
$2y - 4x +$ ___ $>$ ___ $+ 6$	Add ___ to each side.
$2y > 4x + 6$	Simplify.
$\frac{2y}{2} > \frac{4x + 6}{2}$	Divide each side by 2.
$y >$ ___	Simplify.

Step 2 Graph $y = 2x + 3$.
Since $y > 2x + 3$ does not include values when $y = 2x + 3$, the boundary is ______ in the solution set. The boundary should be drawn as a ______.

Step 3 Select a point in one of the half-planes and test it. Let's use (0, 0).

$y > 2x + 3$	Original inequality
$0 > 2(0) + 3$	$x = 0, y = 0$
$0 > 3$	False

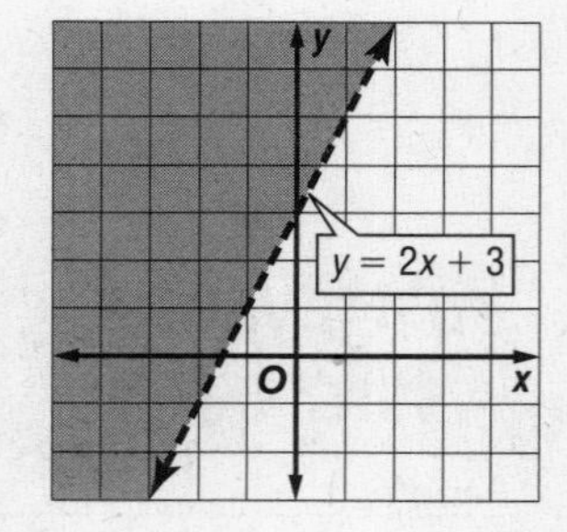

Since the statement is false, the ______ containing the origin is ___ part of the solution. Shade the other half-plane.

Check Test a point in the other half-plane, for example, $(-3, 1)$.

$y > 2x + 3$	Original inequality
$1 > 2(-3) + 3$	$x = -3, y = 1$
$1 > -3$	

Since the statement is true, the half-plane containing $(-3, 1)$ should be ______.

Your Turn Graph $y - 3x < 2$.

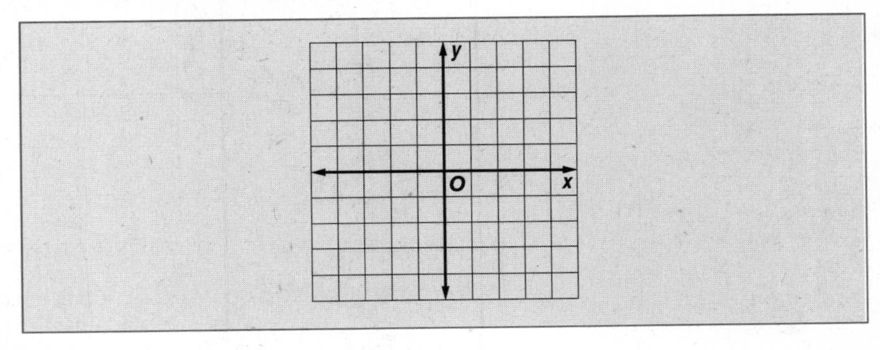

EXAMPLE Write and Solve an Inequality

FOLDABLES™

ORGANIZE IT

In Lesson 6-6 of your journal, explain how to check the solution to an inequality in two variables.

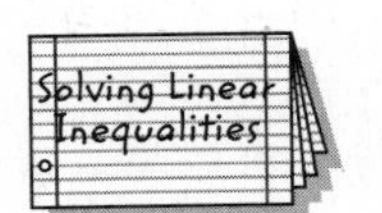

2 JOURNALISM **Lee Cooper writes and edits short articles for a local newspaper. It generally takes her an hour to write an article and about a half-hour to edit an article. If Lee works up to 8 hours a day, how many articles can she write and edit in one day?**

Step 1 Let x equal the number of articles Lee can write. Let y equal the number of articles that Lee can edit. Write an open sentence representing the situation.

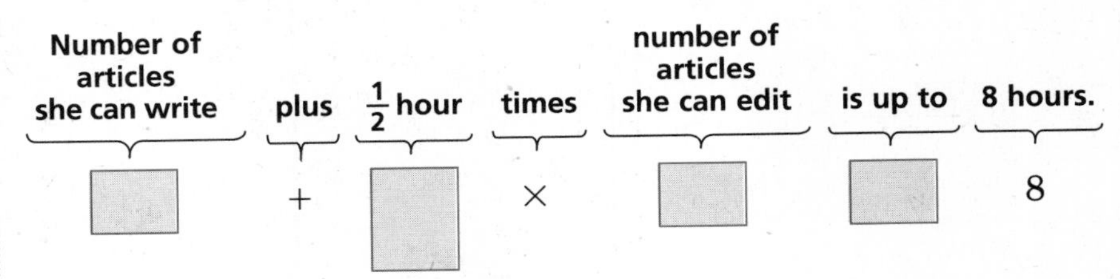

Step 2 Solve for y in terms of x.

$x + \frac{1}{2}y \leq 8$	Original inequality
$x + \frac{1}{2}y - \square \leq \square + 8$	Subtract ______ from each side.
$\square \leq -x + 8$	Simplify.
$(2)\frac{1}{2}y \leq 2(-x + 8)$	Multiply each side by 2.
$y \leq \square$	Simplify.

Step 3 Since the open sentence includes the equation, graph $y = -2x + 16$ as a [ ] line. Test a [ ] in one of the half-planes, for example, (0, 0). Shade the half-plane containing (0, 0) since $0 \leq -2(0) + 16$ is true.

Step 4 Examine the situation

- Lee cannot work a negative number of hours. Therefore, the domain and range contain only [ ] numbers.
- Lee only wants to count articles that are completely written or completely edited. Thus, only points in the half-plane whose x- and y-coordinates are [ ] numbers are possible solutions.
- One solution is (2, 3). This represents [ ] written articles and [ ] edited articles.

Your Turn You offer to go to the local deli and pick up sandwiches for lunch. You have $30 to spend. Chicken sandwiches cost $3.00 and tuna sandwiches are $1.50 each. How may sandwiches can you purchase for $30?

HOMEWORK ASSIGNMENT

Page(s):

Exercises:

CHAPTER 6

BRINGING IT ALL TOGETHER

STUDY GUIDE

FOLDABLES	VOCABULARY PUZZLEMAKER	BUILD YOUR VOCABULARY
Use your **Chapter 6 Foldable** to help you study for your chapter test.	To make a crossword puzzle, word search, or jumble puzzle of the vocabulary words in Chapter 6, go to: www.glencoe.com/sec/math/t_resources/free/index.php	You can use your completed **Vocabulary Builder** (pages 136–137) to help you solve the puzzle.

6–1 Solving Inequalities by Addition and Subtraction

Write the letter of the graph that matches each inequality.

1. $x \leq -1$

2. $x > -1$

3. $x < -1$

a. 3 2 1 0 1 2 3

b. 3 2 1 0 1 2 3

c. 3 2 1 0 1 2 3

4. According to the Subtraction Property of Inequalities, if any number is ______ from each side of a ______ inequality, the resulting inequality is also ______.

6–2 Solving Inequalities by Multiplication and Division

Write an inequality that describes each situation.

5. A number n divided by 8 is greater than 5.

6. Twelve times a number k is at least 7.

Use words to tell what each inequality says.

7. $12 < 6n$

8. $\frac{t}{-3} \geq 14$

6–3

Solving Multi-Step Inequalities

Solve each inequality. Then check your solution.

9. $5 \leq 11 + 3h$

10. $5 - 2n \leq 3 - n$

Define a variable, write an inequality, and solve each problem. Then check your solution.

11. Six plus four times a number is no more than the number.

12. Three times a number plus eight is at least ten less than four times the number.

13. Six times a number is greater than twelve less than 8 times the number.

6–4

Solving Compound Inequalities

14. When is a compound inequality containing *and* true?

15. The graph of a compound inequality containing *and* is the ______ of the graphs of the two inequalities.

16. When is a compound inequality containing *or* true?

17. The graph of a compound inequality containing *or* is the ______ of the graphs of the two inequalities.

6–5

Solving Open Sentences Involving Absolute Value

Complete each compound sentence by writing *and* or *or* in the blank. Use the result to help you graph the absolute value sentence.

	Absolute Value Sentence	Compound Sentence	Graph
18.	$\lvert 2x + 2 \rvert = 8$	$2x + 2 = 8$ ____ $2x + 2 = -8$	−6 −5 −4 −3 −2 −1 0 1 2 3 4
19.	$\lvert x - 5 \rvert \leq 4$	$x - 5 \leq 4$ ____ $x - 5 \geq -4$	0 1 2 3 4 5 6 7 8 9 10
20.	$\lvert 2x - 3 \rvert > 5$	$2x - 3 > 5$ ____ $2x - 3 < -5$	−3 −2 −1 0 1 2 3 4 5 6 7

21. A thermometer is guaranteed to give a temperature no more than 2.1°F from the actual temperature. If the thermometer reads 58°F, what is the range for the actual temperature?

6–6

Graphing Inequalities in Two Variables

22. Complete the chart to show which type of line is needed for each symbol.

Symbol	Type of Line	Boundary Part of Solution?
$<$		
$>$		
$\geq$		
$\geq$		

23. If a test point results in a false statement, what do you know about the graph?

24. If a test point results in a true statement, what do you know about the graph?

ARE YOU READY FOR THE CHAPTER TEST?

Visit **algebra1.com** to access your textbook, more examples, self-check quizzes, and practice tests to help you study the concepts in Chapter 6.

Check the one that applies. Suggestions to help you study are given with each item.

☐ **I completed the review of all or most lessons without using my notes or asking for help.**

- You are probably ready for the Chapter Test.
- You may want take the Chapter 6 Practice Test on page 363 of your textbook as a final check.

☐ **I used my Foldable or Study Notebook to complete the review of all or most lessons.**

- You should complete the Chapter 6 Study Guide and Review on pages 359–362 of your textbook.
- If you are unsure of any concepts or skills, refer to the specific lesson(s).
- You may also want to take the Chapter 6 Practice Test on page 363.

☐ **I asked for help from someone else to complete the review of all or most lessons.**

- You should review the examples and concepts in your Study Notebook and Chapter 6 Foldable.
- Then complete the Chapter 6 Study Guide and Review on pages 359–362 of your textbook.
- If you are unsure of any concepts or skills, refer to the specific lesson(s).
- You may also want to take the Chapter 6 Practice Test on page 363.

Student Signature

Parent/Guardian Signature

Teacher Signature

Solving Systems of Linear Equations and Inequalities

Use the instructions below to make a Foldable to help you organize your notes as you study the chapter. You will see Foldable reminders in the margin of this Interactive Study Notebook to help you in taking notes.

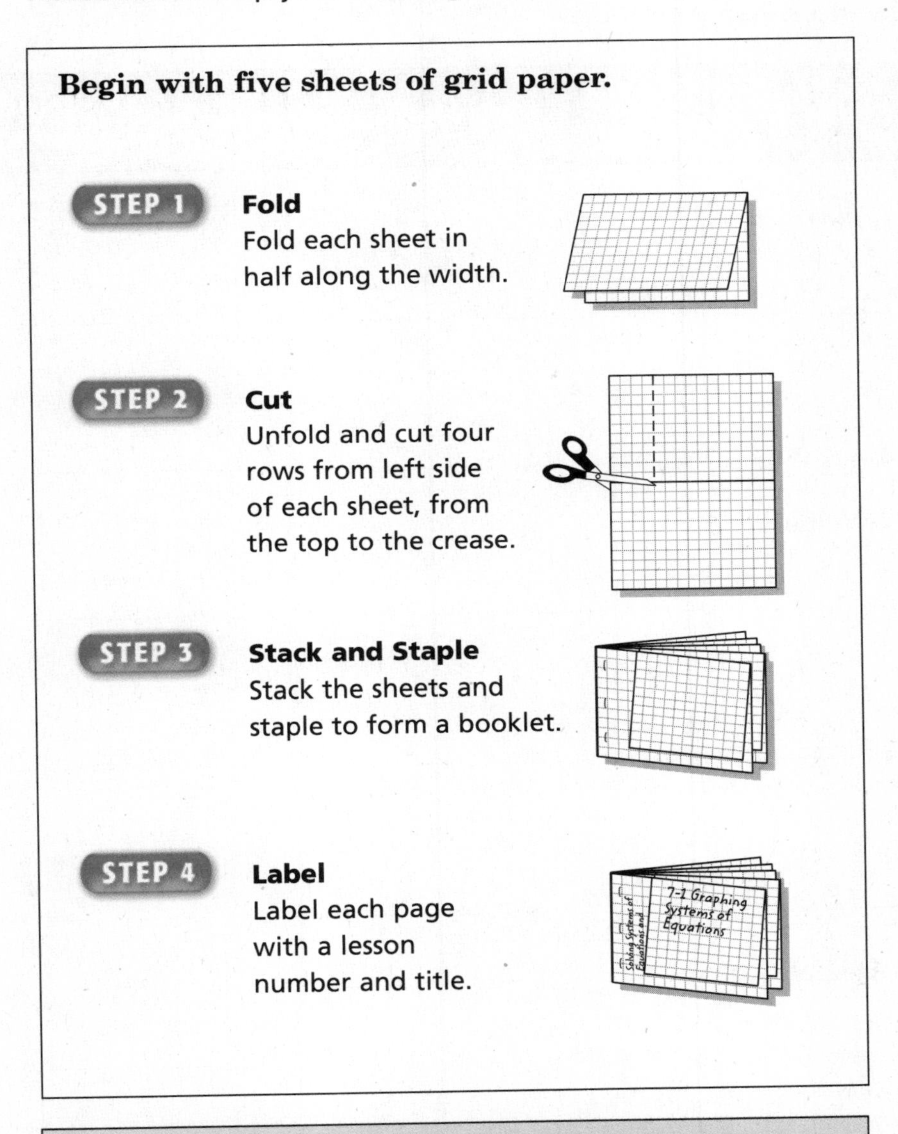

NOTE-TAKING TIP: Before going to class, look over your notes from the previous class, especially if the day's topic builds from the last one.

CHAPTER 7

BUILD YOUR VOCABULARY

This is an alphabetical list of new vocabulary terms you will learn in Chapter 7. As you complete the study notes for the chapter, you will see Build Your Vocabulary reminders to complete each term's definition or description on these pages. Remember to add the textbook page number in the second column for reference when you study.

Vocabulary Term	Found on Page	Definition	Description or Example
consistent [kuhn-SIHS-tuhnt]			
dependent			
elimination [ih-LIH-muh-NAY-shuhn]			
inconsistent			
independent			
substitution [SUHB-stuh-TOO-shuhn]			
system of equations			
system of inequalities			

7–1 Graphing Systems of Equations

What You'll Learn

- Determine whether a system of linear equations has 0, 1, or infinitely many solutions.
- Solve systems of equations by graphing.

BUILD YOUR VOCABULARY (page 158)

Two equations together are called a **system of equations**.

If the graphs intersect or coincide, the system of equations is said to be **consistent**.

If the graphs are ______, the system of equations is said to be **inconsistent**.

If a system has exactly ______ solution, it is **independent**.

If the system has an ______ number of solutions, it is **dependent**.

ORGANIZE IT

In Lesson 7–1 of your booklet, draw the graph of a system of equations that has no solutions.

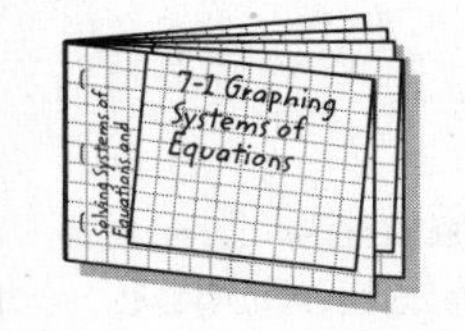

EXAMPLE Number of Solutions

1 **Use each graph to determine whether the system has *no* solution, *one* solution, or *infinitely many* solutions.**

a. $y = -x + 1$
$y = -x + 4$

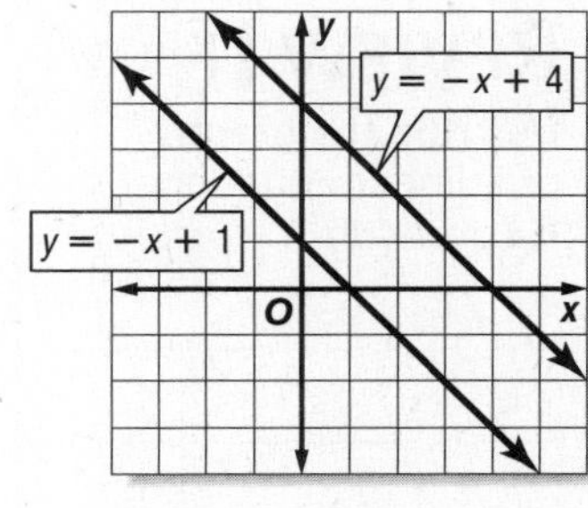

Since the graphs of $y = -x + 1$ and $y = -x + 4$ are ______, there are ______ solutions.

b. $3x - 3y = 9$
$y = -x + 1$

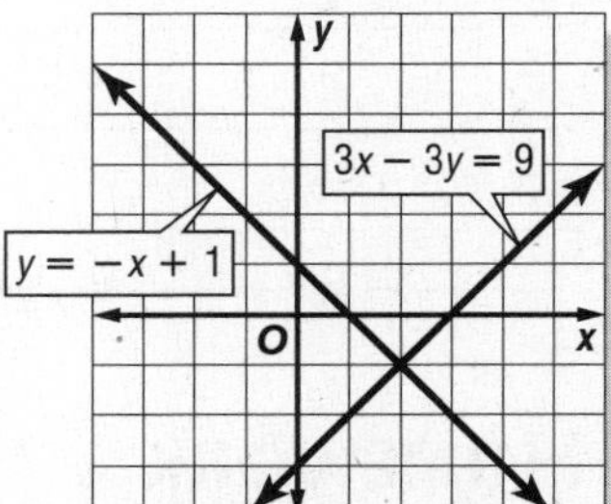

Since the graphs of $3x - 3y = 9$ and $y = -x + 1$ are ______ lines, there is ______ solution.

Your Turn **Use the graph to determine whether each system has *no* solution, *one* solution, or *infinitely many* solutions.**

a. $2y + 3x = 6$
$y = x - 1$

b. $y = x + 4$
$y = x - 1$

c. $y = -\frac{3}{2}x + 3$
$2y + 3x = 6$

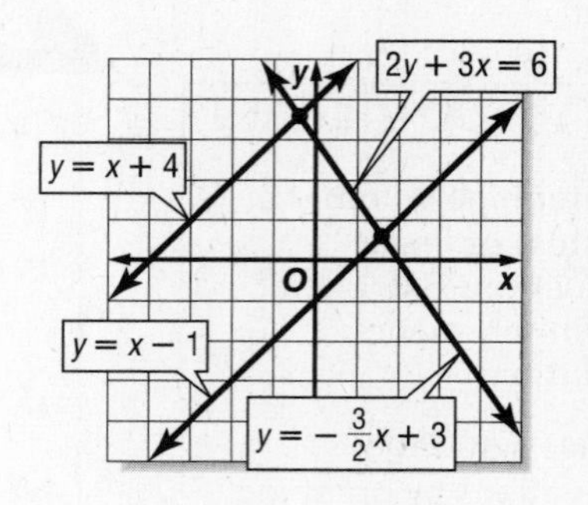

EXAMPLE Solve a System of Equations

2 **Graph the system of equations. Then determine whether the system has *no* solution, *one* solution, or *infinitely many* solutions. If the system has one solution, name it.**

$\mathbf{2x - y = -3}$
$\mathbf{8x - 4y = -12}$

The graphs of the equations ________.

There are ________ solutions of this system of equations.

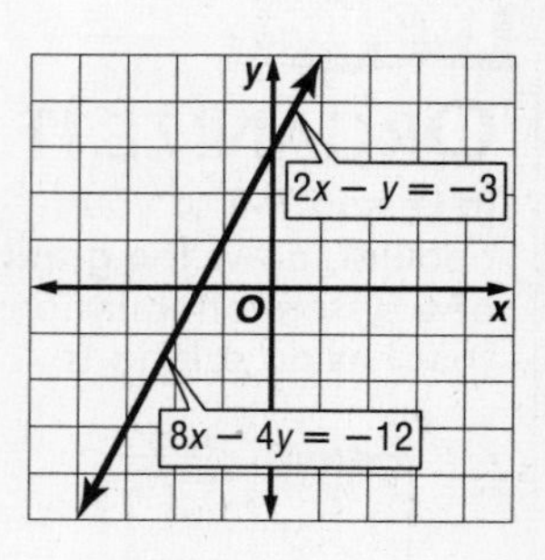

Your Turn **Graph each system of equations. Then determine whether the system has no solution, one solution, or infinitely many solutions. If the system has one solution, name it.**

a. $y = 2x + 3$
$y = \frac{1}{2}x + 3$

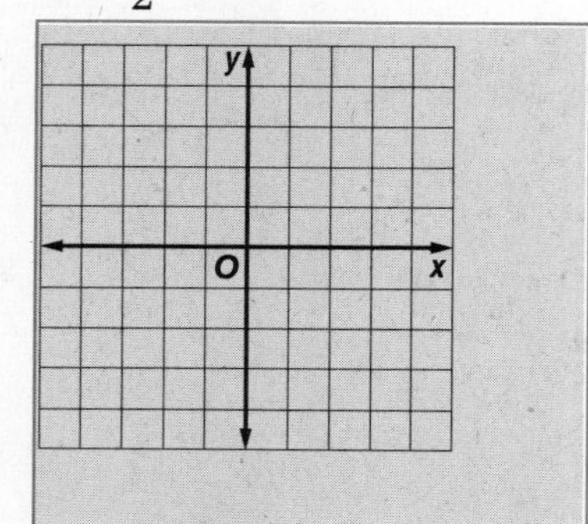

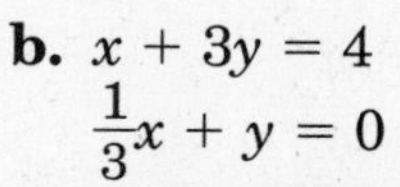

b. $x + 3y = 4$
$\frac{1}{3}x + y = 0$

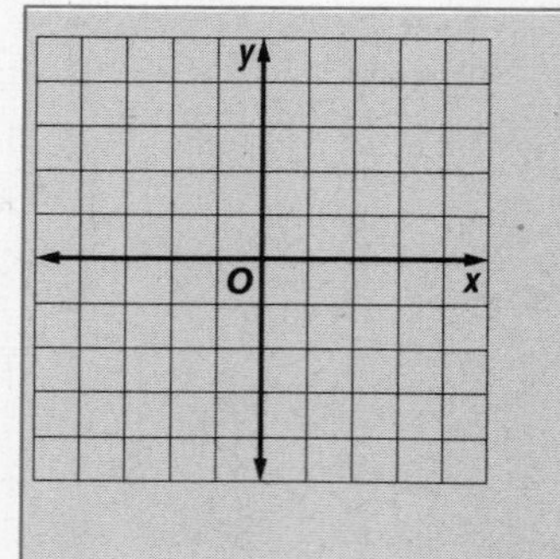

REVIEW IT

Describe the graph of a linear equation. *(Lesson 4-5)*

HOMEWORK ASSIGNMENT

Page(s):
Exercises:

7-2 Substitution

What You'll Learn

- Solve systems of equations by using substitution.
- Solve real-world problems involving systems of equations.

BUILD YOUR VOCABULARY (page 158)

The ____ solution of a system of equations can be found by using algebraic methods. One such method is called **substitution**.

EXAMPLE Solve Using Substitution

1 Use substitution to solve the system of equations.

$\boldsymbol{x = 4y}$

$\boldsymbol{4x - y = 75}$

Since $x = 4y$, substitute $4y$ for x in the second equation.

$4x - y = 75$	Second equation
$4(\ \_\_\_\_\ ) - y = 75$	$x = 4y$
$\_\_\_\_ = 75$	Simplify.
$15y = 75$	Combine like terms.
$\frac{15y}{\_\_\_\_} = \frac{75}{\_\_\_\_}$	Divide each side by ____. 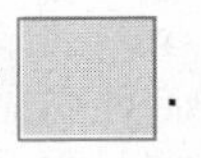
$y = \_\_\_\_$	Simplify.

Use $x = 4y$ to find the value of x.

$x = 4y$	First equation
$x = 4(\ \_\_\_\_\ )$	$y = \_\_\_\_$
$x = \_\_\_\_$	Simplify.

The solution is ____.

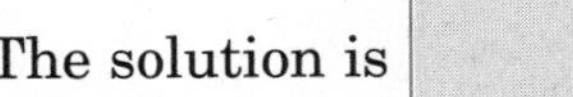

Your Turn Use substitution to solve $y = 2x$ and $3x + 4y = 11$.

____

FOLDABLES

ORGANIZE IT

In Lesson 7-2 of your booklet, explain why it might be easier to solve a system of equations using substitution rather than graphing.

EXAMPLE Solve for One Variable, then Substitute

2 **Use substitution to solve the system of equations.**
$\mathbf{4x + y = 12}$
$\mathbf{-2x - 3y = 14}$

Solve the first equation for y since the coefficient of y is 1.

$4x + y = 12$	First equation
$4x + y$ ___ $= 12$ ___	Subtract $4x$ from each side.
$y =$ ___	Simplify.

Find the value of x by substituting $12 - 4x$ for y in the second equation.

$-2x - 3y = 14$	Second equation
$-2x - 3$ ___ $= 14$	$y = 12 - 4x$
$-2x$ ___ $+$ ___ $= 14$	Distributive Property
$10x - 36 = 14$	___
$10x - 36 + 36 = 14 + 36$	Add 36 to each side.
___ $= 50$	Simplify.
$\frac{10x}{10} = \frac{50}{10}$	Divide each side by ___.
$x =$ ___ 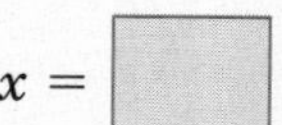	Simplify.

Substitute 5 for x in either equation to find the value of y.

$4x + y = 12$	First equation
$4($ ___ $) + y = 12$	$x =$ ___
___ $+ y = 12$	Simplify.
$y =$ ___	Subtract.

The solution is ___.

The graph verifies the solution.

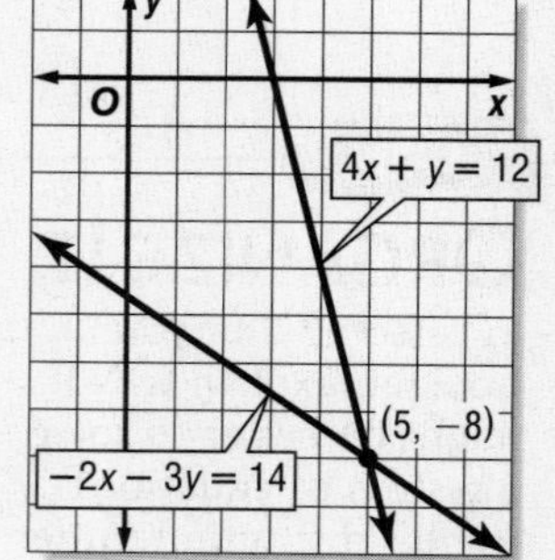

Your Turn Use substitution to solve $x + 2y = 1$ and $5x - 4y = -23$.

___

EXAMPLE Dependent System

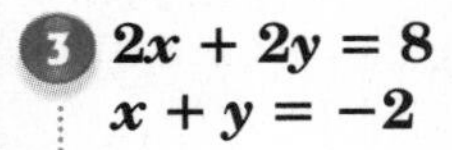

3 $2x + 2y = 8$
$x + y = -2$

REVIEW IT

Describe the first step when using the Distributive Property. *(Lesson 1-5)*

____

____

____

____

____

____

Solve the second equation for y.

$x + y = -2$	Second equation
$x + y$ ____ $= -2$ ____	Subtract x from each side.
$y =$ ____	Simplify.

Substitute ____ for y in the first equation.

$2x + 2y = 8$	First equation
$2x + 2($____$) = 8$	$y = -2 - x$
$2x - 4 - 2x = 8$	Distributive Property
____ $= 8$	Simplify.

The statement ____ $= 8$ is  ____. This means there are ____ solutions of the system of equations. The graphs of the lines are ____.

Your Turn Use substitution to solve the $3x - 2y = 3$ and $-6x + 4y = -6$.

HOMEWORK ASSIGNMENT

Page(s):

Exercises:

7–3 Elimination Using Addition and Subtraction

BUILD YOUR VOCABULARY (page 158)

Sometimes adding two equations together will eliminate one variable. Using this step to solve a system of equations is called **elmination**.

WHAT YOU'LL LEARN

- Solve systems of equations by using elimination with addition.
- Solve systems of equations by using elimination with subtraction.

EXAMPLE Elimination using Addition

1 Use elimination to solve the system of equations.

$$-3x + 4y = 12$$
$$3x - 6y = 18$$

Since the coefficients of the x terms, -3 and 3, are additive inverses, you can eliminate the x terms by adding the equations.

$-3x + 4y = 12$	Write the equation in column
$(+)\ 3x - 6y = 18$	form and add.
$-2y = 30$	Notice that the ____ value is eliminated.
$\frac{-2y}{\_\_\_\_} = \frac{30}{\_\_\_\_}$	Divide each side by .
$y =$ ____	Simplify.

Now substitute ____ for y in either equation to find x.

$-3x + 4y = 12$	First equation
$-3x + 4(\_\_\_\_) = 12$	Replace y with ____.
$-3x - \_\_\_\_ = 12$	Simplify.
$-3x - 60 + \_\_\_\_ = 12 + \_\_\_\_$	Add ____ to each side.
$-3x = 72$	Divide each side by ____.
$x =$ ____	Simplify.

The solution is ____.

Your Turn Use elimination to solve $3x - 5y = 1$ and $2x + 5y = 9$.

EXAMPLE Elimination Using Subtraction

2 Use elimination to solve the system of equations.

$$\mathbf{4x + 2y = 28}$$
$$\mathbf{4x - 3y = 18}$$

Since the coefficients of the x terms, 4 and 4, are the ______, you can eliminate the x terms by subtracting the equations.

$4x + 2y = 28$	Write the equation in column form
$(-)\ 4x - 3y = 18$	and subtract.
______ $= 10$	Notice that the x value is eliminated.
$\frac{5y}{\_\_\_} = \frac{10}{\_\_\_}$	Divide each side by ______.
$y =$ ______	Simplify.

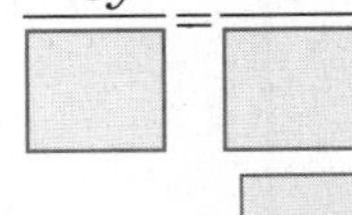

Now substitute ______ for y in either equation.

$4x - 3y = 18$	Second equation
$4x - 3(\_\_\_) = 18$	Replace y with ______.
$4x - 6 = 18$	Simplify.
$4x - 6 + 6 = 18 + 6$	Add 6 to each side.
$4x = 24$	Simplify.
______ = ______	Divide each side 4. Simplify.

The solution is ______.

FOLDABLES™

ORGANIZE IT

In Lesson 7-3 of your booklet, write an example of a system that can be solved by subtracting the equations. Then solve your system.

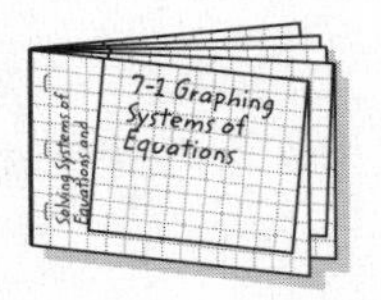

Your Turn **Use elimination to solve each system of equations.**

a. $3x - 5y = 1$
$2x - 5y = 8$

b. $9x - 2y = 30$
$x - 2y = 14$

HOMEWORK ASSIGNMENT

Page(s):

Exercises:

7–4 Elimination Using Multiplication

What You'll Learn

- Solve systems of equations by using elimination with multiplication.
- Determine the best method for solving systems of equations.

EXAMPLE Multiply One Equation to Eliminate

1 **Use elimination to solve the system of equations.**

$2x + y = 23$

$3x + 2y = 37$

Multiply the first equation by ____ so the coefficients of the y terms are additive inverses. Then add the equations.

$2x + y = 23 \rightarrow$ ____ $- 2y =$ ____ Multiply by ____.

$3x + 2y = 37$ $(+)\ 3x + 2y = 37$

____ $=$ ____ Add the equations.

$\frac{x}{\square} = \frac{-9}{\square}$ Divide.

$x =$ ____ Simplify.

Now substitute ____ for x in either equation to find the value of y.

$2x + y = 23$ First equation

$2(\ \ ) + y = 23$ $x =$ ____

____ $+ y = 23$ Simplify.

$18 + y -$ ____ $= 23 -$ ____ Subtract ____ from each side.

$y = 5$ Simplify.

The solution is ____.

Your Turn Use elimination to solve $x + 7y = 12$ and $3x - 5y = 10$.

Foldables™ Organize It

In Lesson 7-4 of your booklet, list the 5 different methods for solving a system of equations. Be sure to tell when it is best to use each one.

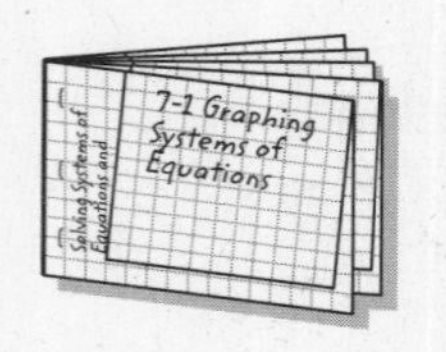

Remember It

When solving a system of equations by elimination, you can choose to eliminate either variable. See Example 2 on page 388 of your textbook.

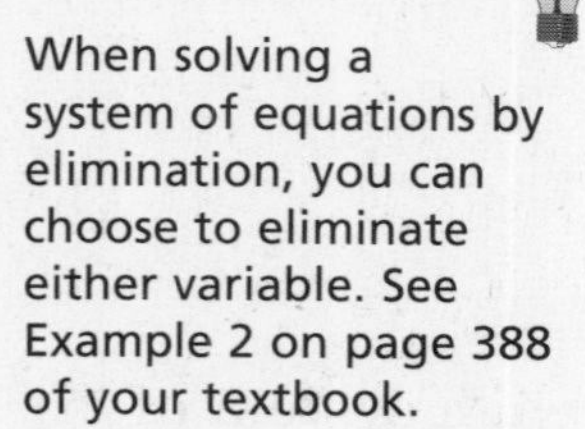

EXAMPLE Multiply Both Equations to Eliminate

2 **Use elimination to solve the system of equations.**

$\mathbf{4x + 3y = 8}$

$\mathbf{3x - 5y = -23}$

Choose either variable to eliminate. Let's eliminate x.

$4x + 3y = 8 \rightarrow$ ______ $+ 9y = 24$ Multiply by ______.

$3x - 5y = -23 \rightarrow (+) -12x +$ ______ $= 92$ Multiply by ______.

______ $=$ ______ Add the equations.

$\frac{29y}{29} = \frac{116}{29}$ Divide each side by ______.

$y =$ ______ Simplify.

Now substitute ______ for y in either equation to find x.

$4x + 3y = 8$ First equation

$4x + 3$ ______ $= 8$ $y =$ ______

______ $+$ ______ $= 8$ Simplify.

$4x + 12 -$ ______ $= 8 -$ ______ Subtract ______ from each side.

$4x =$ ______ Simplify.

$\frac{4x}{4} = \frac{-4}{4}$ Divide each side by 4.

$x =$ ______ Simplify.

The solution is ______.

Your Turn Use elimination to solve $3x + 2y = 10$ and $2x + 5y = 3$.

HOMEWORK ASSIGNMENT

Page(s):

Exercises:

7–5 Graphing Systems of Inequalities

WHAT YOU'LL LEARN

- Solve systems of inequalities by graphing.
- Solve real-world problems involving systems of inequalities.

BUILD YOUR VOCABULARY (page 158)

To solve a **system of inequalities**, you need to find ______ that satisfy ______ the inequalities involved.

FOLDABLES™

ORGANIZE IT

Under the tab for Lesson 7–5, write a description of how to graph the solution of a system of inequalities.

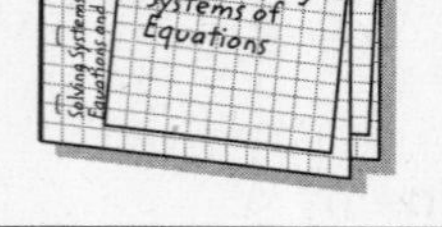

EXAMPLE Solve by Graphing

1 Solve the system of inequalities by graphing.

$y < 2x + 2$
$y \geq -x - 3$

The solution includes the ordered pairs in the intersection of the graphs of $y < 2x + 2$ and $y \geq -x - 3$. The region is shaded in dark grey. The graphs $y = 2x + 2$ and $y = -x - 3$ are ______ of this region. The graph ______ is dashed and is not ______ in the graph of $y < 2x + 2$. The graph of $y = -x - 3$ is included in the graph of $y \geq -x - 3$.

EXAMPLE No Solution

2 Solve the system of inequalities by graphing.

$y \geq -3x + 1$
$y \leq -3x - 2$

The graphs of $y = -3x + 1$ and $y = -3x - 2$ are ______ lines.
Because the two regions have no points in common, the system of inequalities has ______.

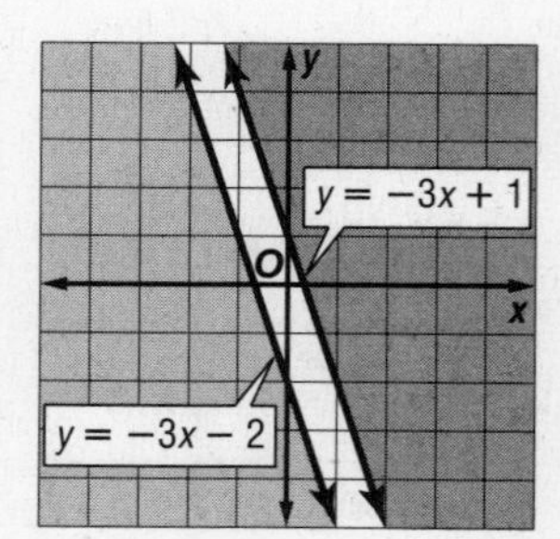

Your Turn **Solve each system of inequalities by graphing.**

a. $2x + y \leq 4$
$x + 2y < -4$

b. $y > 4x$
$y < 4x - 3$

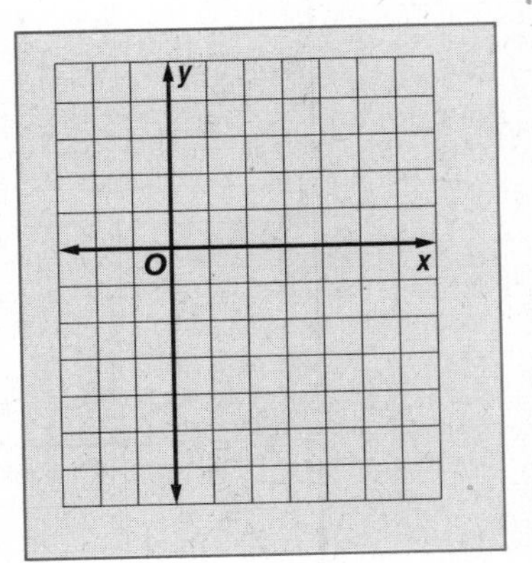

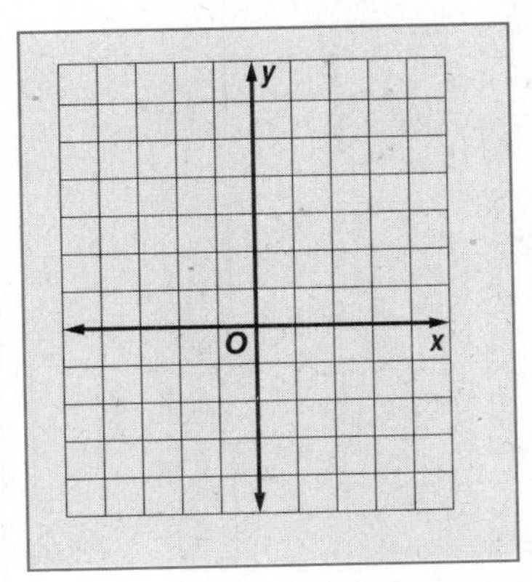

EXAMPLE Use a System of Inequalities to Solve a Problem

3 SERVICE **A college service organization require that it's members maintain at least a 3.0 grade point average, and volunteer at least 10 hours a week. Graph these requirements.**

If g = the grade point average and v = the number of volunteers, the following inequalities represent these requirements.

The grade point average is at least 3.0.

The number of volunteer hours is at least 10.

Volunteer Hours
14
12
10
8
6
4
2
0
1.0 2.0 3.0 4.0
Requirements for Membership

The solution is the set of ordered pairs whose graphs are in the of the graphs of these inequalities.

WRITE IT

Describe the graph of a system of inequalities that has no solution.

Your Turn The senior class is sponsoring a blood drive. Anyone who wishes to give blood must be at least 17 years old and weigh at least 110 pounds. Graph these requirements.

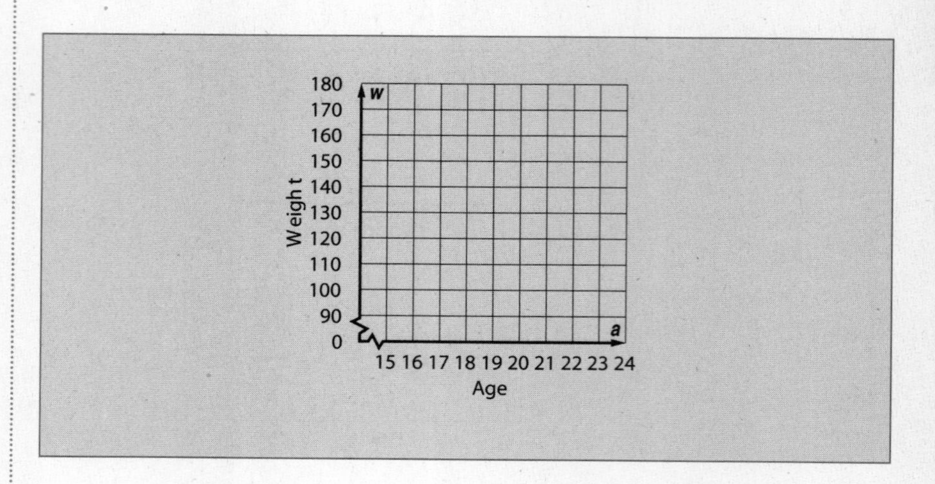

HOMEWORK ASSIGNMENT

Page(s):

Exercises:

CHAPTER 7

BRINGING IT ALL TOGETHER

STUDY GUIDE

FOLDABLES™	VOCABULARY PUZZLEMAKER	BUILD YOUR VOCABULARY
Use your **Chapter 7 Foldable** to help you study for your chapter test.	To make a crossword puzzle, word search, or jumble puzzle of the vocabulary words in Chapter 7, go to: www.glencoe.com/sec/math/t_resources/free/index.php	You can use your completed **Vocabulary Builder** (page 158) to help you solve the puzzle.

7-1

Graphing Systems of Equations

Each figure shows the graph of a system of two equations. Write the letter(s) of the figures that illustrate each statement.

A.

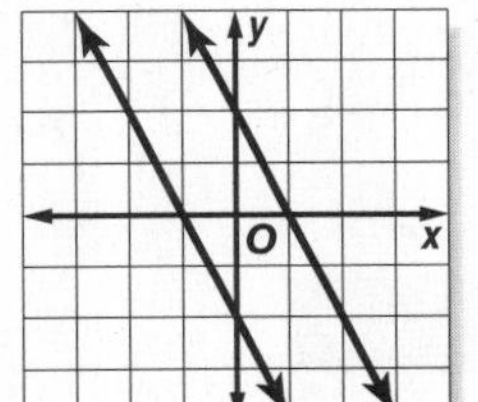

B.

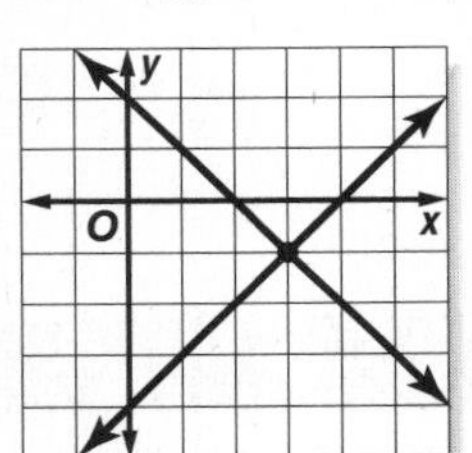

C.

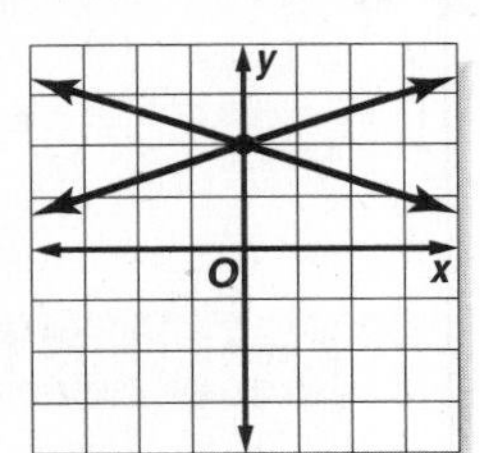

D.

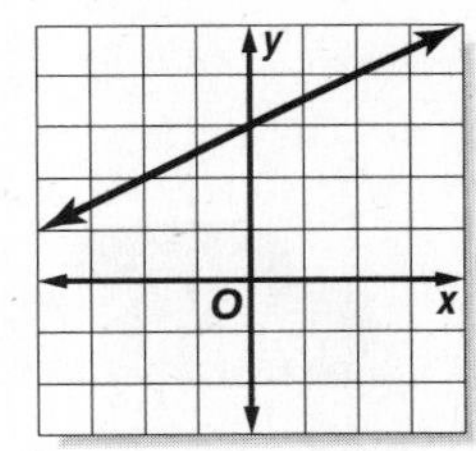

1. A system of two linear equations can have an infinite number of solutions.

2. If two graphs are parallel, there are no ordered pairs that satisfy both equations.

3. If a system of equations has exactly one solution, it is independent.

4. If a system of equations has an infinite number of solutions, it is dependent.

7-2

Substitution

Solve each system using substitution.

5. $y = -2x$
$x + 3y = 15$

6. $3x - 2y = 12$
$x = 2y$

7. $-3x + 5y = 81$
$2x + y = 24$

7-3

Elimination Using Addition and Subtraction

Write *addition* or *subtraction* to tell which operation it would be easiest to use to eliminate a variable of the system. Explain your choice.

	System of Equations	Operation	Explanation
8.	$3x + 5y = 12$ $-3x + 2y = 6$		
9.	$3x + 5y = 7$ $3x - 2y = 8$		

Use elimination to solve each system of equations.

10. $7x + 2y = 10$
$-7x + y = -16$

11. $2x + 5y = -22$
$10x + 3y = 22$

7-4

Elimination Using Multiplication

Three methods for solving systems of linear equations are summarized below. Complete the table.

	Method	The Best Time to Use
12.	Graphing	to ______ the solution, since graphing usually does not give an ______ solution
13.		if one of the variables in either equation has a coefficient of 1 or ______
14.	Elimination Using Multiplication	if none of the coefficients are ______ or -1 and neither of the variables can be eliminated by simply adding or subtracting the equations

7-5

Graphing Systems of Inequalities

Write the inequality symbols that you need to get a system whose graph looks like the one shown. Use <, ≤, >, or ≥.

15.

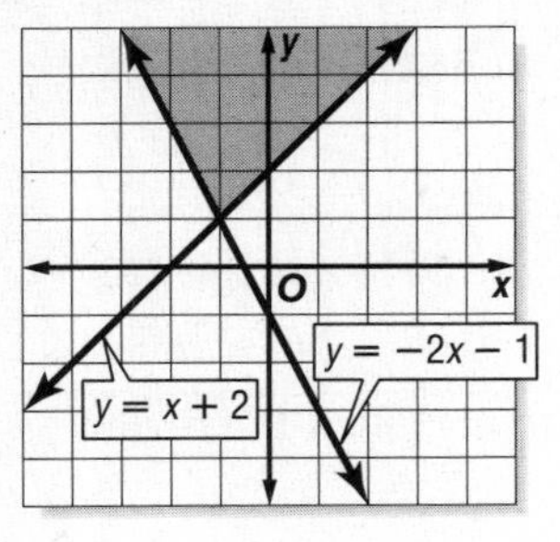

y ☐ $x + 2$

y ☐ $-2x - 1$

16.

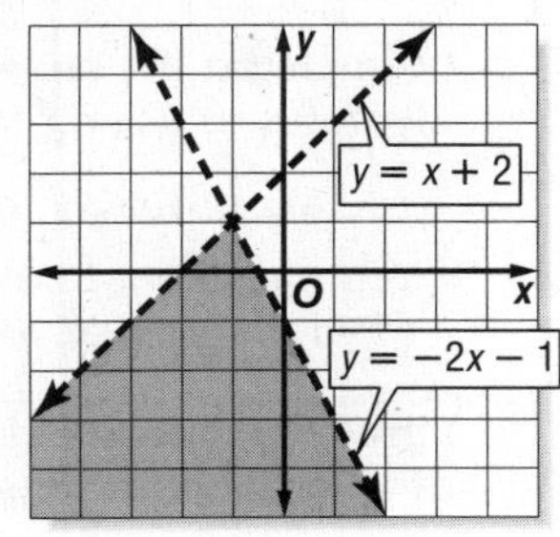

y ☐ $x + 2$

y ☐ $-2x - 1$

17.

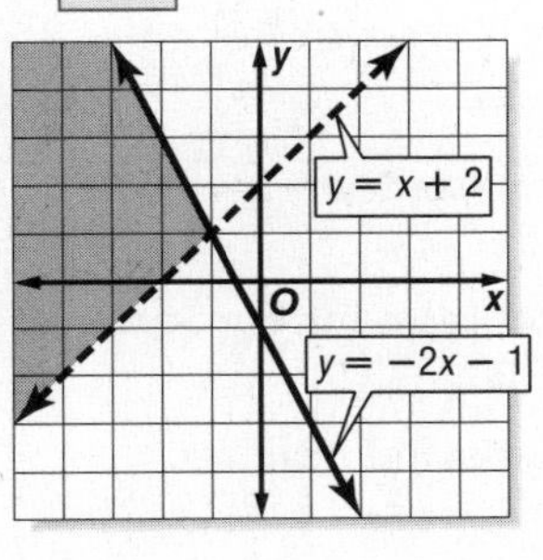

y ☐ $x + 2$

y ☐ $-2x - 1$

18.

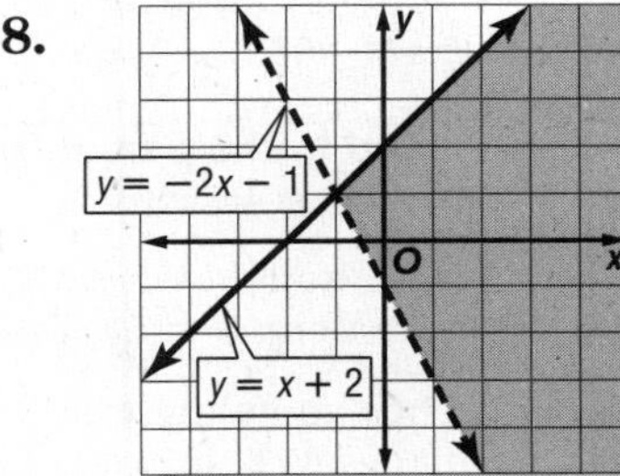

y ☐ $x + 2$

y ☐ $-2x - 1$

19. The solution of a ______ is the set of all ordered pairs that satisfy both inequalities. If you graph the inequalities in the same coordinate plane, the ______ is the region where the graphs ______.

20. Describe how you would explain the process of using a graph to solve a system of inequalities to a friend who missed Lesson 7-5.

ARE YOU READY FOR THE CHAPTER TEST?

Math Online

Visit **www.algebra1.com** to access your textbook, more examples, self-check quizzes, and practice tests to help you study the concepts in Chapter 7.

Check the one that applies. Suggestions to help you study are given with each item.

☐ **I completed the review of all or most lessons without using my notes or asking for help.**

- You are probably ready for the Chapter Test.
- You may want take the Chapter 7 Practice Test on page 403 of your textbook as a final check.

☐ **I used my Foldable or Study Notebook to complete the review of all or most lessons.**

- You should complete the Chapter 7 Study Guide and Review on pages 399–402 of your textbook.
- If you are unsure of any concepts or skills, refer back to the specific lesson(s).
- You may also want to take the Chapter 7 Practice Test on page 403.

☐ **I asked for help from someone else to complete the review of all or most lessons.**

- You should review the examples and concepts in your Study Notebook and Chapter 7 Foldable.
- Then complete the Chapter 7 Study Guide and Review on pages 399–402 of your textbook.
- If you are unsure of any concepts or skills, refer back to the specific lesson(s).
- You may also want to take the Chapter 7 Practice Test on page 403.

Student Signature

Parent/Guardian Signature

Teacher Signature

Polynomials

Use the instructions below to make a Foldable to help you organize your note as you study the chapter. You will see Foldable reminders in the margin of this Interactive Study Notebook to help you in taking notes.

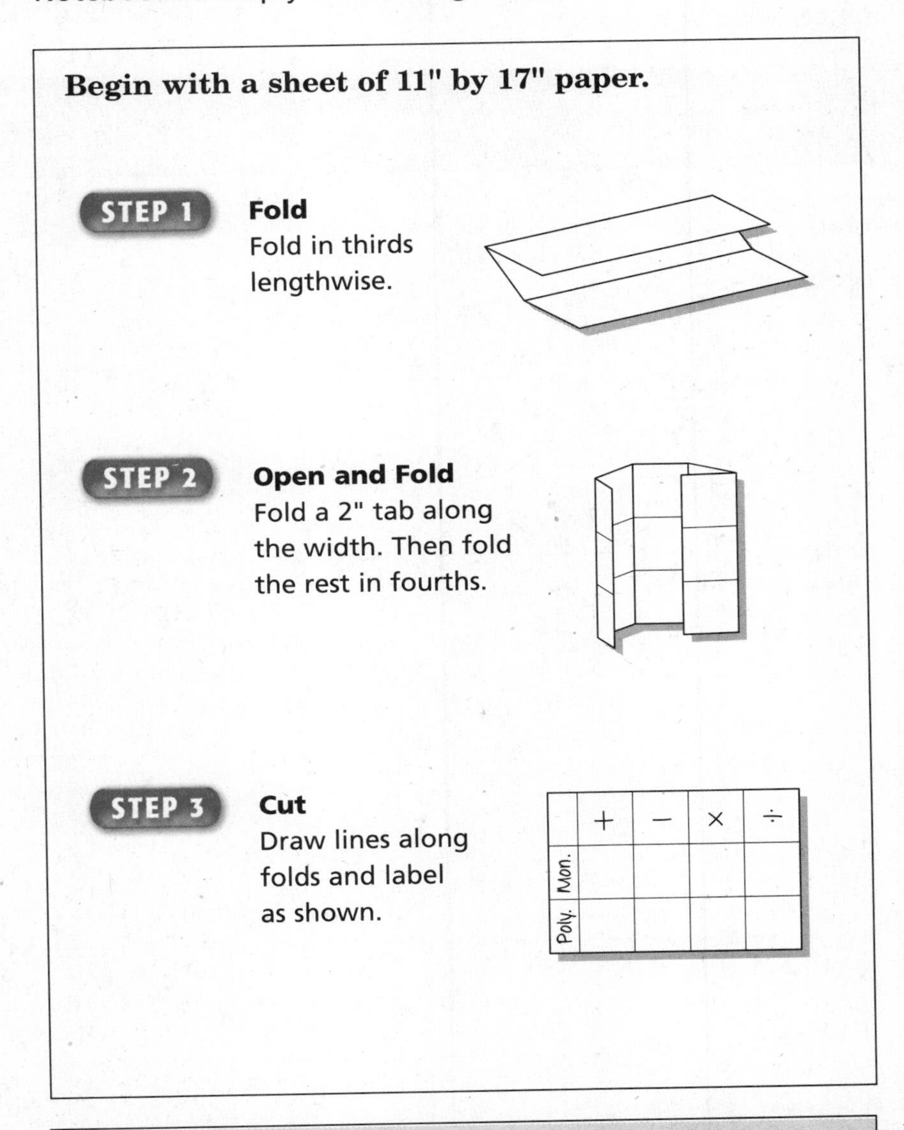

NOTE-TAKING TIP: It is helpful to read through your notes before beginning your homework. Look over any page referenced material.

BUILD YOUR VOCABULARY

This is an alphabetical list of new vocabulary terms you will learn in Chapter 8. As you complete the study notes for the chapter, you will see Build You Vocabulary reminders to complete each term's definition or description on these pages. Remember to add the textbook page number in the second column for reference when you study.

Vocabulary Term	Found on Page	Definition	Description or Example
binomial [by-NOH-mee-uhl]			
constant			
degree of monomial			
degree of polynomial			
FOIL method			
monomial [mah-NOH-mee-uhl]			

Vocabulary Term	Found on Page	Definition	Description or Example
negative exponent			
polynomial [PAH-luh-NOH-mee-uhl]			
Power of a Power			
Power of a Product			
Product of Powers			
Power of a Quotient			
Quotient of Powers			
scientific notation			
trinomial [try-NOH-mee-uhl]			
zero exponent			

8–1 Multiply Monomials

WHAT YOU'LL LEARN

- Multiply monomials.
- Simplify expressions involving powers of monomials.

BUILD YOUR VOCABULARY (page 176)

A **monomial** is a number, a ________, or a product of a number and one or more variables.

Monomials that are ________ numbers are called **constants.**

EXAMPLE Identify Monomials

1 **Determine whether each expression is a monomial. Explain your reasoning.**

	Expression	Monomial?	Reason
a.	$17 - x$	no	The expression involves subtraction, not the product, of two variables.
b.	$8f^2g$		The expression is the product of a number and two variables.
c.	$\frac{3}{4}$	yes	$\frac{3}{4}$ is a real number and an example of a constant.
d.	xy		The expression is the product of two variables.

Your Turn Determine whether each expression is a monomial. Explain your reasoning.

	Expression	Monomial?	Reason
a.	x^5		
b.	$3p - 1$		
c.	$\frac{9x}{y}$		
d.	$\frac{cd}{8}$		

EXAMPLE Product of Powers

2 a. Simplify $(r^4)(-12r^7)$.

$(r^4)(-12r^7) = (1)(-12)(r^4)(r^7)$ Commutative and Associative Properties

= Product of Powers

= Simplify.

b. Simplify $(6cd^5)(5c^5d^2)$.

$(6cd^5)(5c^5d^2) = (6)(5)(c \cdot c^5)(d^5 \cdot d^2)$ Commutative and Associative Properties

$= 30(c$ 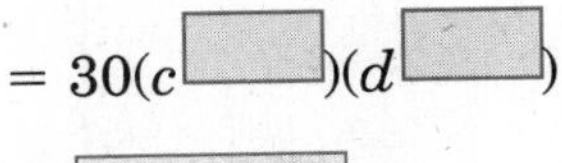 $)(d$ $)$ Product of Powers

=  Simplify.

Your Turn **Simplify each expression.**

a. $(5x^2)(4x^3)$

b. $3xy^2(-2x^2y^3)$

KEY CONCEPTS

Product of Powers To multiply two powers that have the same base, add the exponents.

Power of a Power To find the power of a power, multiply the exponents.

Power of a Product To find the power of a product, find the power of each factor and multiply.

EXAMPLE Power of a Power

3 Simplify $((2^3)^3)^2$.

$((2^3)^3)^2 = (2^{3 \cdot 3})^2$ of a Power

= Simplify.

$= 2$  of a Power

$= 2$ or Simplify.

Your Turn Simplify $((4^2)^2)^3$.

EXAMPLE Power of a Product

4 GEOMETRY **Find the volume of a cube with a side length $s = 5xyz$.**

Volume $= s^3$	Formula for volume of a cube
$= (5xyz)^3$	$s =$ ______
$= 5^3x^3y^3z^3$	Power of a Product
$=$ ______	Simplify.

Your Turn Express the surface area of the cube as a monomial.

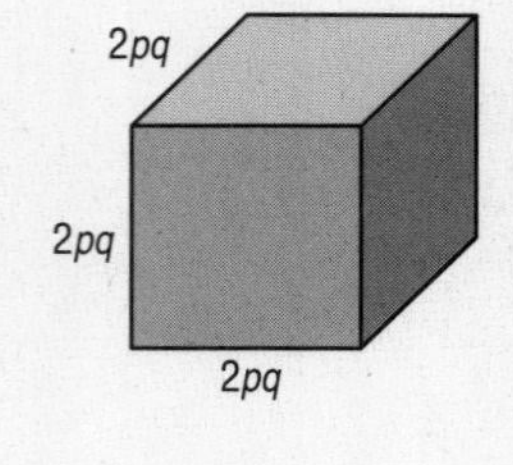

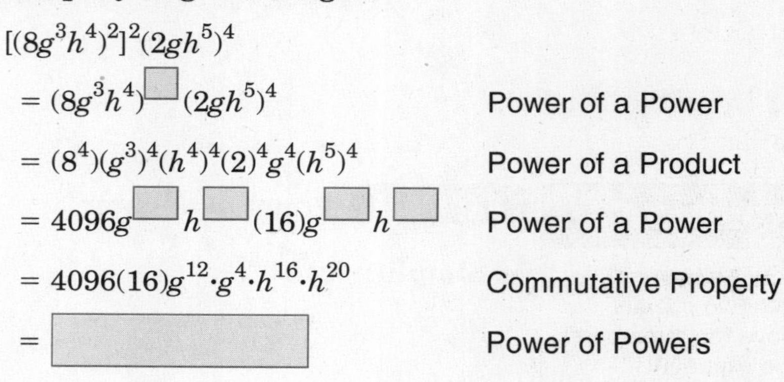

EXAMPLE Simplify Expressions

FOLDABLES

ORGANIZE IT

In your Foldable, in the box for monomial multiplication, write the name of each exponent rule and an example illustrating the rule.

	+	−	×	÷
Mon.				
Poly.				

5 **Simplify $[(8g^3h^4)^2]^2(2gh^5)^4$.**

$[(8g^3h^4)^2]^2(2gh^5)^4$

$= (8g^3h^4)^{\square}(2gh^5)^4$	Power of a Power
$= (8^4)(g^3)^4(h^4)^4(2)^4g^4(h^5)^4$	Power of a Product
$= 4096g^{\square}h^{\square}(16)g^{\square}h^{\square}$	Power of a Power
$= 4096(16)g^{12}\cdot g^4\cdot h^{16}\cdot h^{20}$	Commutative Property
$=$ ______	Power of Powers

Your Turn Simplify $[(2c^2d^3)^2]^3(3c^5d^2)^3$.

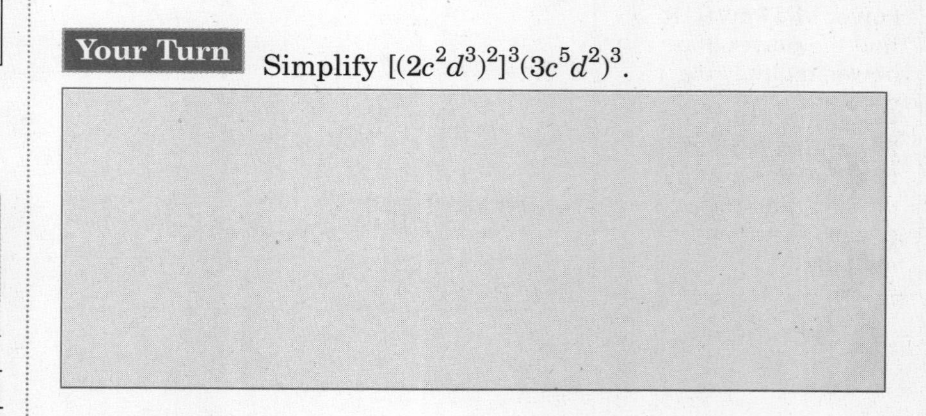

HOMEWORK ASSIGNMENT

Page(s):

Exercises:

8–2 Dividing Monomials

What You'll Learn

- Simplify expressions involving the quotient of monomials.
- Simplify expressions containing negative exponents.

Key Concepts

Quotient of Powers To divide two powers that have the same base, subtract the exponents.

Power of a Quotient To find the power of a quotient, find the power of the numerator and the power of the denominator.

Zero Exponent Any nonzero number raised to the zero power is 1.

EXAMPLE Quotient of Powers

1 **Simplify $\frac{x^7y^{12}}{x^6y^3}$. Assume that x and y are not equal to zero.**

$\frac{x^7y^{12}}{x^6y^3} = \left(\frac{x^7}{x^6}\right)\left(\frac{y^{12}}{y^3}\right)$ Group powers that have the same base.

$= (x^{7-6})(y^{12-3})$ Quotient of Powers

$= x^{\square}y^{\square}$ Simplify.

EXAMPLE Power of a Quotient

2 **Simplify $\left(\frac{4c^3d^2}{5e^4f^7}\right)^3$. Assume that e and f are not equal to zero.**

$\left(\frac{4c^3d^2}{5e^4f^7}\right)^3 = \frac{(4c^3d^2)^3}{(5e^4f^7)^3}$ Power of a ______

$= \frac{4^3(c^3)^3(d^2)^3}{5^3(e^4)^3(f^7)^3}$ Power of a ______

$= \frac{64c^{\square}d^{\square}}{125e^{\square}f^{\square}}$ Power of a ______

Your Turn **Simplify each expression. Assume that a, b, p and q are not equal to zero.**

a. $\frac{a^3b^9}{ab^2}$

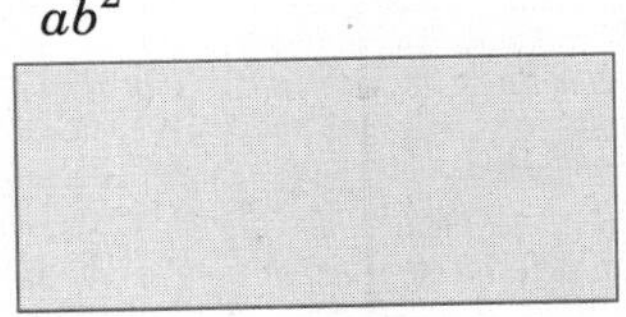

b. $\left(\frac{3m^3n^2}{4p^5q}\right)^3$

EXAMPLE Zero Exponent

3 **Simplify $\left(\frac{12m^8n^7}{8m^5n^{10}}\right)^0$. Assume that m and n are not equal to zero.**

$\left(\frac{12m^8n^7}{8m^5n^{10}}\right)^0 = 1$ $\quad a^0 = 1$

EXAMPLE Negative Exponents

4 a. Simplify $\frac{x^{-6}}{y^{-4}z^{9}}$. Assume that y and z are not equal to zero.

$\frac{x^{-6}}{y^{-4}z^{9}} = \left(\frac{x^{-6}}{1}\right)\left(\frac{1}{y^{-4}}\right)\left(\frac{1}{z^{9}}\right)$ Write as a product of fractions.

$= \left(\frac{1}{x^{6}}\right)\left(\frac{y^{4}}{1}\right)\left(\frac{1}{z^{9}}\right)$

$= \frac{y^{\square}}{x^{\square}z^{\square}}$ Multiply fractions.

b. Simplify $\frac{75p^{3}q^{-5}}{15p^{5}q^{-4}r^{-8}}$. Assume that p, q, and r are not equal to zero.

$\frac{75p^{3}q^{-5}}{15p^{5}q^{-4}r^{-8}} = \left(\frac{75}{15}\right)\left(\frac{p^{3}}{p^{5}}\right)\left(\frac{q^{-5}}{q^{-4}}\right)\left(\frac{1}{r^{-8}}\right)$ Group powers with the same base.

$=$ 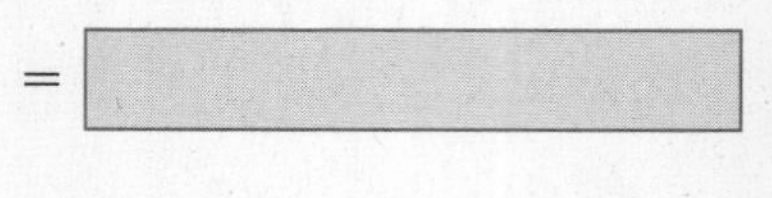 Quotient of Powers and Negative Exponent Properties

$= 5p^{\square}q^{\square}r^{\square}$ Simplify.

$= 5\left(\frac{1}{p^{2}}\right)q^{-1}r^{8}$ Negative Exponent Property

$=$ 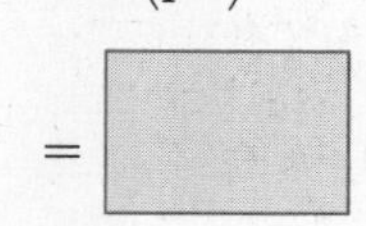 Multiply Fractions.

KEY CONCEPT

Negative Exponent For any nonzero number a and any integer n, a^{-n} is the reciprocal of a^{n}. In addition, the reciprocal of a^{-n} is a^{n}.

FOLDABLES In your foldable, in the monomial division box, write the name of each exponent rule in the lesson and an example illustrating the rule.

Your Turn **Simplify each expression. Assume that no denominator is equal to zero.**

a. $\left(\frac{3x^{2}y^{9}}{5z^{12}}\right)^{0}$

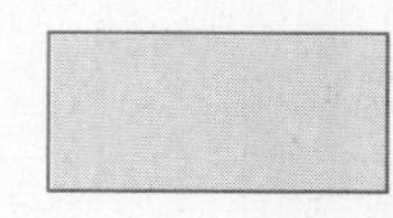

b. $\frac{x^{0}k^{5}}{k^{3}}$

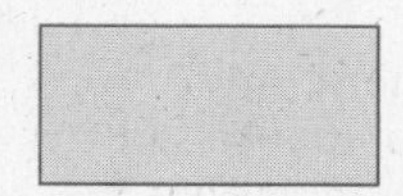

c. $\frac{a^{-2}b^{3}}{c^{-5}}$

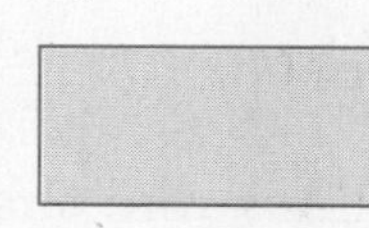

d. $\frac{36x^{5}y^{8}z^{2}}{9x^{4}y^{2}z^{6}}$

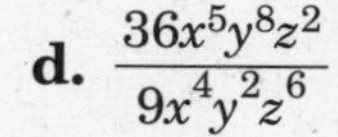

HOMEWORK ASSIGNMENT

Page(s):

Exercises:

8–3 Scientific Notation

What You'll Learn

- Express numbers in scientific notation and standard notation.
- Find products and quotients of numbers expressed in scientific notation.

Key Concept

Scientific Notation A number is expressed in scientific notation when it is written as a product of a factor and a power of 10. The factor must be greater than or equal to 1 and less then 10.

EXAMPLE Scientific to Standard Notation

1 **a. Express 7.48×10^{-3} in standard notation.**

$7.48 \times 10^{-3} = 0.00748$ $\quad$ $n = -3$ move decimal point ____ places to the ____.

b. Express 2.19×10^{5} in standard notation.

$2.19 \times 10^{5} = 219000$ $\quad$ $n =$ ____ move decimal point ____ places to the ____.

Your Turn **Express each number in standard notation.**

a. 3.16×10^{-2} ____

b. 7.61×10^{3} ____

EXAMPLE Standard to Scientific Notation

2 **a. Express 0.000000672 in scientific notation.**

$0.000000672 \rightarrow 0.000000672 \times 10^{n}$ Move decimal point ____ places to the right.

$0.000000672 =$ ____ $\times$ ____ $\quad$ $a =$ ____ and $n =$ ____

b. Express 3,022,000,000,000 in scientific notation.

$3{,}022{,}000{,}000{,}000 \rightarrow 3{,}022{,}000{,}000{,}000 \times 10^{n}$

Move decimal point ____ places to the ____.

$3{,}022{,}000{,}000{,}000 =$ ____ $\times$ ____ $\quad$ $a =$ ____ and $n =$ ____

REMEMBER IT

Count decimal places carefully when converting numbers between scientific and standard notation.

Your Turn **Express each number in scientific notation**

a. 458,000,000

b. 0.0000452

EXAMPLE **Multiplication with Scientific Notation**

3 **Evaluate. Express each result in scientific and standard notation.**

$(7 \times 10^{-6})(4.3 \times 10^{12})$

$= (7 \times 4.3)(10^{-6} \times 10^{12})$ Commutative and Association Properties

$= 30.1 \times 10^{6}$ Product of ______

$= (3.01 \times 10^{1}) \times 10^{6}$ 30.1 = ______

= ______ ______ Property

$= 3.01 \times 10^{7}$ or ______ Product of ______

EXAMPLE **Division with Scientific Notation**

4 **Evaluate** $\frac{6.4 \times 10^4}{1.6 \times 10^7}$.

$\frac{6.4 \times 10^4}{1.6 \times 10^7} = \left(\frac{\square}{\square}\right)\left(\frac{\square}{\square}\right)$ Associative Property

$= 4 \times 10^{\square}$ or ______ Product of ______

Your Turn **Evaluate. Express each result in scientific and standard notation.**

a. $(3 \times 10^{5})(2.1 \times 10^{-3})$

b. $\frac{1.36 \times 10^7}{2.5 \times 10^4}$

HOMEWORK ASSIGNMENT

Page(s):

Exercises:

8–4 Polynomials

BUILD YOUR VOCABULARY (pages 176–177)

A **polynomial** is a monomial or a sum of monomials. A **binomial** is the sum of ______ monomials, and a **trinomial** is the sum of ______ monomials.

The **degree of a monomial** is the ______ of the exponents of all its variables. The **degree of a polynomial** is the greatest ______ of any term in the polynomial.

WHAT YOU'LL LEARN

- Find the degree of a polynomial.
- Arrange the terms of a polynomial in ascending or descending order.

REVIEW IT

Define like terms. *(Lesson 1-5)*

______

______

______

______

______

EXAMPLE Identify Polynomials

1 State whether each expression is a polynomial. If it is a polynomial, identify it as a *monomial*, *binomial*, or *trinomial*.

a. $6 - 4$

Yes, $6 - 4$ is the difference of two real numbers.

It is a ______.

b. $x^2 + 2xy - 7$

Yes , $x^2 + 2xy - 7$ is the sum and difference of three monomials.

It is a ______.

c. $\frac{14d + 19c^2}{5d^4}$

No, $\frac{14d}{5d^4}$ and $\frac{19c^2}{5d^4}$ are not monomials.

Your Turn **State whether each expression is a polynomial. If it is a polynomial, identify it as a *monomial*, *binomial*, or *trinomial*.**

a. $3x^2 + 2y + z$ ______

b. $4a^2 - b^{-2}$ ______

c. $8r - 5s$ ______

EXAMPLE Degree of a Polynomial

2 **Find the degree of each polynomial.**

	Polynomial	Terms	Degree of Each Term	Degree of Polynomial
a.	$12 + 5b + 6bc + 8bc^2$	$12, 5b, 6bc, 8bc^2$	0, 1, 2, 3	
b.	$9x^2 - 2x - 4$	$9x^2, -2x, -4$		
c.	$14g^2h^5i$			

Your Turn **Find the degree of each polynomial.**

	Polynomial	Terms	Degree of Each Term	Degree of Polynomial
a.	$11ab + 6b + 2ac^2 - 7$			
b.	$3r^3 + 5r^2s^2 - s^3$			
c.	$2x^5yz - x^2yz^2$			

EXAMPLE Arrange Polynomials in Ascending Order

3 **Arrange the terms of each polynomial so that the powers of x are in ascending order.**

a. $16 + 14x^3 + 2x - x^2$

$= 16x^0 + \_\_\_ + \_\_\_ - \_\_\_$ $x^0 = \_\_\_$

$= \_\_\_$

b. $7y^2 + 4x^3 + 2xy^3 - x^2y^2$

$= 7y^2 + 4y^4 + 2x^1y^2 - x^2y^2$ $x = x^1$

$= \_\_\_$

EXAMPLE Arrange Polynomials in Descending Order

4 **Arrange the terms of each polynomial so that the powers of x are in descending order.**

a. $8 + 7x^2 - 12xy^3 - 4x^3y$

$= 8x^0 + 7x^2 - 12x^1y^3 - 4x^3y$ $\quad$ $x^0 = 1$ and $x = x^1$

$=$

b. $a^4 + ax^2 - 2a^3xy^3 - 9x^4y$

$= a^4x^0 + a^1x^2 - 2a^3x^1y^3 - 9x^4y^1$ $\quad$ $x^0 = 1$ and $x = x^1$

$=$

Your Turn **Arrange the terms of each polynomial so that the powers of x are in descending order.**

a. $6x^2 - 3x^4 - 2x + 1$

b. $3 - 2xy^4 + 4x^3yz - x^2$

c. $3x^3 + 4x^4 - x^2 + 2$

d. $2y^5 - 7y^3x^2 - 8x^3y^2 - 3x^5$

HOMEWORK ASSIGNMENT

Page(s):

Exercises:

8–5 Adding and Subtracting Polynomials

WHAT YOU'LL LEARN

- Add polynomials.
- Subtract polynomials.

FOLDABLES™

ORGANIZE IT

In your Foldable, write examples that involve adding and subtracting polynomials.

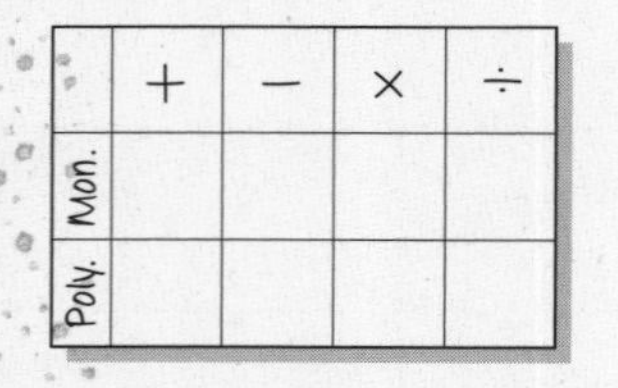

EXAMPLE Add Polynomials

1 **Find $(7y^2 + 2y - 3) + (2 - 4y + 5y^2)$.**

Method 1 Horizontal
Group like terms together.

$(7y^2 + 2y - 3) + (2 - 4y + 5y^2)$

$= (7y^2 + 5y^2) +$ ______ $+ [(-3) + 2]$ Associative and Commutative Properties.

$=$ ______ Add like terms.

Method 2 Vertical
Align the like terms in columns and add.

$7y^2 +$ ___ $- 3$

$(+)$ ___ $- 4y +$ ___

______

Notice that terms are in descending order with like terms aligned.

EXAMPLE Subtract Polynomials

2 **Find $(6y^2 + 8y^4 - 5y) - (9y^4 - 7y + 2y^2)$.**

Method 1 Horizontal
Subtract $9y^4 - 7y + 2y^2$ by adding its additive inverse.

$(6y^2 + 8y^4 - 5y) - (9y^4 - 7y + 2y^2)$

$= (6y^2 + 8y^4 - 5y) +$ ______

The additive inverse of $9y^4 - 7y + 2y^2$ is ______

$= [8y^4 + (-9y^4)] +$ ______ $+ (-5y + 7y)$ Group like terms.

$=$ ______ Add like terms.

Method 2 Vertical

Align like terms in columns and subtract by adding the additive inverse.

$$\begin{array}{r} 6y^2 + 8y^4 - 5y \\ (-)\ 2y^2 + 9y^4 - 7y \\ \hline \end{array}$$

Add the opposite.

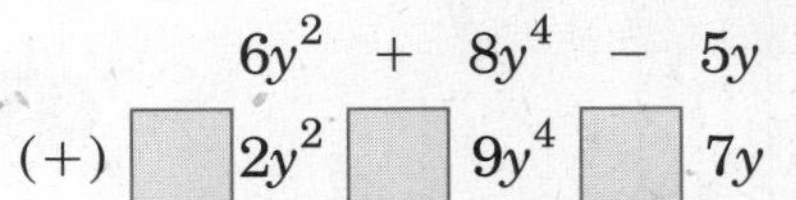

or

Your Turn

a. Find $(3x^2 + 2x - 1) + (-5x^2 + 3x + 4)$.

b. Find $(3x^3 + 2x^2 - x^4) - (x^2 + 5x^3 - 2x^4)$.

HOMEWORK ASSIGNMENT

Page(s):

Exercises:

8–6 Multiplying a Polynomial by a Monomial

WHAT YOU'LL LEARN

- Find the product of a monomial and a polynomial.
- Solve equations involving polynomials.

REVIEW IT

Explain how to write an expression in simplest form. *(Lesson 1-5)*

____

____

____

____

____

____

EXAMPLE Multiply a Polynomial by a Monomial

1 Find $6y(4y^2 - 9y - 7)$.

$6y(4y^2 - 9y - 7)$

$= 6y(\quad) - 6y(\quad) - 6y(\quad)$ Distributive Property

$=$ ______ Multiply.

Your Turn Find $3x(2x^2 + 3x + 5)$.

EXAMPLE Simplify Expressions

2 Simplify $3(2t^2 - 4t - 15) + 6t(5t + 2)$.

$3(2t^2 - 4t - 15) + 6t(5t + 2)$

$= 3(2t^2) - 3(4t) - 3(15) + 6t(5t) + 6t(2)$ Distributive Property

$=$ ______ Product of Powers

$= (6t^2 + 30t^2) + [(-12t) + 12t] - 45$ Commutative and Associative Properties

$=$ ______ Combine like terms.

Your Turn Simplify $5(4y^2 + 5y - 2) + 2y(4y + 3)$.

EXAMPLE Polynomials on Both Sides

3 Solve $b(12 + b) - 7 = 2b + b(-4 + b)$.

$b(12 + b) - 7 = 2b + b(-4 + b)$	Original equation
$12b + b^2 - 7 =$ ____	Distributive Property
$12b + b^2 - 7 =$ ____	Combine like terms.
____ = ____	Subtract b^2 from each side.
$12b = -2b + 7$	Add ____ to each side.
____ = ____	Add ____ to each side.
$b =$ ____	Divide each side by ____.

Your Turn Solve $x(x + 2) + 2x(x - 3) + 7 = 3x(x - 5) - 12$.

HOMEWORK ASSIGNMENT

Page(s):

Exercises:

8–7 Multiplying Polynomials

WHAT YOU'LL LEARN

- Multiply two binomials by using the FOIL method.
- Multiply two polynomials by using the Distributive Property.

EXAMPLE The Distributive Property

1 **Find $(y + 8)(y - 4)$**

$(y + 8)(y - 4) = y(y - 4) + 8(y - 4)$ Distributive Property

$=$ ______ Distributive Property

$=$ ______ Multiply.

$=$ ______ Combine like terms.

Your Turn Find $(c + 2)(c - 4)$.

______

BUILD YOUR VOCABULARY (page 176)

The shortcut of the Distributive Property is called the **FOIL method,** which can be used when multiplying two binomials.

KEY CONCEPT

FOIL Method for Multiplying Binomials

To multiply two binomials, find the sum of the products of

F the *First* terms,

O the *Outer* terms,

I the *Inner* terms, and

L the *Last* terms.

EXAMPLE FOIL Method

2 **a. Find $(z - 6)(z - 12)$.**

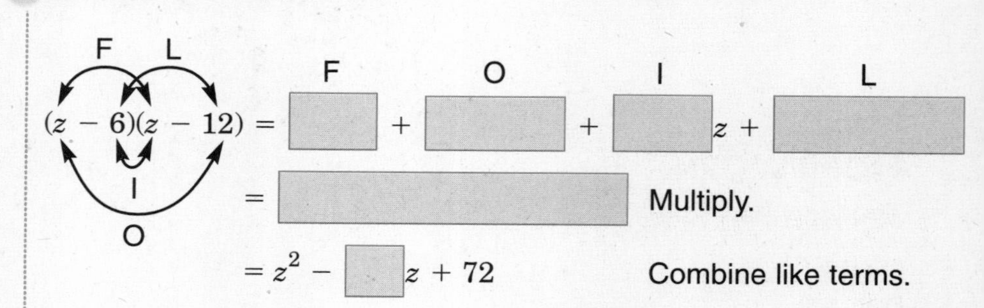

$(z - 6)(z - 12) =$ ______ (F) $+$ ______ (O) $+$ ______ z (I) $+$ ______ (L)

$=$ ______ Multiply.

$= z^2 -$ ______ $z + 72$ Combine like terms.

REMEMBER IT

When multiplying binomials, you can check your answer by reworking the problem using the Distributive Property.

b. Find $(5x - 4)(2x + 8)$.

$(5x - 4)(2x + 8)$

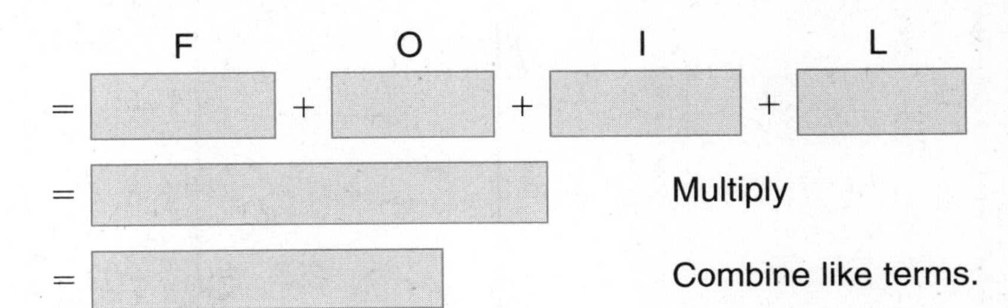

$= \quad + \quad + \quad + \quad$

$=$ Multiply

$=$ Combine like terms.

Your Turn **Find each product.**

a. $(x + 2)(x - 3)$

b. $(3x + 5)(2x - 6)$

FOLDABLES™

ORGANIZE IT

In your Foldable, in the box for polynomial multiplication, write examples of multiplying binomials using the Distributive Property and the FOIL method.

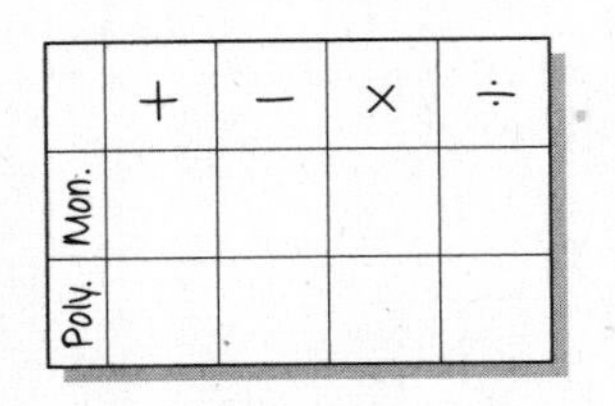

EXAMPLE Foil Method

3 GEOMETRY **The area A of a triangle is one-half the height h times the base b. Write an expression for the area of the triangle.**

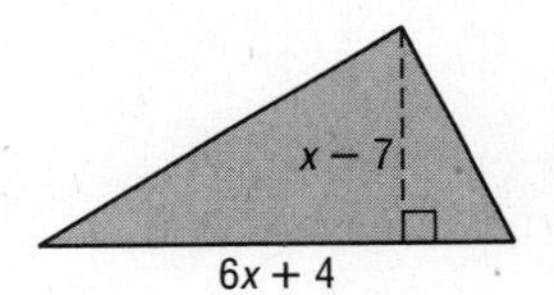

The height is $x - 7$ and the base is $6x + 4$. Write and apply the formula.

$A = \frac{1}{2}hb$ Original formula

$A = \frac{1}{2}$ Substitution

$A = \frac{1}{2}[x(6x) + x(4) - 7(6x) - 7(4)]$ FOIL method

$A = \frac{1}{2}$ Multiply.

$A = \frac{1}{2}$ Combine like terms.

$A =$ Distributive Property

The area of the triangle is  square units.

Your Turn The area of a rectangle is the measure of the base times the height. Write an expression for the area of the rectangle.

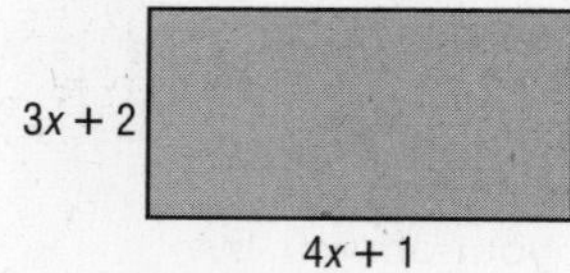

EXAMPLE The Distributive Property

4 **Find $(3a + 4)(a^2 - 12a + 1)$.**

$(3a + 4)(a^2 - 12a + 1)$

$= 3a$ ______ $+ 4$ ______ Distributive Property

$= 3a^3 -$ ______ $+$ ______ Distributive Property

$=$ ______ Combine like terms.

Your Turn **Find each product.**

a. $(3z + 2)(4z^2 + 3z + 5)$

b. $(3x^2 + 2x + 1)(4x^2 - 3x - 2)$

HOMEWORK ASSIGNMENT

Page(s):

Exercises:

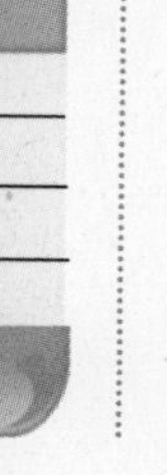

8–8 Special Products

WHAT YOU'LL LEARN

- Find squares of sums and differences.
- Find the product of a sum and a difference.

KEY CONCEPTS

Square of a Sum The square of $a + b$ is the square of a plus twice the product of a and b plus the square of b.

Square of a Difference The square of $a - b$ is the square of a minus twice the product of a and b plus the square of b.

EXAMPLE Square of a Sum

1 Find each product.

a. $(7z + 2)^2$

$(a + b)^2 =$ ______ Square of a Sum

$(7z + 2)^2 =$ ______ $a = 7z$ and $b = 2$

$=$ ______ Simplify.

b. $(5q + 9r)^2$

$(a+b)^2 =$ ______ Square of a Sum

$(5q + 9r)^2 =$ ______ $a = 5q$ and $b = 9r$

$=$ ______ Simplify.

EXAMPLE Square of a Difference

2 Find each product.

a. $(3c - 4)^2$

$(a - b)^2 =$ ______ Square of a Difference

$(3c - 4)^2 =$ ______ $-$ ______ $+$ ______ $a = 3c$ and $b = 4$

$=$ ______ Simplify.

b. $(6e - 6f)^2$

$(a - b)^2 =$ ______ Square of a Difference

$(6e - 6f)^2 =$ ______ $a = 6e$ and $b = 6f$

$=$ ______ Simplify.

EXAMPLE Product of a Sum and a Difference

3 Find each product

a. (9d – 4)(9d + 4).

$(a + b)(a - b) = a^2 - b^2$ Product of a Sum and a Difference

$(9d - 4)(9d + 4) =$ ______ $a = 9d$ and $b = 4$

$=$ ______ Simplify.

b. $(10g + 13h^3)(10g - 13\ h^3)$

$(a + b)(a - b) = a^2 - b^2$ Product of a Sum and a Difference

$(10g + 13h^3)(10g - 13h^3) = (10g)^2 - (13h^3)^2$ $a = 10g$ and $b = 13h^3$

$=$ ______ Simplify.

KEY CONCEPTS

Product of a Sum and a Difference The product of $a + b$ and $a - b$ is the square of a minus the square of b.

Special Products

Square of a Sum
$(a + b)^2 = a^2 + 2ab + b^2$

Square of a Difference
$(a - b)^2 = a^2 - 2ab + b^2$

Product of a Sum and a Difference
$(a - b)(a + b) = a^2 - b^2$

Your Turn **Find each product.**

a. $(3x + 2)^2$

b. $(4x + 2y)^2$

c. $(3m - 2)^2$

d. $(2p - 2q)^2$

e. $(3y + 2)(3y - 2)$

f. $(4y^2 + 5z)(4y^2 - 5z)$

HOMEWORK ASSIGNMENT

Page(s):

Exercises:

CHAPTER 8

BRINGING IT ALL TOGETHER

STUDY GUIDE

FOLDABLES™	VOCABULARY PUZZLEMAKER	BUILD YOUR VOCABULARY
Use your **Chapter 8 Foldable** to help you study for your chapter test.	To make a crossword puzzle, word search, or jumble puzzle of the vocabulary words in Chapter 8, go to: www.glencoe.com/sec/math/t_resources/free/index.php	You can use your completed **Vocabulary Builder** (pages 176–177) to help you solve the puzzle.

8–1 Multiplying Monomials

Simplify.

1. $3^5 \cdot 3^2$

2. $(a^3)^4$

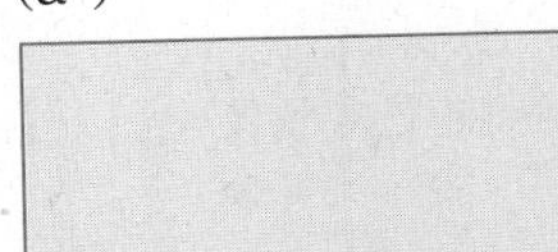

3. $(-4xy)^5$

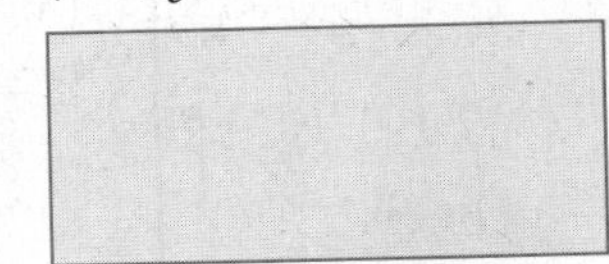

4. $y(y^3)(y^5)$

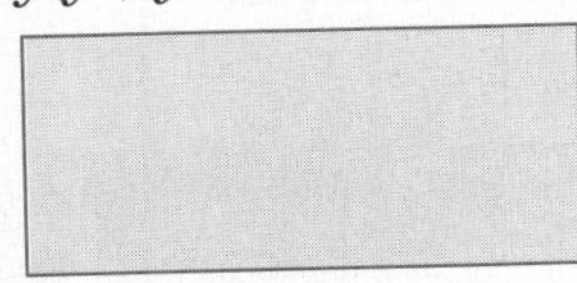

5. $(3c^2d^5)(cd^2)$

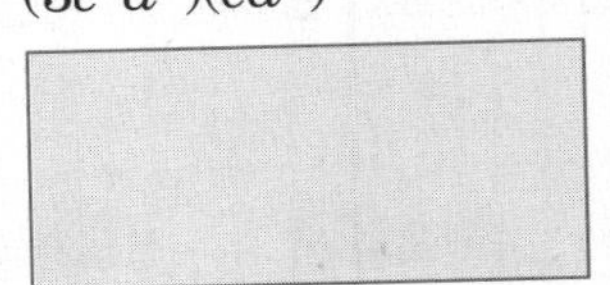

6. $(3m^5n^3)^2$

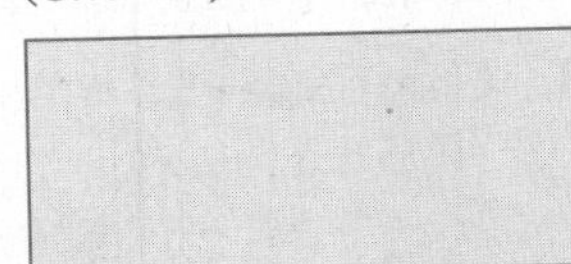

8–2 Dividing Monomials

Simplify. Assume that no denominator is equal to zero.

7. $\dfrac{12x^5}{36x}$

8. 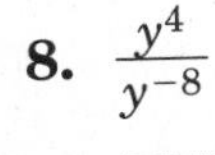 $\dfrac{y^4}{y^{-8}}$

9. $\dfrac{-5w^2}{25w^7}$

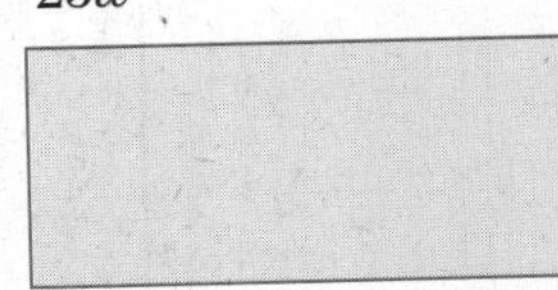

10. $\dfrac{m^{-2}n^{-5}}{(m^4n^3)^{-1}}$

11. $\left(\dfrac{x^{-5}y^4}{5^{-2}}\right)$

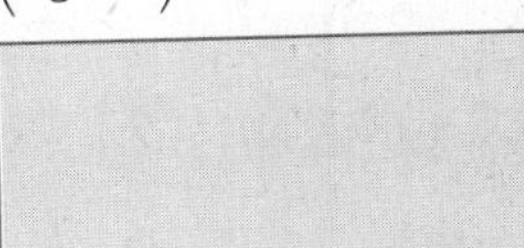

12. $\dfrac{(3q)^3}{q^4}$

8–3

Scientific Notation

13. Is the number 0.0543×10^4 in scientific notation? Explain.

Complete each sentence to change from scientific notation to standard notation.

14. To express 3.64×10^6 in standard notation, move the decimal point ____ places to the ____.

15. To express 7.825×10^{-3} in standard notation, move the decimal point ____ places to the ____.

Complete each sentence to change from standard notation to scientific notation.

16. To express 0.0007865 in scientific notation, move the decimal point ____ places to the right and write ____.

17. To express 54,000,000,000 in scientific notation, move the decimal point ____ places to the left and write ____.

8–4

Polynomials

18. Complete the table.

	monomial	binomial	trinomial	polynomial with more than three terms
Example	$3r^2t$	$2x^2 + 3x$	$5x^2 + 3x + 2$	$7s^2 + s^4 + 2s^3 - s + 5$
Number of Terms				

19. What is the degree of the polynomial $4x^4 + 2x^3y^3 + y^2 + 14$? Explain how you found your answer.

20. Use a dictionary to find the meaning of the terms *ascending* and *descending*. Write their meanings and then describe a situation in your everyday life that relates to them.

8–5 Adding and Subtracting Polynomials

Find each sum or difference.

21. $(3k - 8) + (7k + 12)$

22. $(w^2 + w - 4) + (7w^2 - 4w + 8)$

23. $(7h^2 + 4h - 8) - (3h^2 - 2h + 10)$

24. $(17n^4 + 2n^3) - (10n^4 + n^3)$

8–6 Multiplying a Polynomial by a Monomial

Find each product.

25. $2y^2(3y^2 + 2y - 7)$

26. $-3x^3(x^3 - 2x^2 + 3)$

27. Let n be an integer. What is the product of five times the integer added to two times the next consecutive integer?

8–7 Multiplying Polynomials

Find each product.

28. $(x + 5)(x - 3)$

29. $(3y + 6)(y - 2)$

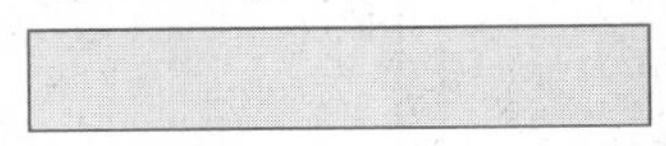

30. $(7x - 4)(7x + 4)$

31. $(3x - 4)(2x^2 + 3x + 3)$

8–8 Special Products

Find each product. Then identify the special product.

33. $(x - 4)^2$

34. $(x + 11)(x - 11)$

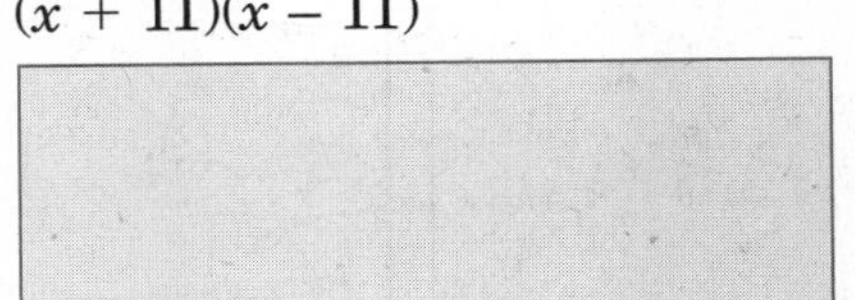

ARE YOU READY FOR THE CHAPTER TEST?

Visit **www.algebra1.com** to access your textbook, more examples, self-check quizzes, and practice tests to help you study the concepts in Chapter 8.

Check the one that applies. Suggestions to help you study are given with each item.

☐ **I completed the review of all or most lessons without using my notes or asking for help.**

- You are probably ready for the Chapter Test.
- You may want take the Chapter 8 Practice Test on page 469 of your textbook as a final check.

☐ **I used my Foldable or Study Notebook to complete the review of all or most lessons.**

- You should complete the Chapter 8 Study Guide and Review on pages 464–468 of your textbook.
- If you are unsure of any concepts or skills, refer back to the specific lesson(s).
- You may also want to take the Chapter 8 Practice Test on page 469.

☐ **I asked for help from someone else to complete the review of all or most lessons.**

- You should review the examples and concepts in your Study Notebook and Chapter 8 Foldable.
- Then complete the Chapter 8 Study Guide and Review on pages 464–468 of your textbook.
- If you are unsure of any concepts or skills, refer back to the specific lesson(s).
- You may also want to take the Chapter 8 Practice Test on page 469.

Student Signature

Parent/Guardian Signature

Teacher Signature

Factoring

Use the instructions below to make a Foldable to help you organize your note as you study the chapter. You will see Foldable reminders in the margin of this Interactive Study Notebook to help you in taking notes.

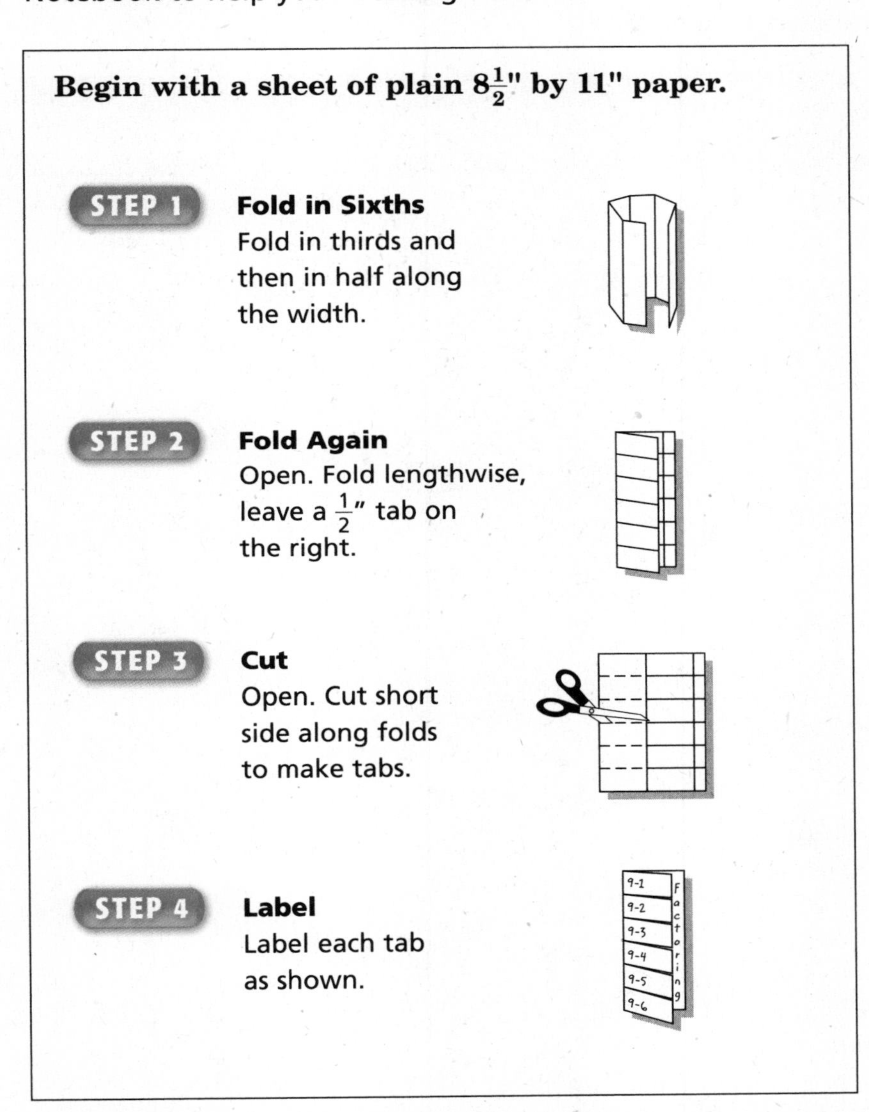

Begin with a sheet of plain $8\frac{1}{2}$" by 11" paper.

STEP 1 **Fold in Sixths**
Fold in thirds and then in half along the width.

STEP 2 **Fold Again**
Open. Fold lengthwise, leave a $\frac{1}{2}$" tab on the right.

STEP 3 **Cut**
Open. Cut short side along folds to make tabs.

STEP 4 **Label**
Label each tab as shown.

NOTE-TAKING TIP: As soon as possible, go over your notes. Clarify any ideas that were not complete.

BUILD YOUR VOCABULARY

This is an alphabetical list of new vocabulary terms you will learn in Chapter 9. As you complete the study notes for the chapter, you will see Build Your Vocabulary reminders to complete each term's definition or description on these pages. Remember to add the textbook page number in the second column for reference when you study.

Vocabulary Term	Found on Page	Definition	Description or Example
composite number [kahm-PAH-zeht]			
factored form			
factoring			
factoring by grouping			
greatest common factor (GCF)			
perfect square trinomial [try-NOH-mee-uhl]			
prime factorization [FAK-tuh-ruh-ZAY-shuhn]			
prime number			
prime polynomial			
Square Root Property			
Zero Product Property			

9–1 Factors and Greatest Common Factors

WHAT YOU'LL LEARN

- Find prime factorizations of integers and monomials.
- Find the greatest common factors of integers and monomials.

KEY CONCEPTS

Prime and Composite Numbers A whole number greater than 1, whose only factors are 1 and itself, is called a **prime number**.

A whole number, greater than 1, that has more than two factors is called a **composite number**.

BUILD YOUR VOCABULARY (page 202)

When a whole number is expressed as a product of ________ that are all ________ numbers, the expression is called the prime factorization of the number.

A monomial is in factored form when it is expressed as the product of ________ numbers and ________ and no variable has an exponent greater than 1.

EXAMPLE Classify Numbers as Prime or Composite

1 **Factor each number. Then classify it as *prime* or *composite*.**

a. 22

To find the factors of 22, list all pairs of whole numbers whose product is 22.

1×22 ________

The factors of 22, in increasing order, are 1, ____, ____, and 22. Since 22 has more than ________ factors, it is a ________ number.

b. 31

The only whole numbers that can be multiplied together to get 31 are ____ and ____. The factors of 31 are ____ and ____. Since the only factors of 31 are ____ and ________, 31 is a ________ number.

Your Turn **Factor each number. Then classify it as prime or composite**

a. 17

____

b. 25

____

WRITE IT

When finding the prime factorization of 36, could you start with 6 6 or 9 4 and still get the same prime factorization? Explain.

____

____

____

____

EXAMPLE Prime Factorization of a Positive Integer

2 **Find the prime factorization of 84. Use least prime factors.**

$84 = 2 \cdot$ ☐ — The least prime factor of 84 is ☐.

$= 2 \cdot$ ☐ — The least prime factor of 42 is ☐.

$= 2 \cdot$ ☐ — The least prime factor of 21 is ☐.

Thus, the prime factorization of 84 is ☐.

EXAMPLE Prime Factorization of a Negative Integer

3 **Find the prime factorization of −132. Use a factor tree.**

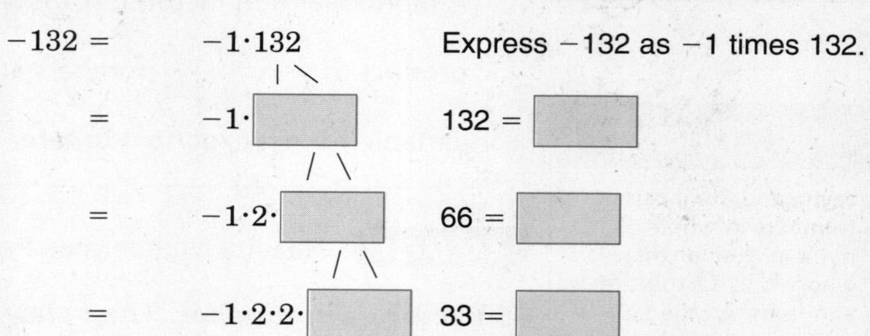

$-132 = -1 \cdot 132$ — Express −132 as −1 times 132.

$= -1 \cdot$ ☐ — 132 = ☐

$= -1 \cdot 2 \cdot$ ☐ — 66 = ☐

$= -1 \cdot 2 \cdot 2 \cdot$ ☐ — 33 = ☐

The prime factorization of −132 is ☐ or $-1 \cdot$ ☐$^2 \cdot 3 \cdot 11$

FOLDABLES

ORGANIZE IT

Under the tab for Lesson 9–1, write a monomial that can be factored. Then factor the monomial.

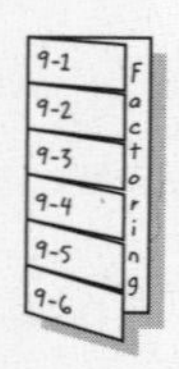

Your Turn

Find the prime factorization of each integer.

a. Find the prime factorization of 60.

☐

b. 24

☐

c. −154

☐

EXAMPLE Prime Factorization of a Monomial

4 **a. Factor $18x^3y^3$ completely.**

$18x^3y^3 = 2 \cdot 9 \cdot x \cdot x \cdot x \cdot y \cdot y \cdot y$ — $18 = 2 \cdot 9$, $x^3 = x \cdot x \cdot x$, and $y^3 = y \cdot y \cdot y$

$= 2 \cdot$ ☐ $\cdot x \cdot x \cdot x \cdot y \cdot y \cdot y$ — 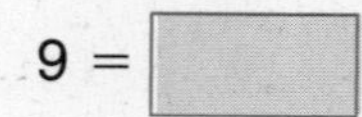 9 = ☐

☐ in factored form is $2 \cdot 3 \cdot 3 \cdot x \cdot x \cdot x \cdot y \cdot y \cdot y$.

b. Factor $-26rst^2$ completely.

$-26rst^2 = -1 \cdot 26 \cdot r \cdot s \cdot t \cdot t$ — Express −26 as −1 times 26.

$= \square \cdot r \cdot s \cdot t \cdot t$ — 26 = 13·2

$\square$ in factored form is $\square$.

Your Turn **Factor each monomial completely.**

a. $15a^3b^2$

b. $-45xy^2$

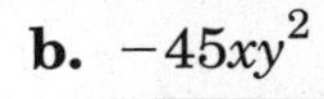

KEY CONCEPT

Greatest Common Factor (GCF)

- The GCF of two or more integers is the product of the prime factors common to the integers.
- The GCF of two or more monomials is the product of their common factors when each monomial is in factored form.
- If two or more integers or monomials have a GCF of 1, then the integers or monomials are said to be *relatively prime.*

EXAMPLE GCF of a set of Monomials

5 a. Find the GCF of 12 and 18.

$12 = ②\ 2\ ③$ — Factor each number.

$18 = ②\ ③\ 3$ — Circle the common prime factors.

The integers 12 and 18 have one 2 and one 3 as common prime factors. The product of these common prime factors, 2·3 or $\square$, is the GCF.

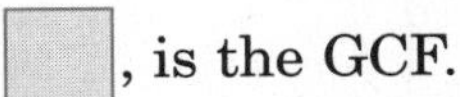

The GCF of 12 and 18 is $\square$.

b. Find the GCF of $27a^2b$ and $15ab^2c$.

$27a^2b = ③\ 3\ 3\ ⓐ\ a\ ⓑ$ — Factor each number.

$15ab^2c = ③\ 5\ ⓐ\ ⓑ\ b\ c$ — Circle the common prime factors.

The GCF of $27a^2b$ and $15ab^2c$ is $\square$.

Your Turn **Find the GCF of each set of monomials.**

a. 15 and 35

b. $39x^2y^3$ and $26xy^4$

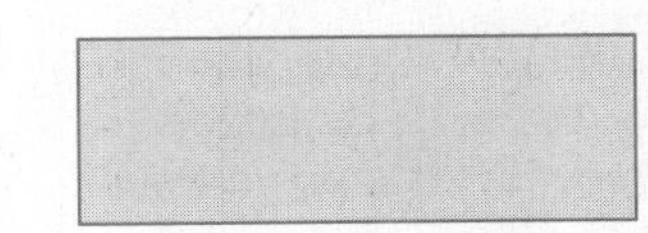

HOMEWORK ASSIGNMENT

Page(s):

Exercises:

9–2 Factoring Using the Distributive Property

What You'll Learn

- Factor polynomials by using the Distributive Property.
- Solve quadratic equations of the form $ax^2 + bx = 0$.

Build Your Vocabulary (page 202)

Factoring a polynomial means to find its ________ factored form. The ________ Property can also be used to factor some polynomials having ________ or more terms. This method is called **factoring by grouping**.

Key Concept

Factoring by Grouping
A polynomial can be factored by grouping if all the following situations exist.

- There are four or more terms.
- Terms with common factors can be grouped together.
- The two common factors are identical or are additive inverses of each other.

Foldables Under the tab for Lesson 9–2, list the steps factor a polynomial grouping.

EXAMPLE Use the Distributive Property

1 Use the Distributive Property of factor $15x + 25x^2$.
First, find the GCF of $15x$ and $25x^2$.

$15x = 3 \cdot 5 \cdot x$ — Factor each number.

$25x^2 = 5 \cdot 5 \cdot x \cdot x$ — Circle the common prime factors.

GCF: $5 \cdot x$ or ________

Then use the Distributive Property to factor out the GCF.

$15x + 25x^2 = 5x(3) + 5x(5 \cdot x)$ — Rewrite each term using the GCF.

$= 5x($____$) + 5x($____$)$ — Simplify remaining factors.

$=$ ____ $(3 + 5x)$ — Distributive Property

EXAMPLE Use Grouping

2 Factor $2xy + 7x - 2y - 7$.

$2xy + 7x - 2y - 7$

$=$ ________ $+ 7x - 7$ — Group terms with common factors.

$= 2y$ ________ $+ 7$ ________ — Factor the GCF from each grouping.

$= (x - 1)$ ________ — Distributive Property

EXAMPLE Use the Additive Inverse Property

3 $15a - 3ab + 4b - 20$

$15a - 3ab + 4b - 20$ — Group terms with common factors.

$= \_\_\_\_ + \_\_\_\_$

$= \_\_\_\_(5 - b) + \_\_\_\_(b - 5)$ — Factor GCF from each grouping.

$= 3a(-1)\_\_\_\_ + 4\_\_\_\_$ — $(5 - b) = -1(b - 5)$

$= \_\_\_\_(b - 5) + \_\_\_\_(b - 5)$ — $3a(-1) = \_\_\_\_$

$= (b - 5)\_\_\_\_$ — Distributive Property

Your Turn **Factor each polynomial.**

a. $3x^2y + 12xy^2$

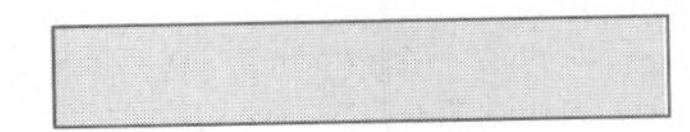

b. $6ab^2 + 15a^2b^2 + 27ab^3$

c. $4xy + 3y - 20x - 15$

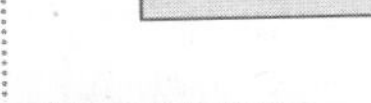

d. $-2xy - 10x + 3y + 15$

EXAMPLE Solve an Equation by Factoring

KEY CONCEPT

Zero Product Property If the product of two factors is 0, then at least one of the factors must be 0.

4 $4y = 12y^2$

Write the equation so that it is of the form $ab = 0$.

$4y = 12y^2$ — Original equation

$4y - \_\_\_\_ = 0$ — Subtract ____ from each side.

$4y\_\_\_\_ = 0$ — Factor the GCF of $4y$ and $12y^2$, which is $4y$.

$4y = 0$ or $1 - 3y = 0$ — Zero Product Property

$y = \_\_\_\_$ $\quad y = \_\_\_\_$ — Solve each equation.

The solution set it ____.

HOMEWORK ASSIGNMENT

Page(s):

Exercises:

Your Turn **Solve each equation. Then check the solutions.**

a. $(s - 3)(3s + 6) = 0$

b. $5x - 40x^2 = 0$

9–3 Factoring Trinomials: $x^2 + bx + c$

WHAT YOU'LL LEARN

- Factor trinomials of the form $x^2 + bx + c$.
- Solve equations of the form $x^2 + bx + c = 0$.

EXAMPLE *b* is Negative and *c* is Positive

1 **Factor $x^2 - 12x + 27$.**

In this trinomial, $b =$ ____ and $c =$ ____. This means $m + n$ is negative and mn is positive. So m and n must both be negative. Make a list of the negative factors of ____, and look for the pair whose sum is ____.

Factors of 27	Sum of Factors
−1, −27	
−3, −9	

The correct factors are ____ and ____.

$x^2 - 12x + 27 = (x + m)(x + n)$ Write the pattern

$=$ ____ $m =$ ____ and $n =$ ____

EXAMPLE *b* is Positive and *c* is Negative

2 **Factor $x^2 + 3x - 18$.**

In this trinomial, $b = 3$ and $c = -18$. This means $m + n$ is positive and mn is negative, so either m or n is negative, but not both. Make a list of the factors of −18. Look for the pair of factors whose sum is 3.

Factors of −18		Sum of Factors
1,		−17
−1,		17
2,		−7
−2,	9	
3,	−6	
−3,	6	

The correct factors are ____ and ____.

$x^2 + 3x - 18 = (x + m)(x + n)$ Write the pattern.

$=$ ____ $m =$ ____ and $n =$ ____

KEY CONCEPT

Factoring $x^2 + bx + c$ To factor quadric trinomials of the form $x + bx + c$, find two integers, m and n, whose sum is equal to b and whose product is equal to c. Then write $x^2 + bx + c$ using the pattern $(x + m)(x + n)$.

FOLDABLES™ Take notes explaining how to factor trinomials in the form $x^2 + bx + c$. Include examples.

REMEMBER IT

Before factoring, rewrite the equation so that one side equals 0.

EXAMPLE ***b* is Negative and *c* is Negative**

3 Factor $x^2 - x - 20$.

Since $b =$ ______ and $c =$ ______, $m + n$ is negative and mn is negative. So either m or n is negative, but not both.

Factors of −20	Sum of Factors
1, −20	______
−1, 20	______
2, ______	−8
______, 10	8
______, −5	−1

The correct factors are 4 and −5.

$x^2 - x - 20 = (x + m)(x + n)$ Write the pattern.

$=$ ______ $m = 4$ and $n = -5$

Your Turn **Factor each trinomial.**

a. $x^2 + 3x + 2$

______

b. $x^2 - 10x + 16$

______

c. $x^2 + 4x - 5$

______

d. $x^2 - 5x - 24$

______

EXAMPLE **Solve an Equation by Factoring**

4 Solve $x^2 + 2x = 15$.

$x^2 + 2x = 15$ Original equation

______ $= 0$ Subtract 15 from each side.

 $= 0$ Factor.

______ $= 0$ or ______ $= 0$ Zero Product Property

$x =$ ______ $x =$ ______ Solve each equation.

The solution is ______.

Your Turn Solve $x^2 - 20 = x$.

______

HOMEWORK ASSIGNMENT

Page(s):

Exercises:

9–4 Factoring Trinomials: $ax^2 + bx + c$

EXAMPLE Factor $ax^2 + bx + c$

What You'll Learn

- Factor trinomials of the form $ax^2 + bx + c$.
- Solve equations of the form $ax^2 + bx + c = 0$.

Review It

Which property is applied when factoring by grouping? *(Lesson 9-2)*

____

____

____

____

____

1 a. Factor $5x^2 + 27x + 10$.

In this trinomial, $a =$ ______, $b =$ ______, and $c =$ ______. Find two numbers whose sum is 27 and whose product is $5 \cdot 10$ or 50. Make a list of factors of 50 and look for a pair of factors whose sum is 27.

Factors of 50	Sum of Factors
1, 50	______
2, 25	______

The correct factors are 2 and 25.

$5x^2 + 27x + 10 = 5x^2 + mx + nx + 10$ — Write the pattern.

$= (5x^2 + 2x) + (25x + 10)$ — $m =$ ______ and $n =$ ______

$=$ ______ $+$ ______ — Group terms with common factors.

$=$ ______ $(5x + 2) +$ ______ $(5x + 2)$ — Factor the GCF from each grouping.

$=$ ______ — Distributive Property

b. Factor $24x^2 - 22x + 3$.

In this trinomial, $a =$ ______, $b =$ ______, and $c =$ ______.
Since b is negative, $m + n$ is negative. Since c is positive mn is positive. So m and n must both be negative. Make a list of the negative factors of $24 \cdot 3$ or 72, and look for the pair of factors whose sum is -22.

Factors of 72	Sum of Factors
-1, ______	-73
-2, ______	-38
______, -24	-27
______, -18	-22

The correct factors are -4, -18.

$24x^2 - 22x + 3 = 24^2 + mx + nx + 3$ Write the pattern.

$=$ ______ $m =$ ______ and $n =$ ______

$= ($ ______ $) + (-18x + 3)$ Group terms with common factors.

$=$ ______ $(6x - 1) + ($ ______ $)(6x - 1)$ Factor the GCF from each grouping.

$=$ ______ Distributive Property

Your Turn **Factor each trinomial.**

a. $3x^2 + 26x + 35$

b. $3x^2 - 17x + 10$

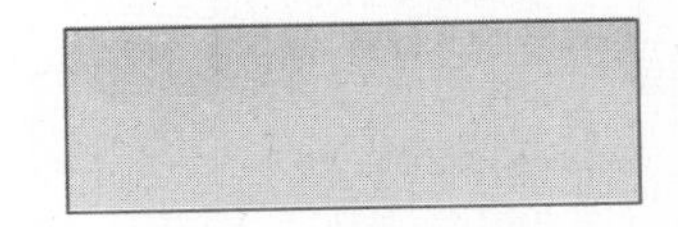

WRITE IT

Before trying to factor a trinomial, what should you check for?

______

______

______

______

EXAMPLE Factor When *a*, *b*, and *c* Have a Common Factor

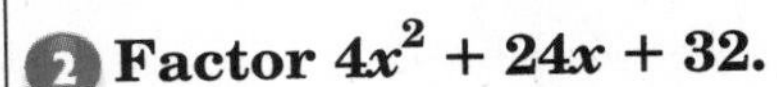

2 Factor $4x^2 + 24x + 32$.

Notice that the GCF of the terms $4x^2$, $24x$, and 32 is ______. When the GCF of the terms of a trinomial is an integer other than 1, you should first factor out this GCF.

$4x^2 + 24x + 32 =$ ______ $(x^2 + 6x + 8)$ Distributive Property

Now factor $x^2 + 6x + 8$. Since the lead coefficient is 1, find the two factors of 8 whose sum is 6.

Factors of 8	Sum of Factors
______	9
______	6

The correct factors are ______ and ______.

So, $x^2 + 6x + 4 = (x + 2)(x + 4)$. Thus, the complete factorization of $4x^2 + 24x + 32$ is ______.

BUILD YOUR VOCABULARY (page 202)

A polynomial that cannot be written as a product of two polynomials with ______ coefficients is called a **prime polynomial**.

EXAMPLE Determine Whether a Polynomial is Prime

3 Factor $3x^2 + 7x - 5$.

In this trinomial, $a = 3$, $b = 7$, and $c = -5$. Since b is positive, $m + n$ is positive. Since c is negative, mn is negative, so either m or n is negative, but not both. Make a list of all the factors of $3(-5)$ or -15. Look for the pair of factors whose sum is 7.

Factors of −15	Sum of Factors	Factors of −15	Sum of Factors
−1, 15		−3, 5	
1, −15		3, −5	

There are no integral factors whose sum is 7. Therefore,

$3x^2 + 7x - 5$ is a ______.

FOLDABLES™

ORGANIZE IT

Under the tab for Lesson 9-4, list the steps you use to solve equations by factoring.

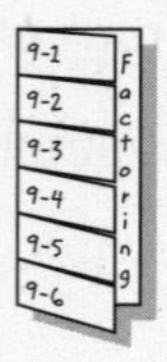

Your Turn **Factor each trinomial. If the trinomial cannot be factored using integers, write *prime*.**

a. $2x^2 + 14x + 20$

b. $3x^2 - 5x + 3$

EXAMPLE Solve Equations by Factoring

4 Solve $18b^2 - 19b - 8 = 3b^2 - 5b$.

$18b^2 - 19b - 8 = 3b^2 - 5b$	Original equation
______ $= 0$	Rewrite so one side equals 0.
______ $(3b - 4) = 0$	Factor the left side.
______ $= 0$ or ______ $= 0$	Zero Product Property
$5b =$ ______ $\quad 3b =$ ______	Solve each equation.
$b =$ ______ $\quad b =$ ______	

The solution set is ______.

HOMEWORK ASSIGNMENT

Page(s):

Exercises:

Your Turn Solve $12x^2 + 19x + 5 = 0$.

9–5 Factoring Differences of Squares

What You'll Learn

- Factor binomials that are the difference of squares.
- Solve equations involving the differences of squares.

EXAMPLE Factor the Difference of Squares

1 **Factor $16y^2 - 81z^2$.**

Write in the form $a^2 - b^2$.

$16y^2 - 81z^2 = \square^2 - \square^2$ — $16y^2 = 4y \cdot 4y$ and $81z^2 = 9z \cdot 9z$

$= \square\square$ — Factor the difference of squares.

EXAMPLE Factor Out a Common Factor

2 **Factor $3b^2 - 27b$.**

$3b^2 - 27b = 3b(b^2 - 9)$ — The GCF is $3b$.

$= 3b[b^2 - 3^2]$ — $b^2 = b \cdot b$ and $9 = 3 \cdot 3$.

$= \square$ — Factor the difference of squares.

Key Concept

Difference of Squares
$a^2 - b^2 = (a + b)(a - b)$ or $(a - b)(a + b)$

Foldables

Under the tab for Lesson 9-5, write a binomial that is the difference of squares. Then factor the binomial.

EXAMPLE Apply a Factoring Technique More Than Once

3 **Factor $4y^4 - 2500$.**

$4y^4 - 2500 = 4(y^4 - 625)$ — The GCF is 4.

$= 4[(\square)^2 - \square]$ — $y^4 = y^2 \cdot y^2$ and $625 = 25 \cdot 25$

$= 4\square\square$ — Factor the difference of squares.

$= 4(y^2 + 25)(y^2 - 5^2)$ — $y^2 = y \cdot y$ and $25 = 5 \cdot 5$

$= \square$ — Factor the difference of squares.

Your Turn **Factor each polynomial.**

a. $b^2 - 9$

b. $36k^2 - 144m^2$

c. $5x^3 - 20x$

d. $3y^4 - 48$

EXAMPLE Apply Several Different Factoring Techniques

4 **Factor $6x^3 + 30x^2 - 24x - 120$.**

$6x^3 + 30x^2 - 24x - 120$	Original polynomial
$= \square(x^3 + 5x^2 - 4x - 20)$	Factor out the GCF.
$= \square\,\square + (5x^2 - 20)$	Group terms with common factors.
$= \square[\square + \square]$	Factor each grouping.
$= 6(x^2 - 4)(x + 5)$	$x^2 - 4$ is the common factor.
$= \square$	Factor the difference of squares.

Your Turn Factor $5x^3 + 25x^2 - 45x - 225$.

EXAMPLE Solve Equations by Factoring

5 **Solve $48y^3 = 3y$ by factoring.**

$48y^3 = 3y$	Original equation
$\square = 0$	Subtract $3y$ from each side.
$\square(16y^2 - 1) = 0$	The GCF of $48y^2$ and $3y$ is $\square$.
$\square = 0$	$16y^2 = 4y \cdot 4y$ and $1 = 1 \cdot 1$

Applying the Zero Product Property, set each factor equal to zero and solve the resulting three equations.

$3y = 0$ or $4y + 1 = 0$ or $4y - 1 = 0$

$y = 0$ $\quad 4y = -1 \quad 4y = 1$

$y = -\frac{1}{4} \quad y = \frac{1}{4}$

The solution set is $\{-\frac{1}{4}, 0, \frac{1}{4}\}$.

Your Turn **Solve each equation by factoring. Check your solutions.**

a. $m^2 - 81 = 0$

b. $8x^3 = 2x$

HOMEWORK ASSIGNMENT

Page(s):

Exercises:

9–6 Perfect Squares and Factoring

BUILD YOUR VOCABULARY (page 202)

Perfect Square Trinomials are trinomials that are the ______ of a ______.

WHAT YOU'LL LEARN

- Factor perfect square trinomials.
- Solve equations involving perfect squares.

KEY CONCEPT

Factoring Perfect Square Trinomials
If a trinomial can be written in the form $a^2 + 2ab + b^2$ or $a^2 - 2ab + b^2$, then it can be factored as $(a + b)^2$ or as $(a - b)^2$, respectively.

EXAMPLE Factor Perfect Square Trinomials

1 Determine whether each trinomial is a perfect square trinomial. If so, factor it.

a. $25x^2 - 30x + 9$

1. Is the first term a perfect square? ______, $25x^2 = (5x)^2$.

2. Is the last term a perfect square? ______, $9 = 3^2$.

3. Is the middle term equal to ______$(5x)(3)$? Yes, ______.

$25x^2 - 30x + 9$ ______ a perfect square trinomial.

$25x^2 - 30x + 9 = (5x)^2 - 2(5x)(3) + 3^2$ Write as $a^2 - 2ab + b^2$

$=$ ______ Factor using the pattern.

b. $49y^2 + 42y + 36$

1. Is the first term a perfect square? ______, $49y^2 = (7y)^2$.

2. Is the last term a perfect square? ______, $36 = 6^2$

3. Is the middle term equal to $2(7y)(6)$? ______, $42y \neq 2(7y)(6)$

$49y^2 + 42y + 36$ ______ a perfect square trinomial.

Your Turn **Determine whether each trinomial is a perfect square trinomial. If so, factor it.**

a. $9x^2 - 12x + 16$

b. $16x^2 + 16x + 4$

EXAMPLE Factor Completely

2 **Factor each polynomial.**

a. $\mathbf{6x^2 - 96}$

First check for a GCF. Then, since the polynomial has two terms, check for the difference of squares.

$6x^2 - 96 = 6$ ______ 6 is the GCF.

$= 6$ ______ $x^2 = x \cdot x$ and $16 = 4 \cdot 4$

$=$ ______ Factor the difference of squares.

FOLDABLES

ORGANIZE IT

Under the tab for Lesson 9-6, use your own words to explain why equations with perfect square trinomials only have one solution.

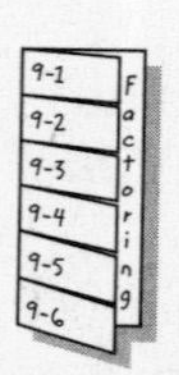

b. $\mathbf{16y^2 + 8y - 15}$

This polynomial has three terms that have a GCF of 1. While the first term is a perfect square, $16y^2 =$ ______, the last term is not. Therefore, this is not a perfect square trinomial.

This trinomial is in the form $ax^2 + bx + c$. Are there two numbers m and n whose product is $16 \cdot -15$ or -240 and whose sum is 8? Yes, the product of ______ and ______ is -240 and their sum is 8.

$16y^2 + 8y - 15 = 16y^2 + mx + nx - 15$ Write the pattern.

$= 16y^2$ ______ $- 15$

$=$ ______ $+$ ______

$=$ ______ $(4y + 5) -$ ______ $(4y + 5)$

$= (4y + 5)$ ______

Your Turn **Factor each polynomial.**

a. $3x^2 - 3$

b. $4x^2 + 10x + 6$

EXAMPLE Solve Equations with Repeated Factors

3 **Solve $4x^2 + 36x + 81 = 0$.**

$4x^2 + 36x + 81 = 0$	Original equation
$(\_\_\_)^2 + 2\_\_\_ + \_\_\_^2 = 0$	Recognize $4x^2 + 36x + 81$ as a perfect square trinomial.
$\_\_\_ = 0$	Factor the perfect square trinomial.
$\_\_\_ = 0$	Set the repeated factor equal to zero.
$x = \_\_\_$	Solve for x.

Thus, the solution set is .

Your Turn Solve $9x^2 - 30x + 25 = 0$.

BUILD YOUR VOCABULARY (page 202)

The **square root property** states that for any number $n > 0$, if $x^2 = n$, then $x = \_\_\_\sqrt{n}$.

EXAMPLE Use the Square Root Property To Solve Equations

4 **a. Solve $y^2 + 12y + 36 = 100$.**

$y^2 + 12y + 36 = 100$	Original equation
$\_\_\_^2 + \_\_\_ + 6^2 = 100$	Recognize perfect square trinomial.
$\_\_\_^2 = 100$	Factor perfect square trinomial.
$y + 6 = \_\_\_$	Square Root Property
$y + 6 = \_\_\_$	$100 = 10 \cdot 10$
$y = \_\_\_$	Subtract 6 from each side.
$y = -6 + 10$ or $y = -6 - 10$	Separate into two equations.
$= \_\_\_ \qquad = \_\_\_$	Simplify.

The solution set is ____.

b. Solve $(x + 9)^2 = 8$.

$(x + 9)^2 = 8$	Original equation
$\_\_\_ = \_\_\_$	Square Root Property
$x = \_\_\_$	Subtract 9 from each side.

Since 8 is not a perfect square, the solution set is ____. Using a calculator, the approximate solutions are $-9 + \sqrt{8}$ or about ____ and $-9 - \sqrt{8}$ or about ____.

HOMEWORK ASSIGNMENT

Page(s):

Exercises:

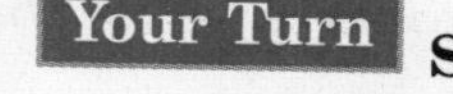

Solve each equation. Check your solutions.

a. $(x - 4)^2 = 25$ **b.** $y^2 + 6y + 9 = 64$ **c.** $(x - 5)^2 = 15$

CHAPTER 9

BRINGING IT ALL TOGETHER

STUDY GUIDE

FOLDABLES™	VOCABULARY PUZZLEMAKER	BUILD YOUR VOCABULARY
Use your **Chapter 9 Foldable** to help you study for your chapter test.	To make a crossword puzzle, word search, or jumble puzzle of the vocabulary words in Chapter 9, go to: www.glencoe.com/sec/math/t_resources/free/index.php	You can use your completed **Vocabulary Builder** (page 202) to help you solve the puzzle.

9–1 Factors and Greatest Common Factors

Choose the letter of the term that best matches each phrase.

a. composite number
b. prime number
c. factored form

1. the number 14 ____
2. a monomial that is expressed as the product of prime numbers and variables, with no variable having an exponent greater than 1 ____

3. the number 5 ____

Find the GCF of each set of monomials.

4. 12, 30, 114 ____
5. $6a^2$, $8a$ ____
6. $24xy^5$, $56x^3y$ ____

9–2 Factoring Using the Distributive Property

7. Complete.

$d^2 = -2d$	Original equation
$d^2 + 2d = 0$	____ $2d$ to each side.
____ $(d + 2) = 0$	Factor the GCF.
$d =$ ____ or $d + 2 = 0$	____ Product Property
$d =$ ____	Solve each equation.

The solution set is ____

9–3

Factoring Trinomials: $x^2 + bx + c$

Tell what sum and product m and n must have in order to use the pattern $(x + m)(x + n)$ to factor the given trinomial.

8. $x^2 + 10x + 24$ sum: product:

9. $x^2 - 12x + 20$ sum: product:

10. $x^2 - 4x - 21$ sum: product:

11. $x^2 + 6x - 16$ sum: product:

12. Find two consecutive even integers whose product is 168.

9–4

Factoring Trinomials: $ax^2 + bx + c$

Factor each trinomial, if possible. If the trinomial cannot be factored using integers, write *prime*.

13. $2b^2 + 10b + 12$

14. $4y^2 + 4y - 3$

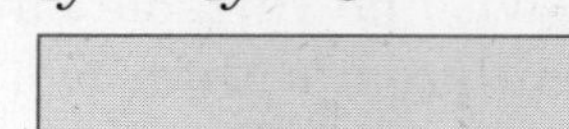

15. $12x^2 - 4y - 5$

16. $10x^2 - 9x + 6$

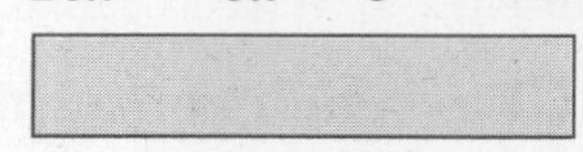

17. Explain how you know that the trinomial $2x^2 - 7x + 4$ is a prime polynomial.

9–5

Factoring Differences of Squares

Factor each polynomial.

18. $4x^2 - 25$

19. $49a^2 - 64b^2$

20. Explain what is done in each step to factor $4x^4 - 64$.

$4x^4 - 64$

$= 4(x^4 - 16)$

$= 4[(x^2)^2 - 4^2)]$

$= 4(x^2 + 4)(x^2 - 4)$

$= 4(x^2 + 4)(x^2 - 2^2)$

$= 4(x^2 + 4)(x + 2)(x - 2)$

9–6

Perfect Squares and Factoring

Match each polynomial from the first column with a factoring technique in the second column. Some of the techniqes may be used more than once. If none of the techniques can be used to factor the polynomial, write *none*.

21. $9x^2 - 64$

22. $9x^2 + 12x + 4$

23. $x^2 - 5x + 6$

24. $4x^2 + 13x + 9$

a. factor as $x^2 + bx + c$

b. factor as $ax^2 + bx + c$

c. difference of squares

d. perfect square trinomial

25. The area of a circle is given by the formula $A = \pi r^2$, where r is the radius. If increasing the radius of a circle by 3 inches gives the resulting circle an area of 81π square inches, what is the radius of the original circle?

ARE YOU READY FOR THE CHAPTER TEST?

Visit **www.algebra1.com** to access your textbook, more examples, self-check quizzes, and practice tests to help you study the concepts in Chapter 9.

Check the one that applies. Suggestions to help you study are given with each item.

☐ **I completed the review of all or most lessons without using my notes or asking for help.**

- You are probably ready for the Chapter Test.
- You may want take the Chapter 9 Practice Test on page 519 of your textbook as a final check.

☐ **I used my Foldable or Study Notebook to complete the review of all or most lessons.**

- You should complete the Chapter 9 Study Guide and Review on pages 515–518 of your textbook.
- If you are unsure of any concepts or skills, refer back to the specific lesson(s).
- You may also want to take the Chapter 9 Practice Test on page 519.

☐ **I asked for help from someone else to complete the review of all or most lessons.**

- You should review the examples and concepts in your Study Notebook and Chapter 9 Foldable.
- Then complete the Chapter 9 Study Guide and Review on pages 515–518 of your textbook.
- If you are unsure of any concepts or skills, refer back to the specific lesson(s).
- You may also want to take the Chapter 9 Practice Test on page 519.

Student Signature

Parent/Guardian Signature

Teacher Signature

Quadratic and Exponential Functions

Use the instructions below to make a Foldable to help you organize your notes as you study the chapter. You will see Foldable reminders in the margin of this Interactive Study Notebook to help you in taking notes.

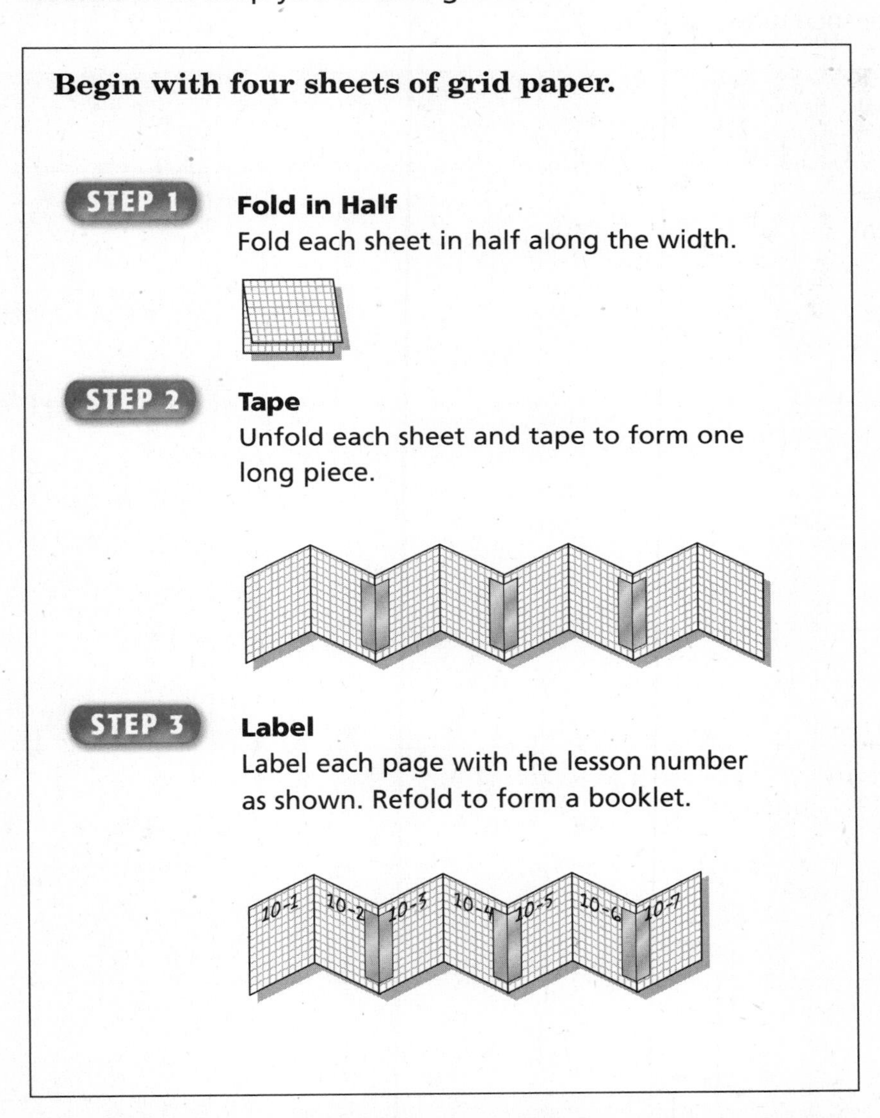

NOTE-TAKING TIP: If you find it difficult to write and pay attention at the same time, ask your instructor if you may record the classes with a tape recorder.

BUILD YOUR VOCABULARY

This is an alphabetical list of new vocabulary terms you will learn in Chapter 10. As you complete the study notes for the chapter, you will see Build Your Vocabulary reminders to complete each term's definition or description on these pages. Remember to add the textbook page number in the second column for reference when you study.

Vocabulary Term	Found on Page	Definition	Description or Example
axis of symmetry [SIH-muh-tree]			
common ratio			
completing the square			
compound interest			
discriminant			
exponential decay [EHK-spuh-NEHN-chuhl]			
exponential function			
exponential growth			
geometric means			
geometric sequence			

Vocabulary Term	Found on Page	Definition	Description or Example
maximum			
minimum			
parabola [puh-RA-buh-lh]			
quadratic equation [kwah-DRA-tihk]			
Quadratic Formula			
quadratic function			
roots			
symmetry			
vertex			
zeros			

10–1 Graphing Quadratic Functions

WHAT YOU'LL LEARN

- Graph quadratic functions.
- Find the equation of the axis of symmetry and the coordinates of the vertex of a parabola.

BUILD YOUR VOCABULARY (page 225)

The graph of a ________ function is called a **parabola**.

When graphing a parabola the ________ point is called the **minimum** and the ________ point is called the **maximum**. The ________ or ________ point of a parabola is called the **vertex**.

KEY CONCEPT

Quadratic Function A **quadratic function** can be described by an equation of the form $y = ax^2 + bx + c$, where $a \neq 0$.

FOLDABLES On the page for Lesson 10–1, write an example of a quadratic function that opens upward. Then write an example of a quadratic function that opens downward.

EXAMPLE Graph Opens Upward

1 Use a table of values to graph $y = x^2 - x - 2$.

x	y
−2	
−1	0
0	
1	
2	0
3	

Graph these ordered pairs and connect them with a smooth curve.

EXAMPLE Graph Opens Downward

2 Use a table of values to graph $y = -2x^2 + 2x + 4$.

x	y
−2	−8
−1	
0	4
1	
2	0
3	

Graph these ordered pairs and connect them with a smooth curve.

Your Turn **Use a table of values to graph each function.**

a. $y = x^2 + 2x + 3.$

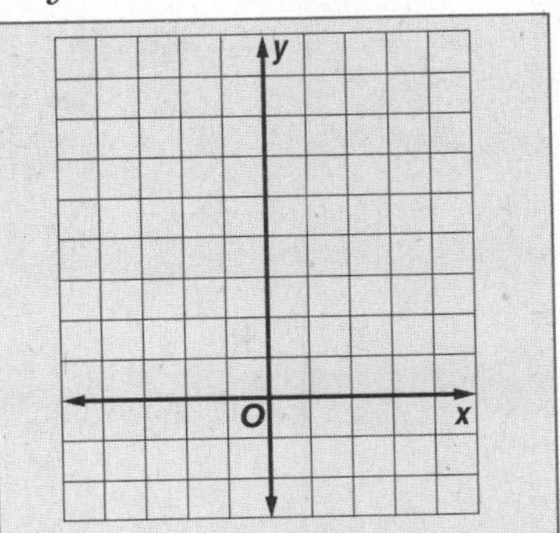

b. $y = -x^2 + 4.$

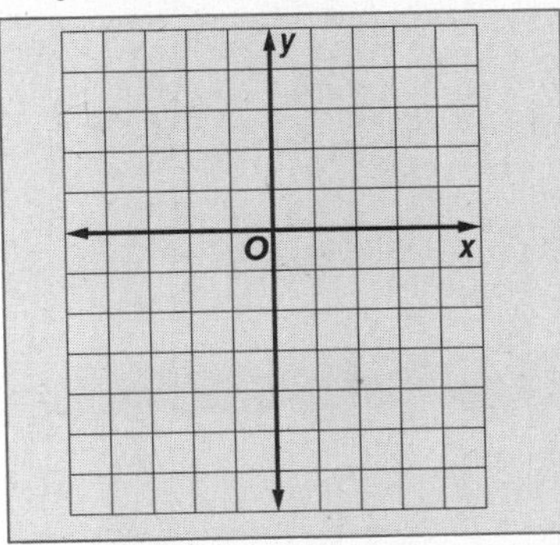

BUILD YOUR VOCABULARY (pages 224–225)

Parabolas possess a geometric property called **symmetry**.

The **axis of symmetry** ______ the parabola into two ______ halves.

EXAMPLE Vertex and Axis of Symmetry

3 **Consider the graph of $y = -2x^2 - 8x - 2$.**

a. Write the equation of the axis of symmetry.

In $y = -2x^2 - 8x - 2$, $a =$ ______ and $b =$ ______.

$x = -\frac{b}{2a}$ Equation for the axis of symmetry of a parabola

$x = -\frac{\square}{2\square}$ $a =$ ______ and $b =$ ______

$= \square$

The equation of the axis of symmetry is $x =$ ______.

KEY CONCEPT

Equation of the Axis of Symmetry of a Parabola The equation of the axis of symmetry for the graph of $y = ax^2 + bx + c$, where $a \neq 0$, is $x = -\frac{b}{2a}$.

b. Find the coordinates of the vertex.

Since the equation of the axis of symmetry is $x = -2$ and the vertex lies on the axis, the x-coordinate for the vertex is -2.

$y = -2x^2 - 8x - 2$	Original Equation
$y = -2(\_\_\_)^2 - 8(\_\_\_) - 2$	$x = -2$
$y =$ ___	Simplify.
$y =$ ___	Add.

The vertex is at ___.

REMEMBER IT

Functions can be graphed using the symmetry of the parabola. See page 526 of your textbook.

c. Identify the vertex as a maximum or minimum.

Since the coefficient of the x^2 term is ___, the parabola opens ___ and the vertex is a ___ point.

Your Turn **Consider the graph of $y = 3x^2 - 6x + 1$.**

Write the equation of the axis of symmetry.

Find the coordinates of the vertex.

Identify the vertex as a maximum or minimum.

HOMEWORK ASSIGNMENT

Page(s):

Exercises:

10–2 Solving Quadratic Equations by Graphing

BUILD YOUR VOCABULARY (page 225)

In a **quadratic equation**, the value of the related quadratic function is ______.

The ______ of a quadratic equation are called the **roots** of the equation. They can be found by the ______ or **zeros** of the related quadratic function.

WHAT YOU'LL LEARN

- Solve quadratic equations by graphing.
- Estimate solutions of quadratic equations by graphing.

EXAMPLE Two Roots

1 **Solve $x^2 - 3x - 10 = 0$ by graphing.**

Graph the related function $f(x) = x^2 - 3x - 10$.

The equation of the axis of symmetry is $x = -\frac{-3}{2(1)}$ or $x = \frac{3}{2}$.

When $x = \frac{3}{2}$, $f(x)$ equals $\left(\frac{3}{2}\right)^2 - 3\left(\frac{3}{2}\right) - 10$ or ______. So the coordinates of the vertex are ______.

Make a table of values to find other points to sketch the graph.

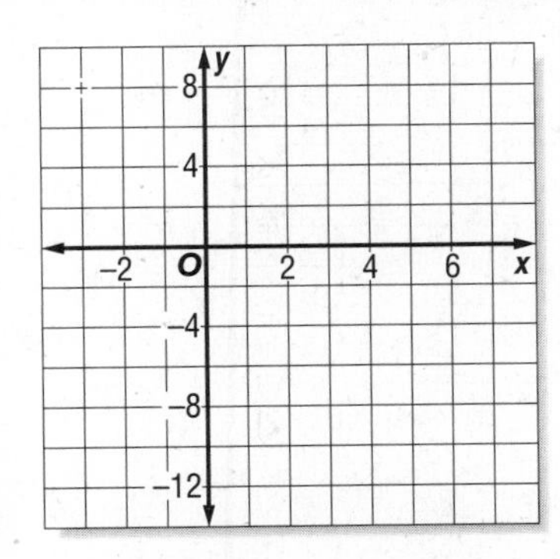

x	$f(x)$
−3	8
−1	
0	−10
1	−12
2	
3	−10
4	
6	

To solve $x^2 - 3x - 10 = 0$, you need to know where the value of $f(x)$ is ______. This occurs at the x-intercepts. The x-intercepts of the parabola appear to be ______ and ______.

EXAMPLE A Double Root

2 $x^2 - 6x = -9$

First rewrite the equation so one side is equal to zero.

$x^2 - 6x = -9$	Original equation
$x^2 - 6x$ ___ $= -9$ ___	Add ___ to each side.
$x^2 - 6x + 9 =$ ___	Simplify.

Graph the related function $f(x) = x^2 - 6x + 9$.

x	f(x)
1	4
2	
3	
4	
5	

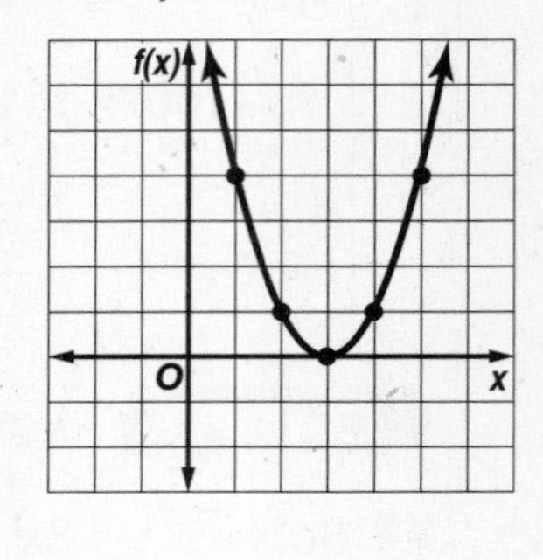

Notice that the vertex of the parabola is the x-intercept. Thus, one solution is ___. What is the other solution?

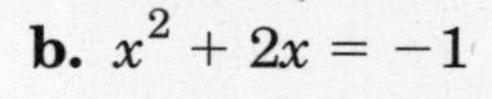

Your Turn **Solve each equation by graphing.**

a. $x^2 - 2x - 8 = 0$

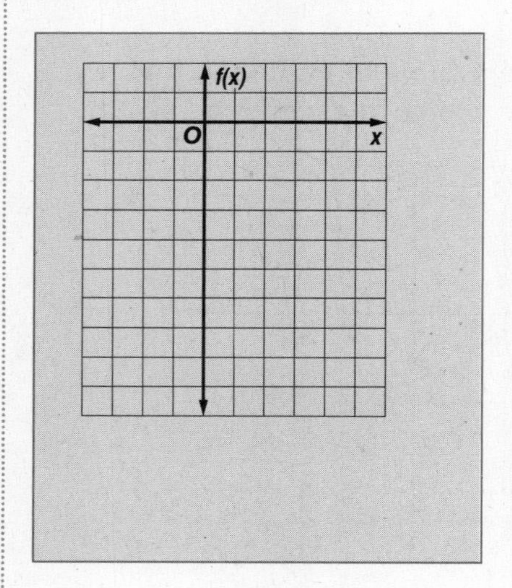

b. $x^2 + 2x = -1$

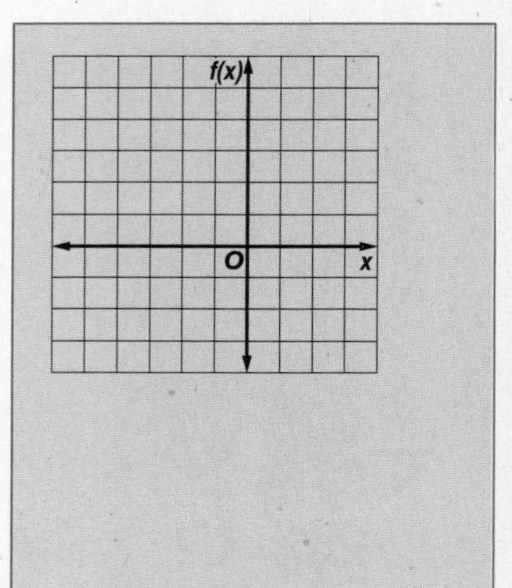

EXAMPLE Rational Roots

3 **Solve $x^2 - 4x + 2 = 0$ by graphing. If integral roots cannot be found, estimate the roots by stating the consecutive integers between which the roots lie.**

Graph the related function $f(x) = x^2 - 4x + 2$.

x	f(x)
0	2
1	
2	
3	
4	2

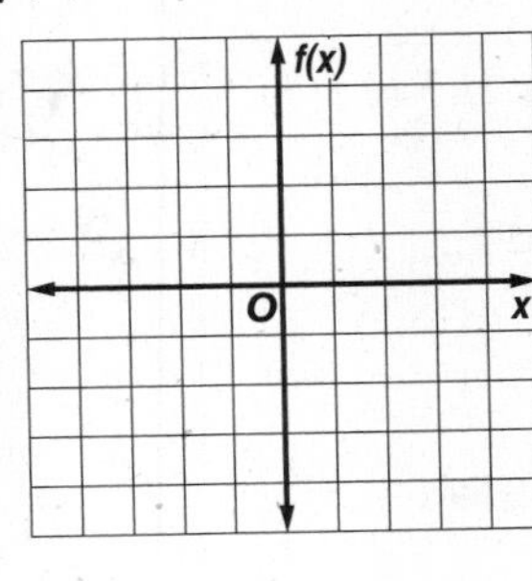

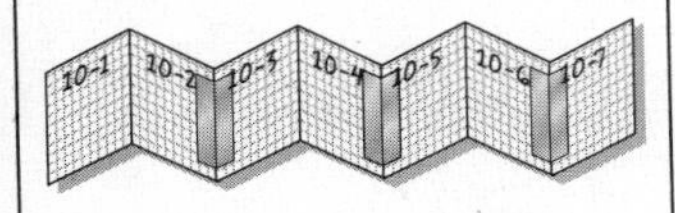

ORGANIZE IT

On the page for Lesson 10-2, write how you solve a quadratic equation by graphing.

The x-intercepts of the graph are between 0 and 1 and between 3 and 4. One root is between ______ and ______, and the other root is between ______ and ______.

Your Turn Solve $x^2 - 2x - 5$ by graphing. If integral roots cannot be found, estimate the roots by stating the consecutive integers between which the roots lie.

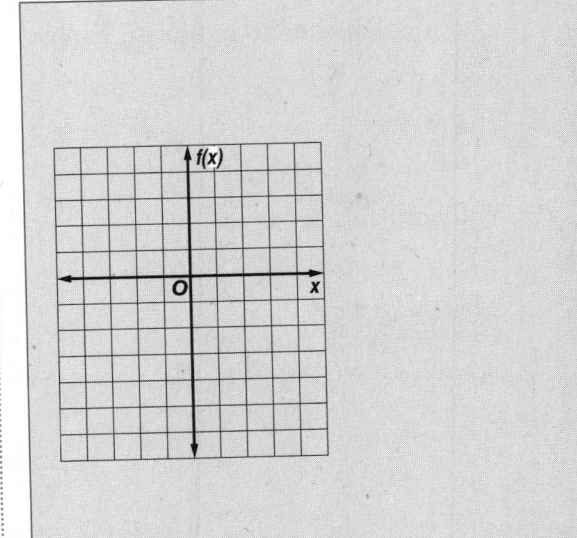

HOMEWORK ASSIGNMENT

Page(s):

Exercises:

10–3 Solving Quadratic Equations by Completing the Square

EXAMPLE Irrational Roots

WHAT YOU'LL LEARN

- Solve quadratic equations by finding the square root.
- Solve quadratic equations by completing the square.

1 **Solve $x^2 + 6x + 9 = 5$ by taking the square roots of each side. Round the nearest tenth if necessary.**

$x^2 + 6x + 9 = 5$	Original equation
$(x + 3)^2 = 5$	$x^2 + 6x + 9$ is a ______ ______ trinomial. 
______ = ______	Take the square root of each side.
$\lvert x + 3 \rvert = \sqrt{5}$	Simplify.
$x + 3 =$ ______	Definition of absolute value
$x + 3 -$ ______ $= \pm\sqrt{5} -$ ______	Subtract ______ from each side.
$x =$ ______	Simplify.

REMEMBER IT

When taking a square root of a positive number, there are two roots, one positive and one negative.

Use a calculator to evaluate each value of x.

$x = -3 + \sqrt{5}$ or $x = -3 - \sqrt{5}$

$\approx$ ______ $\approx$ ______

The solution set is ______.

Your Turn Solve $x^2 + 8x + 16 = 3$ by taking the square root of each side. Round to the nearest tenth if necessary.

BUILD YOUR VOCABULARY (page 224)

To add a ______ term to a binomial of the form $ax^2 + bx$ so that the resulting trinomial is a ______ is referred to as **completing the square.**

EXAMPLE Complete the Square

2 Find the value of c that makes $x^2 - 12x\ c$ a perfect square.

Complete the square.

Step 1 Find $\frac{1}{2}$ of -12. $-\frac{12}{2} =$ ______

Step 2 Square the result of Step 1. $(\ \_\_\_\ )^2 =$ ______

Step 3 Add the result of Step 2 to $x^2 - 12x$. $x^2 - 12x$ ______

Thus, $c =$ ______. Notice that $x^2 - 12x + 36 =$ ______.

Your Turn Find the value of c that makes $x^2 + 14x + c$ a perfect square.

______

KEY CONCEPT

Completing the Square To complete the square for a quadratic expression of the form $x^2 + bx$, you can follow the steps below.

Step 1 Find $\frac{1}{2}$ of b, the coefficient of x.

Step 2 Square the result of Step 1.

Step 3 Add the result of Step 2 to $x^2 + bx$, the original expression.

FOLDABLES

On the page for Lesson 10-3, write the steps for completing the square.

EXAMPLE Solve an Equation by Completing the Square

3 Solve $x^2 - 18x + 5 = -12$ by completing the square.

Step 1 Isolate the x^2 and x terms.

$x^2 - 18x + 5 = -12$	Original equation
$x^2 - 18x + 5 - \_\_\_ = -12 - \_\_\_$	Subtract.
$x^2 - 18x = \_\_\_$	Simplify.

Step 2 Complete the square and solve.

$x^2 - 18x +$ ____ $= -17 +$ ____ Add ____ to each side.

____ $= 64$ Factor $x^2 - 18x + 81$.

____ $=$ ____ Take the square root of each side.

$x - 9 +$ ____ $= \pm 8 +$ ____ Add ____ to each side.

$x =$ ____ Simplify.

$x =$ ____ $+$ ____ or $x =$ ____ $-$ ____

$=$ ____ $=$ ____

Check Substitute each value for x in the original equation.

$x^2 - 18x + 5 = -12$	$x^2 - 18x + 5 = -12$
$(17)^2 - 18(17) + 5 \stackrel{?}{=} -12$	$(1)^2 - 18(1) + 5 \stackrel{?}{=} -12$
____ $-$ ____ $+ 5 \stackrel{?}{=} -12$	____ $-$ ____ $+ 5 \stackrel{?}{=} -12$
____ $= -12$	____ $= -12$

The solution set is ____.

Your Turn Solve $x^2 - 8x + 10 = 30$.

HOMEWORK ASSIGNMENT

Page(s):

Exercises:

10–4 Solving Quadratic Equations by Using the Quadratic Formula

What You'll Learn

- Solve quadratic equations by using the Quadratic Formula.
- Use the discriminant to determine the number of solutions for a quadratic equation.

Build Your Vocabulary (page 225)

When solving the standard form of the  equation for ______, the result produces the **Quadratic Formula**.

Key Concept

The Quadratic Formula The solutions of a quadratic equation in the form $ax^2 + bx + c = 0$, where $a \neq 0$, are given by the Quadratic Formula.

$$x = \frac{-b \pm \sqrt{b^2 - 4ac}}{2a}$$

Foldables Write this formula in your Foldable. Be sure to explain the formula.

Example Integral Roots

1 Solve $x^2 - 2x - 35 = 0$ by using the Quadratic Formula.

For this equation, $a = 1$, $b = -2$, and $c = -35$.

$x = \frac{-b \pm \sqrt{b^2 - 4ac}}{2a}$ Quadratic Formula

$= \frac{-(-2) \pm \sqrt{(-2)^2 - 4(1)(-35)}}{2(1)}$ $a = 1$, $b = -2$, and $c = -35$

$= \frac{2 \pm \text{______}}{2}$ Multiply.

$= \frac{2 \pm \sqrt{\text{______}}}{2}$ Add.

$= \frac{2 \pm \text{______}}{2}$ Simplify.

$x =$ ______ or $x =$ ______

$=$ ______ $=$ ______

The solution set is ______.

Your Turn Solve $x^2 + x - 30 = 0$.

______

EXAMPLE Irrational Roots

2 Solve $15x^2 - 8x = 4$ by using the Quadratic Formula. Round to the nearest tenth if necessary.

Step 1 Rewrite the equation in standard form.

$15x^2 - 8x = 4$ Original equation

$15x^2 - 8x -$ ____ $= 4 -$ ____ Subtract ____ from each side.

$15x^2 - 8x - 4 =$ ____ Simplify.

Step 2 Apply the Quadratic Formula.

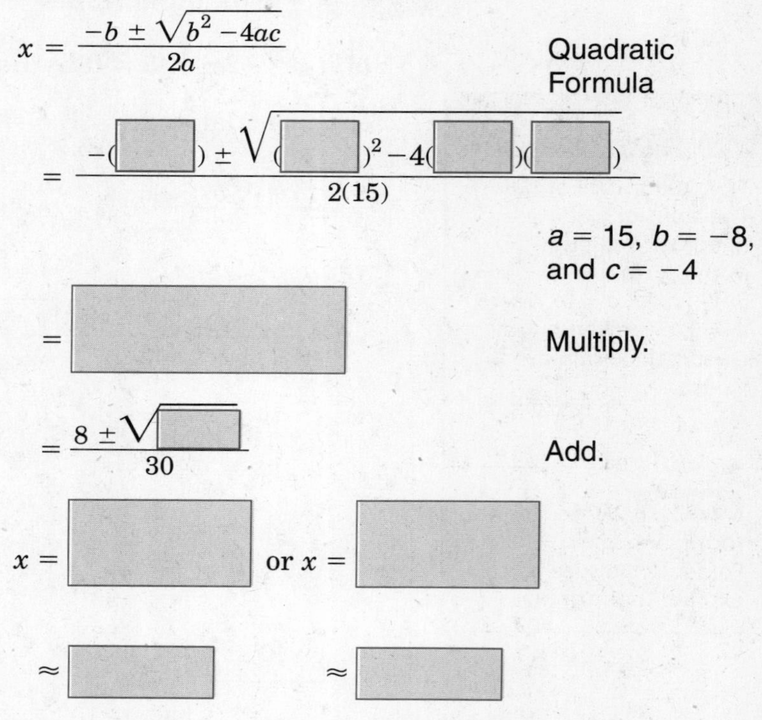

The approximate solution set is ____.

Your Turn Solve $20x^2 - 4x = 8$ by using the Quadratic Formula. Round to the nearest tenth if necessary.

EXAMPLE Use the Discriminant

3 State the value of the discriminant. Then determine the number of real roots of the equation.

KEY CONCEPT

Using the Discriminant

Negative Discriminant: There are no real roots since no real number can be the square root of a negative number.

Zero Discriminant: There is a double root.

Positive Discriminant: There are two roots.

a. $\mathbf{4x^2 - 2x + 14 = 0}$

$b^2 - 4ac = (-2)^2 - 4(4)(14)$ $a = 4$, $b = -2$ and $c = 14$

$= \_\_\_\_$ Simplify.

The discriminant is ____. Since the discriminant is ____, the equation has ____ real roots.

b. $\mathbf{x^2 + 24x = -144.}$

Step 1 Rewrite the equation in standard form.

$x^2 + 24x = -144$ Original equation

$x^2 + 24x + 144 = -144 + 144$ Add 144 to each side.

$x^2 + 24x + 144 = 0$ Simplify.

Step 2 Find the discriminant.

$b^2 - 4ac = (\_\_\_\_)^2 - 4(\_\_\_\_)(\_\_\_\_)$ $a = 1$, $b = 24$, and $c = 144$

$= \_\_\_\_$ Simplify.

The discriminant is ____. Since the discriminant is ____, the equation has ____ real root.

Your Turn State the value of the discriminant for each equation. Then determine the number of real roots for the equation.

a. $x^2 + 2x + 2 = 0$

b. $-5x^2 + 10x = -1$

HOMEWORK ASSIGNMENT

Page(s):

Exercises:

10–5 Exponential Functions

WHAT YOU'LL LEARN

- Graph exponential functions.
- Identify data that displays exponential behavior.

BUILD YOUR VOCABULARY (page 224)

The type of function is which the ______ is the ______, is called an **exponential function.**

KEY CONCEPT

Exponential Function An exponential function is a function that can be described by an equation of the form $y = a^x$, where $a > 0$ and $a \neq 1$.

EXAMPLE Graph an Exponential Function with $a > 1$

1 Graph $y = 3^x$. State the y-intercept.

x	3^x	y
-1		
0		
1		
2		

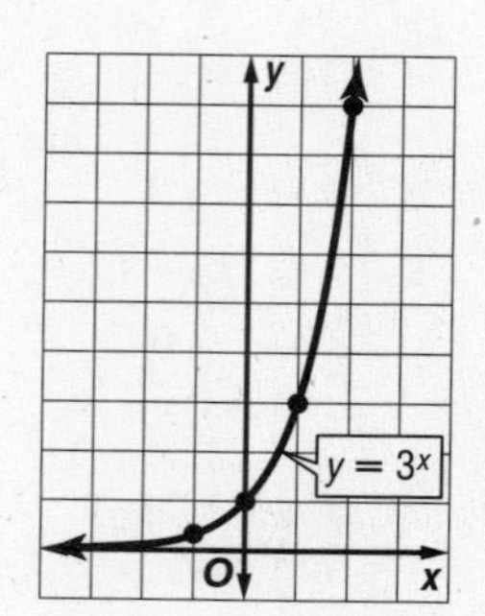

Graph the ordered pairs and connect the points with a smooth curve. The y-intercept is ______.

EXAMPLE Graph Exponential Functions with $0 < a < 1$

2 Graph $y = \left(\frac{1}{4}\right)^x$. State the y-intercept.

x	$\left(\frac{1}{4}\right)^x$	y
-1	$\left(\frac{1}{4}\right)^{-1}$	
0		
1		

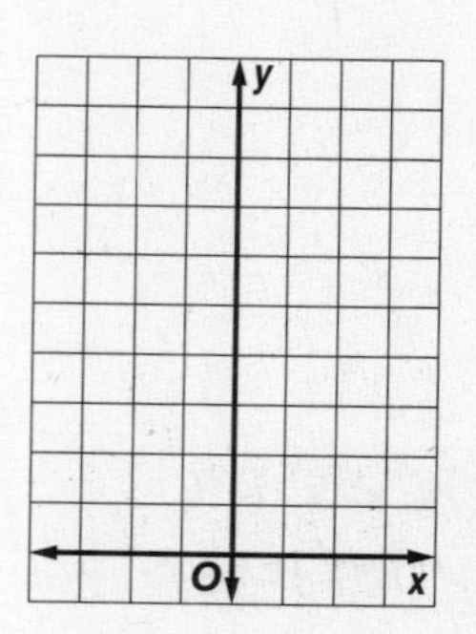

Graph the ordered pairs and connect the points with a smooth curve. The y-intercept is ______.

REMEMBER IT

The graph of $y = a^x$, where $a > 0$ and $a \neq 1$ never has an x-intercept.

Your Turn **Graph each function. State the *y*-intercept.**

a. $y = 5^x$

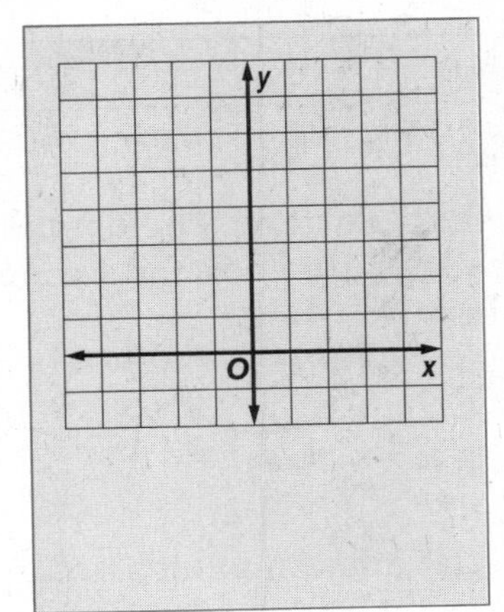

b. $y = \left(\frac{1}{8}\right)^x$

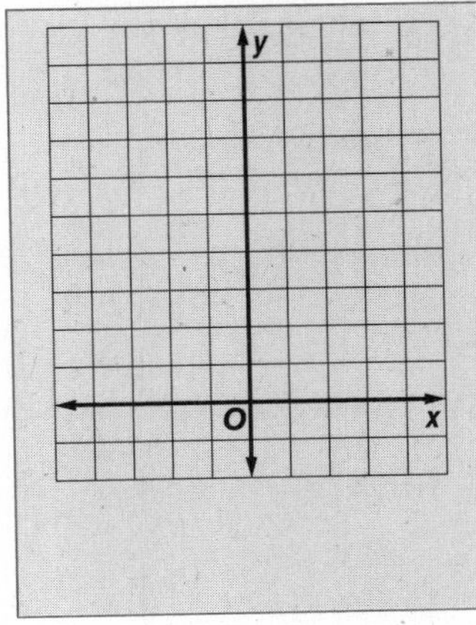

EXAMPLE Use Exponential Functions to Solve Problems

FOLDABLES

ORGANIZE IT

On the page for Lesson 10–5, sketch a graph of an exponential function when $a > 1$. Then sketch a graph of an exponential function when $0 < a < 1$.

3 **The function $V = 25{,}000 \cdot 0.82^t$ models the depreciation of the value of a new car that originally cost \$25,000. *V* represents the value of the car and *t* represents the time in years from the time the car was purchased.**

a. What values of *V* and *t* are meaningful in the function?

Only the values of $V \leq$ ______ and $t \geq$ ______ are meaningful in the context of the problem.

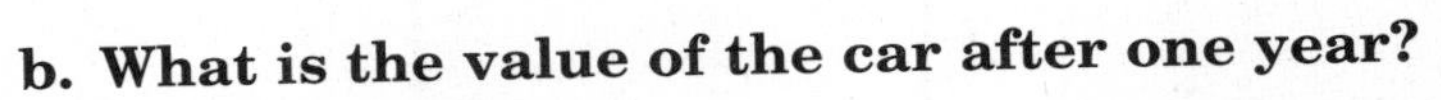

b. What is the value of the car after one year?

$V = 25{,}000 \cdot 0.82^t$ Original equation

$V = 25{,}000 \cdot 0.82^1$ $t = 1$

$V =$ ______ Use a calculator.

After one year, the car's value is about ______.

Your Turn **The function $V = 22{,}000 \cdot 0.82^t$ models the depreciation of the value of a new car that originally cost \$22,000. *V* represents the value of the car and *t* represents the time in years from the time the car was purchased.**

a. What values of *V* and *t* are meaningful in the function?

______

b. What is the value of the car after one year?

______

EXAMPLE Identify the Exponential Behavior

4 Determine whether each set of data displays exponential behavior.

a.

x	0	10	20	30
y	10	25	62.5	156.25

Look for a Pattern The domain values are at regular intervals of 10. Look for a common factor among the range of values.

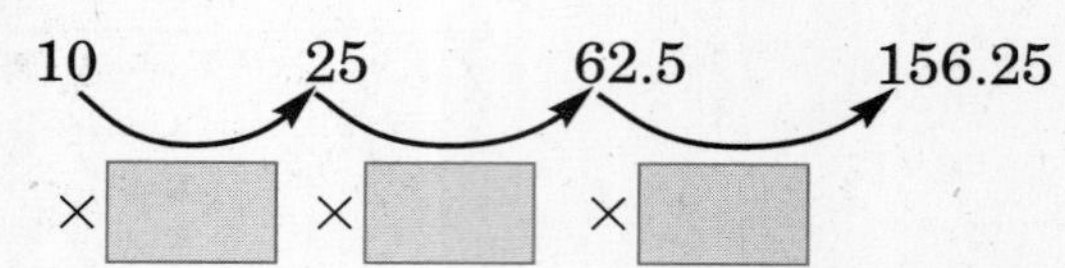

Since the domain values are at regular intervals and the range values have a common factor, the data are probably exponential. The equation for the data may involve $\square^x$.

b.

x	0	10	20	30
y	10	25	40	55

Look for a Pattern The domain values are at regular intervals of ______. The range values have a common difference of ______.

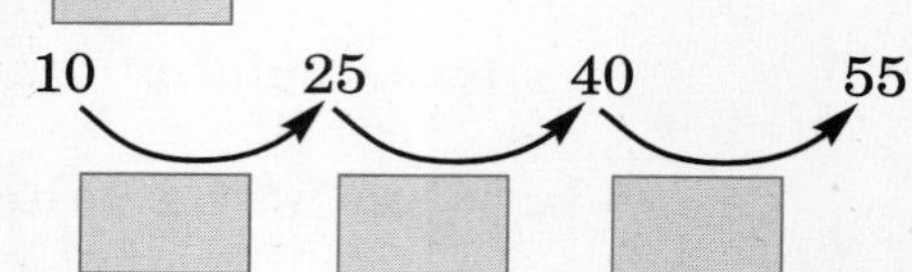

Since the domain values are at regular intervals and there is a common difference of ______, the data display ______ behavior.

Your Turn **Determine whether each set of data displays exponential behavior.**

a.

x	0	10	20	30
y	100	50	25	12.5

b.

x	0	10	20	30
y	−5	0	5	10

HOMEWORK ASSIGNMENT

Page(s):

Exercises:

10–6 Growth and Decay

What You'll Learn

- Solve problems involving exponential growth.
- Solve problems involving exponential decay

Key Concept

General Equation for Exponential Growth The general equation for exponential growth is $y = C(1 + r)^t$ where y represents the final amount, C represents the initial amount, r represents the rate of change expressed as a decimal, and t represents time.

EXAMPLE Exponential Growth

1 POPULATION **In 2000, the United States had a population of about 280 million, and a growth rate of about 0.85% per year.**

a. Write an equation to represent the population of the United States since the year 2000.

$y = C(1 + r)^t$ — General equation the exponential growth

$y =$ ______ $(1 +$ ______ $)^t$ — $C = 280{,}000{,}000$ and $r = 0.85\%$ or ______

$y =$ ______ — Simplify.

b. According to the equation, what will be the population of the United States in the year 2010?

$y = 280{,}000{,}000(1.0085)^t$ — Equation for the population of the U.S.

$y = 280{,}000{,}000(1.0085)^{10}$ — $t = 2010 - 2000$ or ______

$y =$ ______

In 2010, the population will be about ______.

Your Turn **In 2000, Scioto School District had a student population of about 4500 students, and a growth rate of about 0.15% per year.**

a. Write an equation to represent the student population of the Scioto School District since the year 2000.

______

b. According to the equation, what will be the student population of the Scioto School District in the year 2006?

______

BUILD YOUR VOCABULARY (page 224)

The equation $A = P\left(1 + \frac{r}{n}\right)^{nt}$ is used to find **compound interest** which is an application of ______ growth.

EXAMPLE Compound Interest

2 COMPOUND INTEREST **When Jing May was born, her grandparents invested $1000 in a fixed rate savings account at a rate of 7% compounded annually. The money will go to Jing May when she turns 18 to help with her college expenses. What amount of money will Jing May receive from the investment?**

$A = P\left(1 + \frac{r}{n}\right)^{nt}$ Compound interest equation

$A = \_\_\_\left(1 + \frac{\_\_\_}{\_\_\_}\right)^{\_\_\_}$ $P = 1000$, $r = 7\%$ or 0.07, $n = 1$, and $t = 18$

$A = \_\_\_$ Compound interest equation

$A = \_\_\_$ Simplify.

She will receive about ______.

Your Turn When Lucy was 10 years old, her father invested $2500 in a fixed rate savings account at a rate of 8% compounded semiannually. When Lucy turns 18, the money will help to buy her a car. What amount of money will Lucy receive from the investment?

KEY CONCEPT

General Equation for Exponential Decay The general equation for exponential decay is $y = C(1 - r)^t$ where y represents the final amount, C represents the initial amount, r represents the rate of decay expressed as a decimal, and t represents time.

FOLDABLES Write the equations for exponential growth and decay in your Foldable.

EXAMPLE Exponential Decay

3 CHARITY **During an economic recession, a charitable organization found that its donations dropped by 1.1% per year. Before the recession, its donations were $390,000.**

a. Write an equation to represent the charity's donations since the beginning of the recession.

$y = C(1-r)^t$ — General equation for exponential decay

$y =$ ______ $(1 -$ ______ $)^t$ — $C = 390{,}000$ and $r = 1.1\%$ or 0.011

$y =$ ______ — Simplify.

b. Estimate the amount of the donations 5 years after the start of the recession.

$y = 390{,}000(0.989)^t$ — Equation for the amount of donations

$y = 390{,}000(0.989)^{\square}$ — $t =$ ______

$y =$ ______

The amount of donations should be about ______.

Your Turn **A charitable organization found that the value of its clothing donations dropped by 2.5% per year. Before this downturn in donations, the organization received clothing valued at $24,000.**

a. Write an equation to represent the value of the charity's clothing donations since the beginning of the downturn.

b. Estimate the value of the clothing donations 3 years after the start of the downturn.

HOMEWORK ASSIGNMENT

Page(s):

Exercises:

10–7 Geometric Sequences

WHAT YOU'LL LEARN

- Recognize and extend geometric sequences.
- Find geometric means.

KEY CONCEPT

Geometric Sequence
A **geometric sequence** is a sequence in which each term after the nonzero first term is found by multiplying the previous term by a constant called the common ratio *r*, where $r \neq 0, 1$.

EXAMPLE Recognize Geometric Sequence

1 **Determine whether the sequence 1, 4, 16, 64, 256, ... is geometric.**

Determine the pattern.

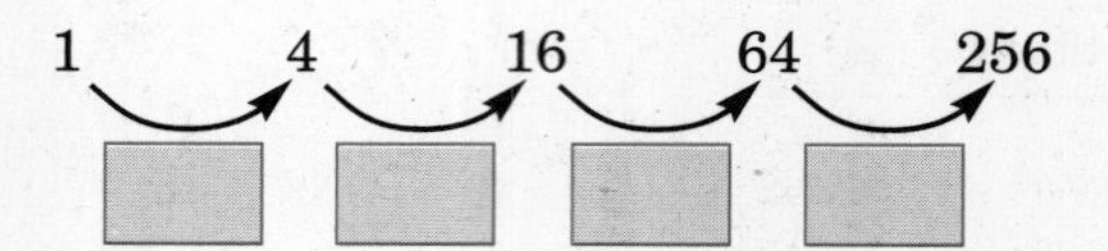

In this sequence, each term is found by multiplying the previous term by ______. This sequence is ______.

Your Turn **Determine whether each sequence is geometric.**

a. $3, 1\frac{1}{2}, \frac{3}{4}, \frac{3}{8}, \ldots$

b. 0, 11, 22, 33, 44, ...

EXAMPLE Continue Geometric Sequences

2 **Find the next three terms in the geometric sequence 64, 48, 36, ...**

To find the common factor, divide the ______ term by the ______.

$\frac{48}{64} =$ ______

Use this information to find the next three terms.

64, 48, 36 ______ ______ ______

The next three terms are ______, ______, and ______.

Your Turn Find the next three terms in the geometric sequence. 42, 25.2, 15.12, ...

EXAMPLE *n*th Term of a Geometric Sequence

3 Find the eighth term of a geometric sequence in which $a_1 = 7$ and $r = 3$.

$a_n = a_1 \times r^{n-1}$	Formula for the *n*th term of a geometric sequence
$a_8 = \_\_\_ \cdot \_\_\_^{8-1}$	$n =$ ___, $a_1 =$ ___, and $r =$ ___
$a_8 = 7 \cdot (3)^{\_\_\_}$	$8 - 1 =$ ___
$a_8 = 7 \cdot$ ___	$(3)^7 =$ ___
$a_8 =$ ___	$7 \cdot (2187) =$ ___

The eighth term in the sequence is ___.

FOLDABLES

ORGANIZE IT

On the page for Lesson 10-7, write your own geometric sequence. Include the common ratio and the next three terms of the sequence.

Your Turn Find the ninth term of a geometric sequence in which $a_1 = 12$ and $r = 4$.

KEY CONCEPT

Formula for the *n*th term of a Geometric Sequence
The *n*th tem of a geometric sequence with the first term a_1 and common ratio r is given by $a_n = a_1 \cdot r^{n-1}$.

BUILD YOUR VOCABULARY (page 224)

Missing term(s) between two nonconsecutive terms in a ___ sequence are called **geometric means**.

EXAMPLE Find Geometric Means

4 Find the geometric mean in the sequence 7, __ , 112.

In the sequence, $a_1 = \square$ and $a_3 = \square$. To find a_2, you must first find r.

$a_n = a_1 \cdot r^{n-1}$	Formula for the *n*th term of a geometric sequence
$a_3 = a_1 \cdot r^{\square - 1}$	$n = \square$
$\square = \square \cdot r^2$	$a_1 = \square$ *and* $a_3 = \square$
$\dfrac{112}{\square} = \dfrac{7 \cdot r^2}{\square}$	Divide each side by $\square$.
$\square = \square$	Simplify.
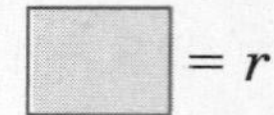 $= r$	Take the square root of each side.

If $r = \square$, the geometric mean is $7(\square)$ or $\square$. If $r = \square$, the geometric mean is $7(\square)$ or $\square$.

The geometric mean is $\square$ or $\square$.

Your Turn Find the geometric mean in the sequence 9, __ , 576.

HOMEWORK ASSIGNMENT

Page(s):

Exercises:

CHAPTER 10

BRINGING IT ALL TOGETHER

STUDY GUIDE

FOLDABLES™	VOCABULARY PUZZLEMAKER	BUILD YOUR VOCABULARY
Use your **Chapter 10 Foldable** to help you study for your chapter test.	To make a crossword puzzle, word search, or jumble puzzle of the vocabulary words in Chapter 10, go to: www.glencoe.com/sec/math/t_resources/free/index.php	You can use your completed **Vocabulary Builder** (pages 224–225) to help you solve the puzzle.

10–1 Graphing Quadratic Functions

The graphs of two quadratic functions are shown below. Complete each statement about the graphs.

A.

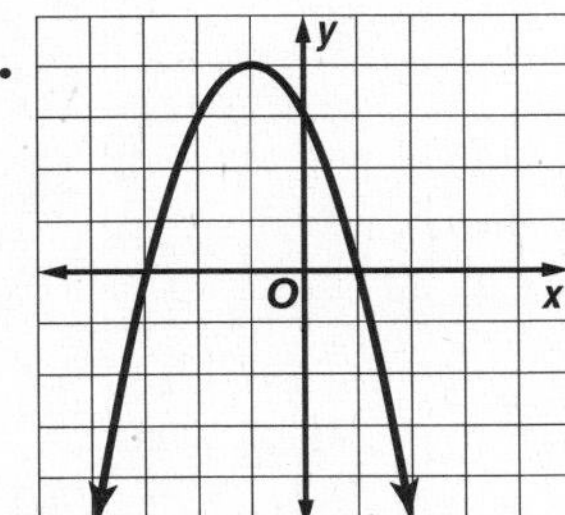

B.

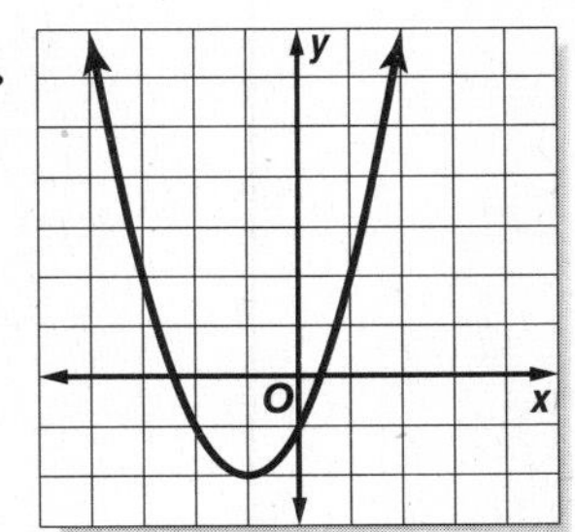

1. Each graph is a curve called a ________.

2. The highest point of graph A is located at ________.

3. The lowest point of graph B is located at ________.

4. The maximum or minimum point of a parabola is called the

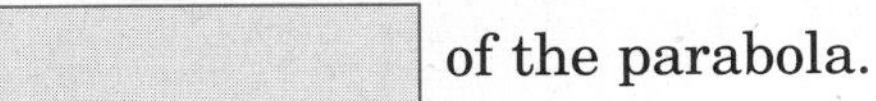

of the parabola.

10–2 Solving Quadratic Equations by Graphing

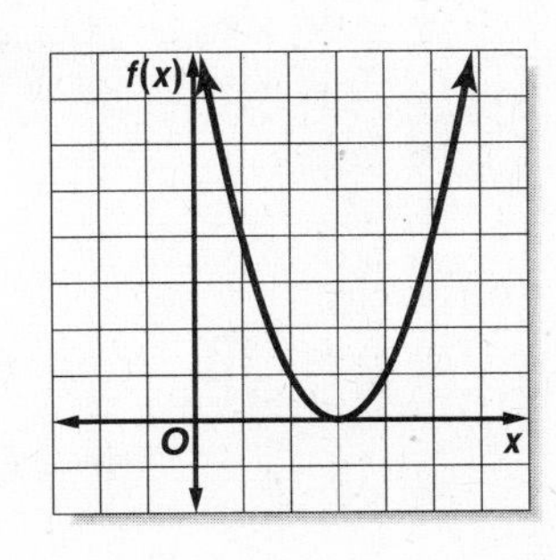

Refer to the graph shown at the right to answer the questions about the related equation $f(x) = x^2 - 6x + 9$.

5. The related quadratic equation is ________.

6. How many real number solutions are there? ________

7. Name one solution.

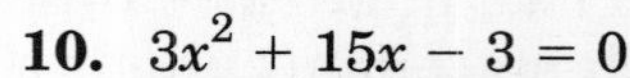

10–3 Solving Quadratic Equations by Completing the Square

8. Draw a line under each quadratic equation that you could solve by taking the square root of each side.

$x^2 + 6x + 9 = 100$ $x^2 - 14x + 40 = 25$ $x^2 - 16x + 64 = 26$

$x^2 - 20x + 80 = 16$ $x^2 + 10x + 36 = 49$ $x^2 - 12x + 36 = 6$

10–4 Solving Quadratic Equations by Using the Quadratic Formula

Solve each equation by completing the square.

9. $x^2 + 18x + 50 = 9$

10. $3x^2 + 15x - 3 = 0$

11. What is the quadratic formula?

Solve each equation by using the quadratic formula. Round to the nearest tenth if necessary.

12. $2a^2 - 3a = -1$

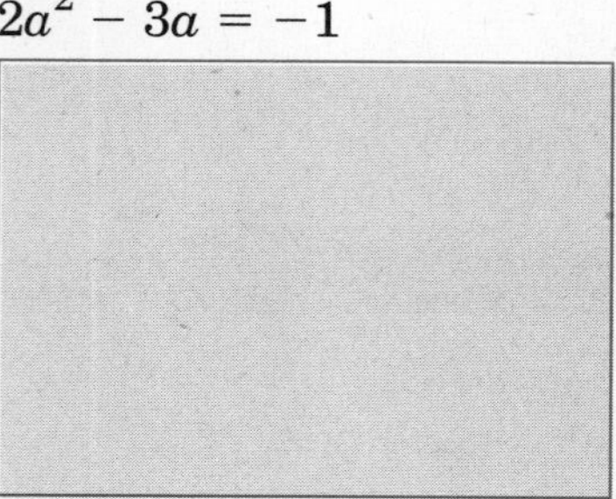

13. $3w^2 - 1 = 8w$

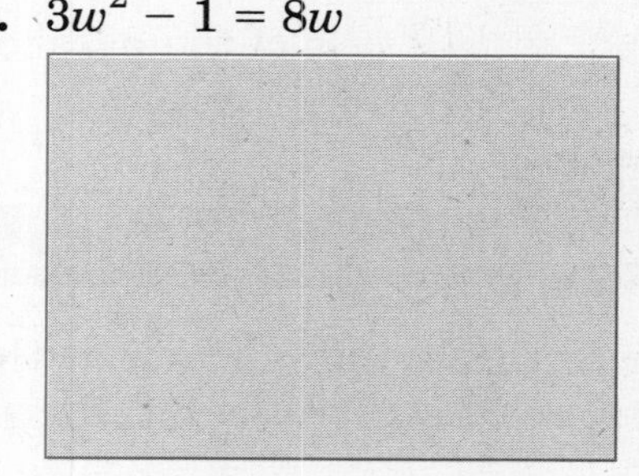

14. You can use the discriminant to determine the number of real roots for a quadratic equation. What is the discriminant?

10–5

Exponential Functions

The graphs of two exponential functions of the form $y = a^x$ are shown below.

A.

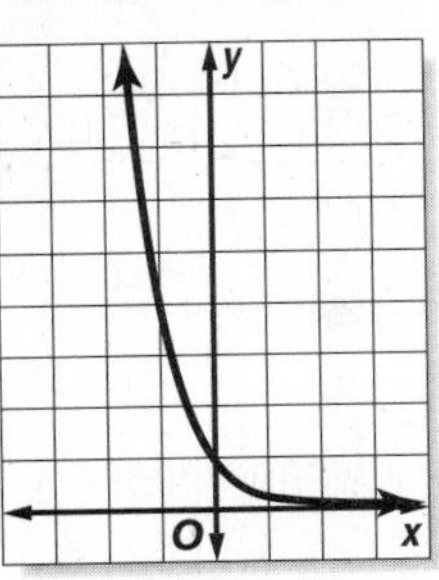

B.

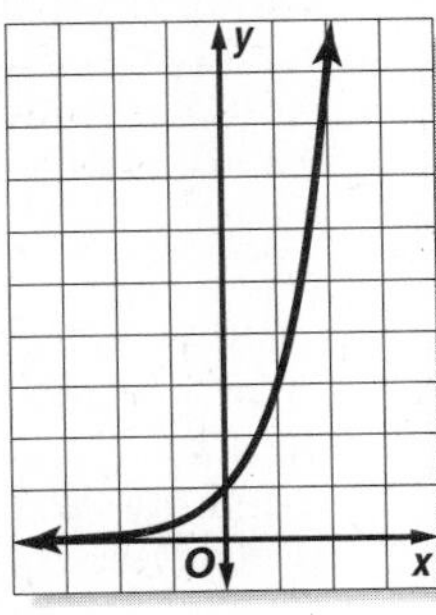

15. In graph A, the value of a is greater than 0 and less than ______. The y values decrease as the x values ______.

16. In graph B, the value of a is greater than ______. The y values increase as the x values ______.

10–6

Growth and Decay

Match an equation to each solution, and then indicate whether the situation is an example of exponential growth or decay.

17. A coin had a value of \$1.17 in 1995. Its value has been increasing at a rate of 9% per year.

A. $y = 1.17(1.09)^t$

B. $y = 1.17(0.91)^t$

18. A business owner has just paid \$6000 for a computer. It depreciates at a rate of 22% per year. How much will it be worth in 5 years?

A. $A = 6000(1.22)^5$

B. $A = 6000(0.78)^5$

10–7

Geometric Sequences

Find the next 3 terms of each geometric sequence.

19. $7, \frac{1}{7}, \frac{1}{49}$

20. $20, -42, 88.2$

ARE YOU READY FOR THE CHAPTER TEST?

Visit **algebra1.com** to access your textbook, more examples, self-check quizzes, and practice tests to help you study the concepts in Chapter 10.

Check the one that applies. Suggestions to help you study are given with each item.

☐ **I completed the review of all or most lessons without using my notes or asking for help.**

- You are probably ready for the Chapter Test.
- You may want take the Chapter 10 Practice Test on page 579 of your textbook as a final check.

☐ **I used my Foldable or Study Notebook to complete the review of all or most lessons.**

- You should complete the Chapter 10 Study Guide and Review on pages 574–578 of your textbook.
- If you are unsure of any concepts or skills, refer back to the specific lesson(s).
- You may also want to take the Chapter 10 Practice Test on page 579.

☐ **I asked for help from someone else to complete the review of all or most lessons.**

- You should review the examples and concepts in your Study Notebook and Chapter 10 Foldable.
- Then complete the Chapter 10 Study Guide and Review on pages 574–578 of your textbook.
- If you are unsure of any concepts or skills, refer back to the specific lesson(s).
- You may also want to take the Chapter 10 Practice Test on page 579.

Student Signature

Parent/Guardian Signature

Teacher Signature

Radical Expressions and Triangles

FOLDABLES™ Use the instructions below to make a Foldable to help you organize your notes as you study the chapter. You will see Foldable reminders in the margin of this Interactive Study Notebook to help you in taking notes.

Begin with a sheet of plain $8\frac{1}{2}$" by 11" paper.

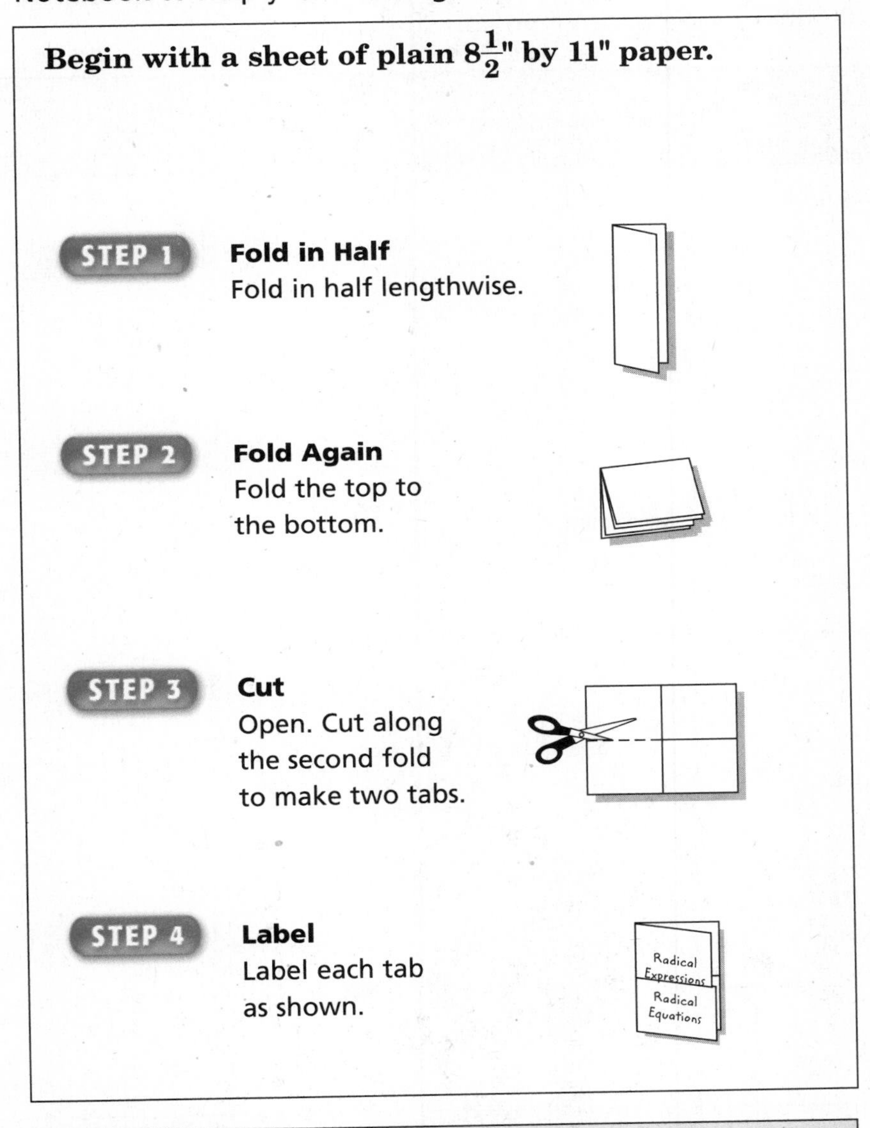

NOTE-TAKING TIP: Remember to study your notes daily. Reviewing small amounts at a time will help you retain the information.

BUILD YOUR VOCABULARY

This is an alphabetical list of new vocabulary terms you will learn in Chapter 11. As you complete the study notes for the chapter, you will see Build Your Vocabulary reminders to complete each term's definition or description on these pages. Remember to add the textbook page number in the second column for reference when you study.

Vocabulary Term	Found on Page	Definition	Description or Example
angle of depression			
angle of elevation			
conjugate [KAHN-jih-guht]			
corollary [KAWR-uh-LEHR-ee]			
cosine			
Distance Formula			
extraneous solution [ehk-STRAY-nee-uhs]			
hypotenuse [hy-PAH-tn-oos]			
leg			
Pythagorean triple [puh-THA-guh-REE-uhn]			

Vocabulary Term	Found on Page	Definition	Description or Example
radical equation			
radical expression			
radicand [RA-duh-KAND]			
rationalizing the denominator			
similar triangles			
sine			
solve a triangle			
tangent			
trigonometric ratios [TRIH-guh-nuh-MEH-trihk]			

11–1 Simplifying Radical Expressions

BUILD YOUR VOCABULARY (page 253)

A **radical expression** is an expression that contains a ______ root.

A **radicand** is the expression under the ______ sign.

WHAT YOU'LL LEARN

- Simplify radical expressions using the Product Property of Square Roots.
- Simplify radical expressions using the Quotient Property of Square Roots.

KEY CONCEPT

Product Property of Square Roots For any numbers *a* and *b*, where $a \geq 0$ and $b \geq 0$, the square root of the product *ab* is equal to the product of each square root.

EXAMPLE Simplify Square Roots

1 **a. Simplify $\sqrt{52}$.**

$\sqrt{52} =$ ______	Prime factorization of 52
$= \sqrt{2^2} \cdot \sqrt{13}$	Product Property of Square Roots
$=$ ______	Simplify.

b. Simplify $\sqrt{72}$.

$\sqrt{72} =$ ______	Prime factorization of 72
$=$ ______ $\cdot \sqrt{3^2}$	Product Property of Square Roots
$= 2\sqrt{2} \cdot$ ______	Simplify.
$=$ ______	Simplify.

EXAMPLE Multiply Square Roots

2 **Find $\sqrt{2} \cdot \sqrt{24}$.**

$\sqrt{2} \cdot \sqrt{24} = \sqrt{2} \cdot \sqrt{2} \cdot \sqrt{2} \cdot \sqrt{2} \cdot \sqrt{3}$	Product Property
$=$ ______ $\cdot$ ______ $\cdot \sqrt{3}$	Product Property
$=$ ______	Simplify.

Your Turn Simplify.

a. $\sqrt{45}$

b. $\sqrt{60}$

c. $\sqrt{5} \cdot \sqrt{35}$

EXAMPLE Simplify a Square Root with Variables

3 **Simplify $\sqrt{45a^4b^5c^6}$.**

$\sqrt{45a^4b^5c^6}$

= ______ Prime factorization

= ______ $\cdot \sqrt{5} \cdot$ ______ $\cdot \sqrt{b^4} \cdot$ ______ $\cdot \sqrt{c^6}$ Product Property

= ______ $\cdot \sqrt{b} \cdot |c^3|$ Simplify.

= ______ The absolute value of $|c^3|$ ensures a nonnegative result.

Your Turn Simplify $\sqrt{32m^2n^3c}$.

BUILD YOUR VOCABULARY (page 253)

Rationalizing the denominator of a radical expression is a method used to eliminate ______ from the ______ of a fraction.

EXAMPLE Rationalizing the Denominator

KEY CONCEPT

Quotient Property of Square Roots For any numbers *a* and *b*, where $a \geq 0$ and $b > 0$, the square root of the quotient $\frac{a}{b}$ is equal to the quotient of each square root.

4 **Simplify each quotient.**

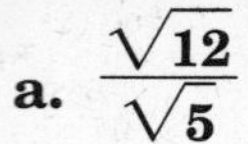

a. $\dfrac{\sqrt{12}}{\sqrt{5}}$

$\dfrac{\sqrt{12}}{\sqrt{5}} = \dfrac{\sqrt{12}}{\sqrt{5}} \cdot$ ☐ — Multiply by ☐.

$= \dfrac{☐}{5}$ — Product Property of Square Roots

$= \dfrac{2\sqrt{15}}{5}$ — Simplify.

b. $\dfrac{\sqrt{11y}}{\sqrt{27}}$

$\dfrac{\sqrt{11y}}{\sqrt{27}} = \dfrac{\sqrt{11y}}{☐}$ — Prime factorization

$= \dfrac{\sqrt{11y}}{3\sqrt{3}} \cdot$ ☐ — Multiply by ☐.

$=$ ☐ — Product Property of Square Roots

FOLDABLES™ ORGANIZE IT

Under the tab for Radical Expressions, write the three steps that must be met for a radical expression to be in simplest form.

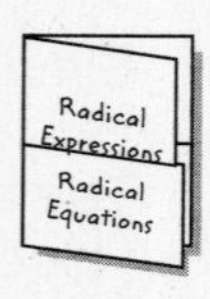

c. $\dfrac{\sqrt{3}}{\sqrt{8}}$

$\dfrac{\sqrt{3}}{\sqrt{8}} = \dfrac{\sqrt{3}}{\sqrt{8}} \cdot$ ☐ — Multiply by ☐.

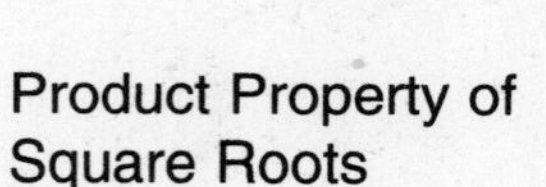

$= \dfrac{\sqrt{24}}{☐}$ — Product Property of Square Roots

$= \dfrac{\sqrt{2 \cdot 2 \cdot 2 \cdot 3}}{8}$ — Prime factorization

$=$ ☐ — $\sqrt{2^2} = 2$

$=$ ☐ — Divide the numerator and denominator by 2.

HOMEWORK ASSIGNMENT

Page(s):

Exercises:

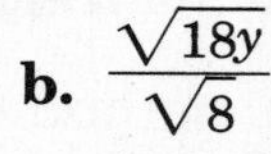

Your Turn **Simplify.**

a. $\dfrac{\sqrt{5}}{\sqrt{2}}$ ☐

b. $\dfrac{\sqrt{18y}}{\sqrt{8}}$ ☐

c. $\dfrac{\sqrt{2}}{\sqrt{27}}$ ☐

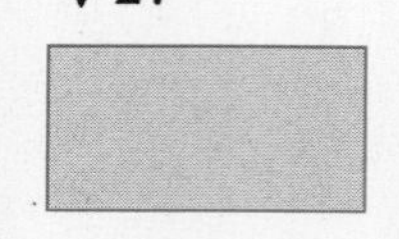

11–2 Operations with Radical Expressions

WHAT YOU'LL LEARN

- Add and subtract radical expressions.
- Multiply radical expressions.

FOLDABLES™

ORGANIZE IT

Under the tab for Radical Expressions, write how simplifying like radicands is similar to simplifying like terms.

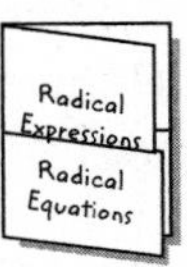

EXAMPLE Expressions with Like Radicands

1 **Simplify $7\sqrt{2} + 8\sqrt{11} - 4\sqrt{11} - 6\sqrt{2}$.**

$7\sqrt{2} + 8\sqrt{11} - 4\sqrt{11} - 6\sqrt{2}$

$= \_\_\_\_ - 6\sqrt{2} + \_\_\_\_ - 4\sqrt{11}$ Commutative Property

$= \_\_\_\_\sqrt{2} + \_\_\_\_\sqrt{11}$ Distributive Property

$= \_\_\_\_$ Simplify.

Your Turn Simplify $4\sqrt{2} + 8\sqrt{3} - 5\sqrt{3} + 2\sqrt{2}$.

EXAMPLE Expressions with Unlike Radicands

2 **Simplify $6\sqrt{27} + 8\sqrt{12} + 2\sqrt{75}$.**

$6\sqrt{27} + 8\sqrt{12} + 2\sqrt{75}$

$= 6\sqrt{3^2 \cdot 3} + \_\_\_\_ + 2\sqrt{5^2 \cdot 3}$

$= 6\_\_\_\_ + 8\_\_\_\_ + 2(\sqrt{5^2} \cdot \sqrt{3})$

$= 6(3\sqrt{3}) + 8(2\sqrt{3}) + 2(5\sqrt{3})$

$= \_\_\_\_$

$= \_\_\_\_$

The simplified form is ____.

Your Turn Simplify $6\sqrt{245} + 3\sqrt{125} + \sqrt{80}$.

REVIEW IT

Give an example of two binomials. Then explain how you multiply them using FOIL method. *(Lesson 8–7)*

____

____

____

____

____

EXAMPLE Multiply Radical Expressions

3 Find the area of a rectangle with a width of $4\sqrt{6} - 2\sqrt{10}$ and a length of $5\sqrt{3} + 7\sqrt{5}$.

To find the area of the rectangle multiply the measures of the length and width.

$$\left(4\sqrt{6} - 2\sqrt{10}\right)\left(5\sqrt{3} + 7\sqrt{5}\right)$$

First terms   *Outer terms*

$= (4\sqrt{6})$ _____ $+ (4\sqrt{6})$ _____ $+$

Inner terms   *Last terms*

$(-2\sqrt{10})$ _____ $+ (-2\sqrt{10})$ _____

$= 20\sqrt{18} +$ _____ $- 10\sqrt{30} -$ _____   Multiply.

$= 20$ _____ $+ 28\sqrt{30} -$ _____ $- 14\sqrt{5^2 \cdot 2}$ Prime factorization

$= 60\sqrt{2} +$ _____ $- 10\sqrt{30} -$ _____   Simplify.

$=$ _____   Combine like terms.

The area of the rectangle is _____ square units.

Your Turn Find the area of a rectangle with a width of $3\sqrt{3} + 5\sqrt{7}$ and a length of $2\sqrt{14} + 4\sqrt{6}$.

HOMEWORK ASSIGNMENT

Page(s):

Exercises:

11–3 Radical Equations

WHAT YOU'LL LEARN

- Solve radical equations.
- Solve radical equations with extraneous solutions.

BUILD YOUR VOCABULARY (page 253)

Equations that contain radicals with variables in the ________ are called **radical equations**.

EXAMPLE Radical Equation with a Variable

1 FREE-FALL HEIGHT **An object is dropped from an unknown height and reaches the ground in 5 seconds. From what height is it dropped?**

Use the equation $t = \frac{\sqrt{h}}{4}$ to replace t with ____.

$t = \frac{\sqrt{h}}{4}$	Original equation
____ $= \frac{\sqrt{h}}{4}$	Replace t with 5.
____ $= \sqrt{h}$	Multiply each side by 4.
____ $= (\sqrt{h})^2$	Square each side.
____ $= h$	Simplify.

The object is dropped from ____ feet.

REMEMBER IT

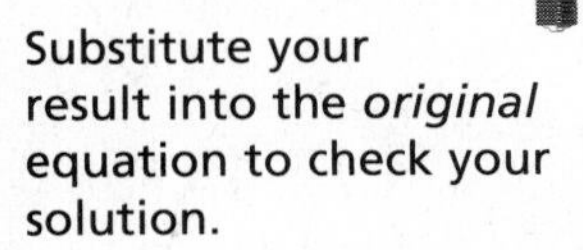

Substitute your result into the *original* equation to check your solution.

EXAMPLE Radical Equation with an Expression

2 **Solve $\sqrt{x-3} + 8 = 15$.**

$\sqrt{x-3} + 8 = 15$	Original equation
$\sqrt{x-3} =$ ____	Subtract ____ from each side.
____ $= 7^2$	Square each side.
____ $= 49$	$(\sqrt{x-3})^2 = x - 3$
$x =$ ____	Add 3 to each side.

The solution is ____.

REVIEW IT

Explain the Zero Product Property in your own words. *(Lesson 9–2)*

Your Turn

a. Refer to Example 1. If an unknown object reaches the ground in 7 seconds, from what height is it dropped?

b. Solve $\sqrt{x + 4} + 6 = 14$.

BUILD YOUR VOCABULARY (page 252)

An **extraneous solution** is a solution derived from an equation that is a solution of the equation.

EXAMPLE Variable on Each Side

3 **Solve $\sqrt{2 - y} = y$.**

$\sqrt{2 - y} = y$ Original equation

$\left(\sqrt{2 - y}\right)^2 = y^2$

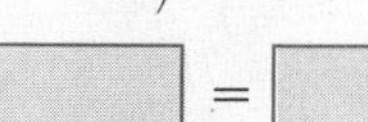

Simplify.

$0 = y^2 + y - 2$ Subtract and add to each side.

$0 =$ Factor.

$= 0$ or $= 0$ Zero Product Property

$y =$ $y =$ Solve.

Check

$\sqrt{2 - y} = y$ | $\sqrt{2 - y} = y$

$\sqrt{2 - (-2)} \stackrel{?}{=} -2$ | $\sqrt{2 - 1} \stackrel{?}{=} 1$

$\sqrt{4} \stackrel{?}{=} -2$ | $\sqrt{1} \stackrel{?}{=} 1$

$2 \neq -2$ ✘ | $1 = 1$ ✔

Since does not satisfy the original equation, is the only solution.

Your Turn Solve each equation.

a. $\sqrt{x + 3} - 1 = 8$

b. $y = \sqrt{2y + 3}$

HOMEWORK ASSIGNMENT

Page(s):

Exercises:

11–4 The Pythagorean Theorem

WHAT YOU'LL LEARN

- Solve problems by using the Pythagorean Theorem.
- Determine whether a triangle is a right triangle.

BUILD YOUR VOCABULARY (page 252)

In a right triangle, the side opposite the ______ angle is called the **hypotenuse**. The other two ______ are called the **legs** of the triangle.

Whole numbers that satisfy the ______ are called **Pythagorean triples**.

KEY CONCEPT

The Pythagorean Theorem If a and b are the lengths of the legs of a right triangle and c is the length of the hypotenuse, then the square of the length of the hypotenuse is equal to the sum of the squares of the lengths of the legs.

EXAMPLE Find the Length of the Hypotenuse

1 **Find the length of the hypotenuse of a right triangle if $a = 18$ and $b = 24$.**

______	Pythagorean Theorem
$c^2 = 18^2 + 24^2$	$a =$ ______ and $b =$ ______
$c^2 =$ ______	Simplify.
$\sqrt{c^2} =$ ______	Take the square root of each side.
$c =$ ______	Use the positive value.

The length of the hypotenuse is ______ units.

Your Turn Find the length of the hypotenuse of a right triangle if $a = 25$ and $b = 60$.

EXAMPLE Find the Length of a Side

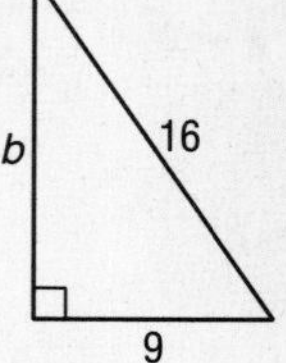

2 Find the length of the missing side.

In the triangle, c = ____ and a = ____ units.

$c^2 = a^2 + b^2$	Pythagorean Theorem
____ = ____ $+ b^2$	$a = 9$ and $c = 16$
____ = ____ $+ b^2$	Evaluate squares.
____ $= b^2$	Subtract 81 from each side.
$\pm\sqrt{175} = b$	Take the square root of each side.
 $\approx b$	Use the positive value. Use a calculator to evaluate $\sqrt{175}$.

To the nearest hundredth, the length of the leg is ____ units.

Your Turn Find the length of the missing side.

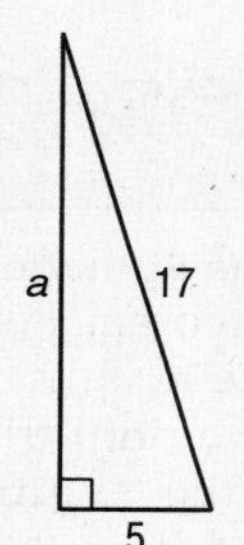

EXAMPLE Pythagorean Triples

3 What is the area of triangle *XYZ*?

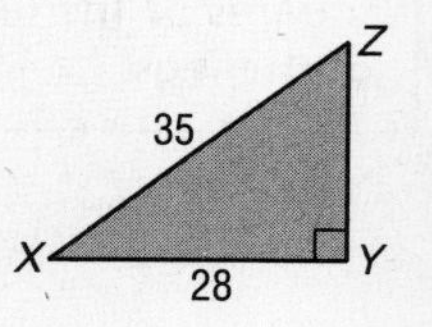

Use the measures of the hypotenuse and the base to find the height of the triangle.

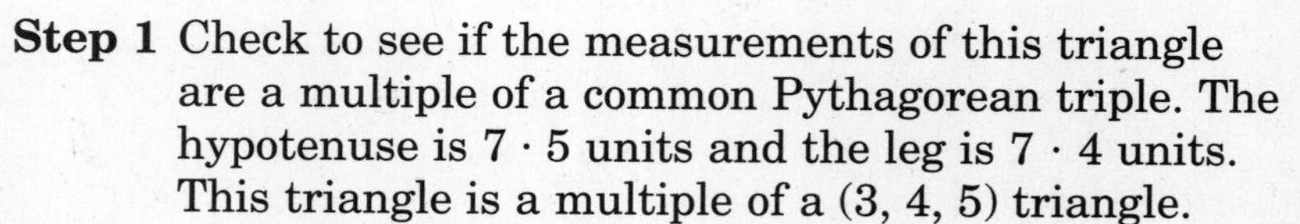

Step 1 Check to see if the measurements of this triangle are a multiple of a common Pythagorean triple. The hypotenuse is $7 \cdot 5$ units and the leg is $7 \cdot 4$ units. This triangle is a multiple of a (3, 4, 5) triangle.

____ $= 21$

____ $= 28$

____ $= 35$

The height of the triangle is ____ units.

WRITE IT

Do the numbers 6, 8, and 10 represent a Pythagorean triple? Explain.

______

______

______

______

Step 2 Find the area of the triangle.

$A = \frac{1}{2}bh$	Area of a triangle
$A = \frac{1}{2}$ ☐ ☐	$b = 28$ and $h = 21$
$A =$ ☐	Simplify.

The area of the triangle is 294 square units.

Your Turn What is the area of triangle *RST*?

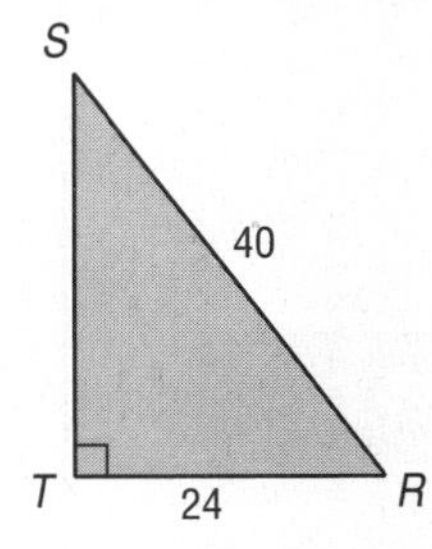

KEY CONCEPT

Corollary to the Pythagorean Theorem If *a* and *b* are measures of the shorter sides of a triangle, *c* is the measure of the longest side, and $c^2 = a^2 + b^2$, then the triangle is a right triangle. If $c^2 \neq a^2 + b^2$, then the triangle is not a right triangle.

BUILD YOUR VOCABULARY (page 252)

A statement that can easily be proved using a ☐ is often called a **corollary**.

EXAMPLE Check for Right Triangles

4 Determine whether the side measures of 27, 36, and 45 form a right triangle.

Since the measure of the longest side is 45, let $c =$ ☐, $a =$ ☐, and $b =$ ☐. Then determine whether $c^2 = a^2 + b^2$.

$c^2 = a^2 + b^2$	Pythagorean Theorem
☐^{2} = ☐$^2 + 36^2$	$a = 27$, $b = 36$, and $c = 45$
$2025 = 729 +$ ☐	Multiply.
☐ = ☐	Add.

Since $c^2 = a^2 + b^2$, the triangle ☐ a right triangle.

HOMEWORK ASSIGNMENT

Page(s):

Exercises:

Your Turn Determine whether the side measures 12, 22, and 40 form a right triangle.

11–5 The Distance Formula

What You'll Learn

- Find the distance between two points on the coordinate plane.
- Find a point that is a given distance from a second point in a plane.

Key Concept

The Distance Formula
The distance d between any two points with coordinates (x_1, y_1) and (x_2, y_2) is given by $d = \sqrt{(x_2 - x_1)^2 + (y_2 - y_1)^2}$.

EXAMPLE Distance Between Two Points

1 **Find the distance between the points at (1, 2) and (−3, 0).**

$d =$ ________ Distance Formula

$= \sqrt{\text{\_\_\_\_}^2 + \text{\_\_\_\_}^2}$ $(x_1, y_1) = (1, 2)$ and $(x_2, y_2) = (-3, 0)$

$= \sqrt{\text{\_\_\_\_}^2 + (-2)^2}$ Simplify.

$=$ ________ Evaluate squares and simplify.

$=$ ________ or about 4.47 units

Your Turn Find the distance between the points at (5, 4) and (0, 22).

____

EXAMPLE Use the Distance Formula

2 BIATHLON **Julianne is sighting her rifle for an upcoming biathlon competition. Her first shot is 2 inches to the right and 7 inches below the bull's-eye. What is the distance between the bull's-eye and where her first shot hit the target?**

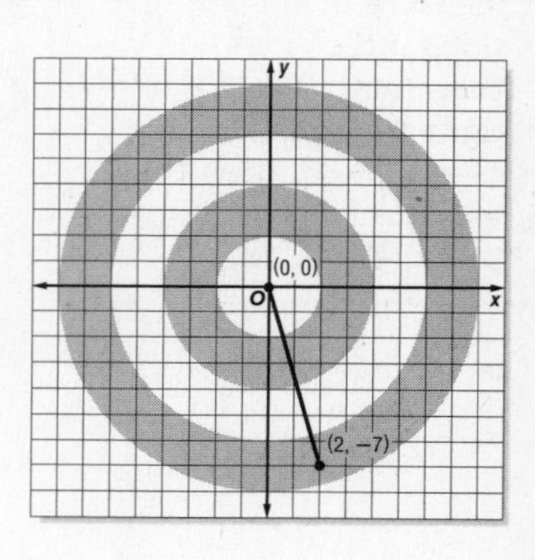

If the bull's-eye is at (0, 0), then the location of the first shot is ________. Use the Distance Formula.

$d =$ ________ Distance Formula

$= \sqrt{\text{\_\_\_\_}^2 + (-7 - 0)^2}$ $(x_1, y_1) = (0, 0)$ and $(x_2, y_2) = (2, -7)$

$= \sqrt{2^2 + (-7)}$ Simplify.

$=$ ________ or about 7.28 inches

The distance is  or about ________ inches.

REMEMBER IT You can choose either point to be (x_1, y_1) when using the Distance Formula.

Your Turn Marcy is pitching a horseshoe in her local park. Her first pitch is 9 inches to the left and 3 inches below the pin. What is the distance between the horseshoe and the pin?

EXAMPLE Finding a Missing Coordinate

3 **Find the value of a if the distance between the points at (2, −1) and (a, −4) is 5 units.**

$d = \sqrt{(x_2 - x_1)^2 + (y_2 - y_1)^2}$ Distance Formula

$\boxed{} = \sqrt{\boxed{}^2 + (-4 - (-1))^2}$ Let $d = 5$, $x_2 = a$, $x_1 = 2$, $y_2 = -4$, and $y_1 = -1$.

$5 = \sqrt{\boxed{}^2 + \boxed{}^2}$ Simplify.

$5 = \sqrt{a^2 - 4a + 4 + 9}$ Evaluate squares.

$5 = \sqrt{\boxed{}}$ Simplify.

$25 = a^2 - 4a + 13$ Square each side.

$0 = \boxed{}$ Subtract $\boxed{}$ from each side.

$0 = \boxed{}(a + 2)$ Factor.

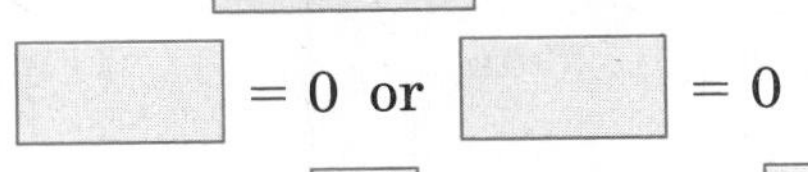

$\boxed{} = 0$ or $\boxed{} = 0$ Zero Product Property

$a = \boxed{}$ $a = \boxed{}$ Solve.

The value of a is $\boxed{}$ or $\boxed{}$.

Your Turn Find the value of a if the distance between the points at (2, 3) and (a, 2) is $\sqrt{37}$ units.

HOMEWORK ASSIGNMENT

Page(s):

Exercises:

11–6 Similar Triangles

BUILD YOUR VOCABULARY (page 253)

Similar triangles have the same ______, but not necessarily the same ______.

WHAT YOU'LL LEARN

- Determine whether two triangles are similar.
- Find the unknown measures of sides of two similar triangles.

KEY CONCEPT

Similar Triangles If two triangles are similar, then the measures of their corresponding sides are proportional, and the measures of their corresponding angles are equal.

EXAMPLE Determine Whether Two Triangles Are Similar

1 **Determine whether the pair of triangles is similar. Justify your answer.**

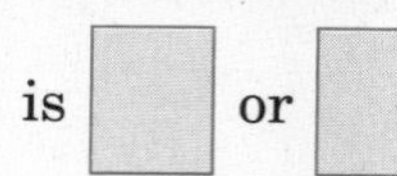
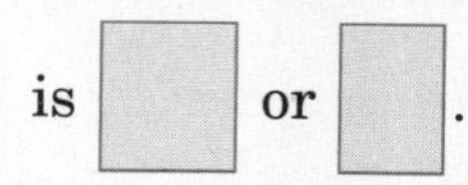

The ratio of sides $\overline{XY}$ to $\overline{AB}$ is ______ or ______.

The ratio of sides $\overline{YZ}$ to $\overline{BC}$ is ______ or ______.

The ratio of sides $\overline{XZ}$ to $\overline{AC}$ is ______ or ______.

Since the measures of the corresponding sides are ______ and the measures of the corresponding angles are equal, triangle ______ is similar to triangle ______.

Your Turn Determine whether the pair of triangles is similar. Justify your answer.

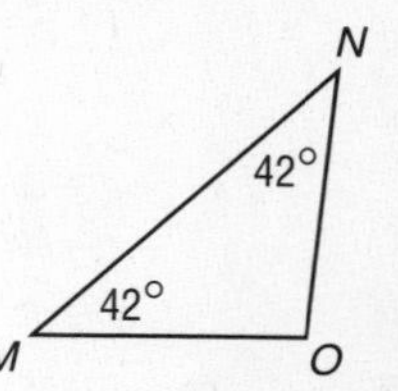

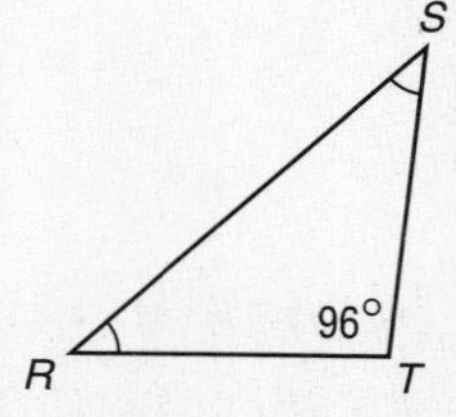

WRITE IT

Find the word *corresponding* in a dictionary, and write its definition below.

____

____

____

____

EXAMPLE Find Missing Measures

2 **a. Find the missing measures if the pair of triangles is similar.**

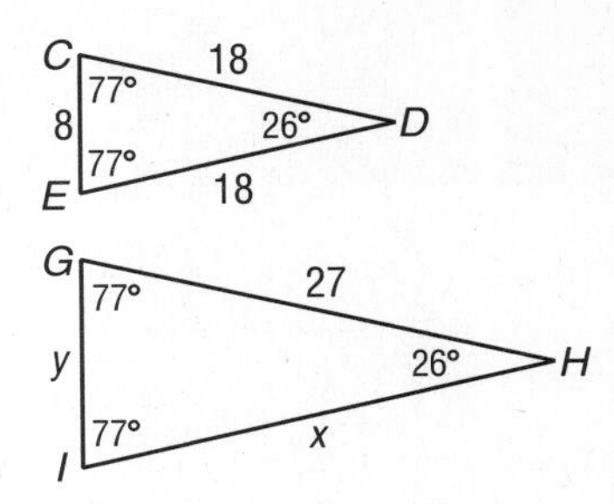

Since the corresponding angles have equal measures, ______. The lengths of the corresponding sides are proportional.

$\frac{CE}{\square} = \frac{\square}{GH}$ Corresponding sides of similar triangles are proportional.

$\frac{\square}{y} = \frac{18}{\square}$ $CE = 8$, $GI = y$, $CD = 18$, and $GH = 27$

$\square = 18y$ Find the cross products.

$\square = y$ Divide each side by 18.

$\frac{\square}{GH} = \frac{\square}{IH}$ Corresponding sides of similar triangles are proportional.

$\frac{\square}{27} = \frac{\square}{x}$ $CD = 18$, $GH = 27$, $ED = 18$, and $IH = x$

$\square = 486$ Find the cross products.

$x = \square$ Divide each side by $\square$.

The missing measures are $\square$ and $\square$.

b. Find the missing measures if the pair of triangles is similar.

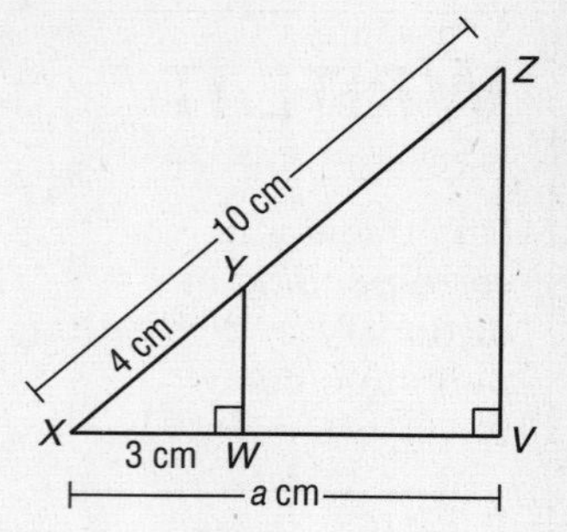

$\triangle XYW \sim$ ______

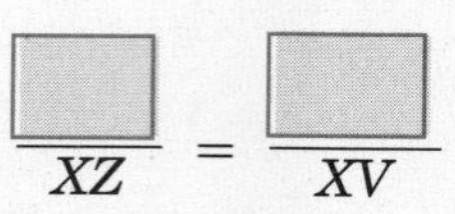

$\dfrac{\square}{XZ} = \dfrac{\square}{XV}$

Corresponding sides of similar triangles are proportional.

$\square = \dfrac{3}{a}$

$XY =$ ____, $XZ =$ ____,

$XW = 3$, and $XV = a$

$\square = 30$

Find the cross products.

$a =$ ____

Divide each side by 4.

The missing measure is ____.

Your Turn **Find the missing measures if each pair of triangles is similar.**

a.

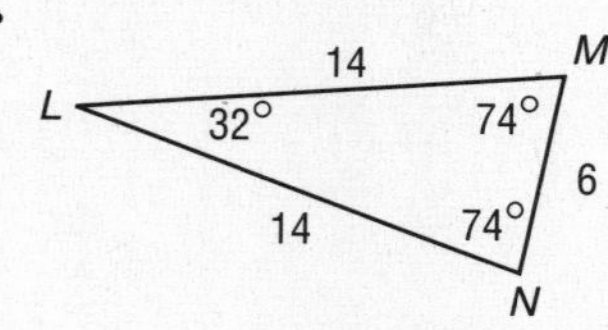

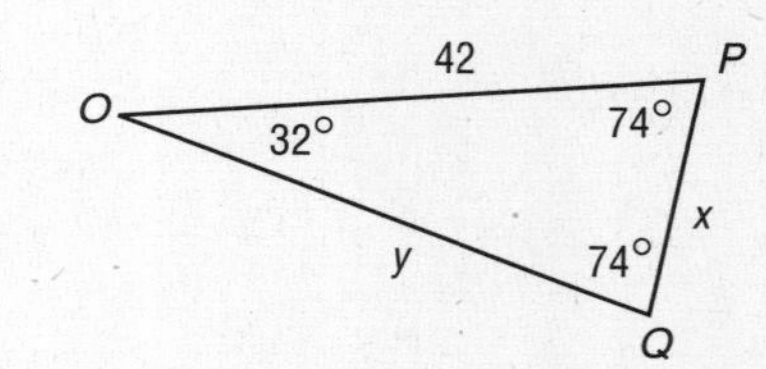

b.

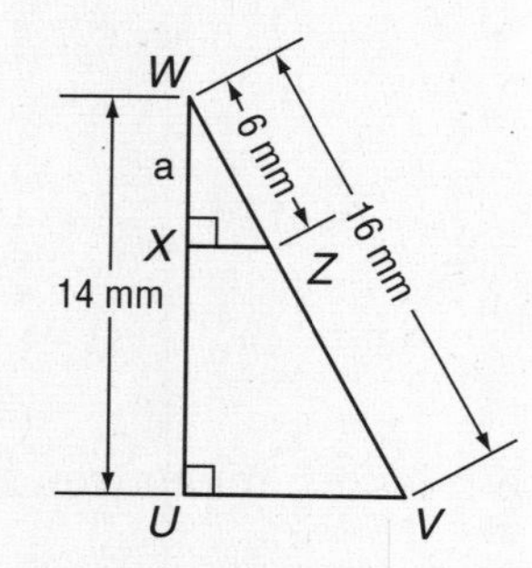

$\triangle WXZ \sim \triangle WUV$

HOMEWORK ASSIGNMENT

Page(s):

Exercises:

11–7 Trigonometric Ratios

WHAT YOU'LL LEARN

- Define the sine, cosine, and tangent ratios.
- Use trigonometric ratios to solve right triangles.

BUILD YOUR VOCABULARY (page 252–253)

Trigonometric ratios are ratios of measures of two ______ of a ______ triangle.

Three common ______ ratios are called **sine, cosine**, and **tangent**.

sine of $\angle A = \dfrac{\text{______}}{\text{measure of hypotenuse}}$

cosine of $\angle A = \dfrac{\text{measure of leg adjacent to } \angle A}{\text{______}}$

tangent of $\angle A = \dfrac{\text{______}}{\text{measure of leg adjacent to } \angle A}$

EXAMPLE Sine, Cosine, and Tangent

1 Find the sine, cosine, and tangent of each acute angle of $\triangle DEF$. Round to the nearest ten-thousandth.

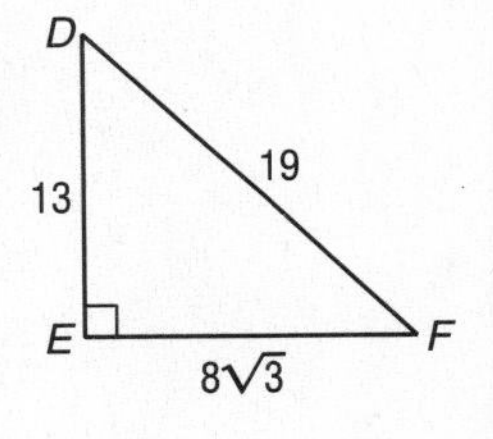

Write each ratio and substitute the measures. Use a calculator to find each value.

$\sin D =$ ______ $= \dfrac{\text{___}}{19}$ or 0.7293

$\cos D =$ ______ $= \dfrac{\text{___}}{19}$ or 0.6842

$\tan D =$ ______ $= \dfrac{\text{___}}{13}$ or 1.0659

$\sin F =$ ______ $=$ ___ or 0.6842

$\cos F = \dfrac{\text{adjacent leg}}{\text{hypotenuse}} =$ ___ or 0.7293

$\tan F = \dfrac{\text{opposite leg}}{\text{adjacent leg}} =$ ___ or 0.9382

Your Turn Find the sine, cosine, and tangent of each acute angle of $\triangle XYZ$. Round to the nearest ten-thousandth.

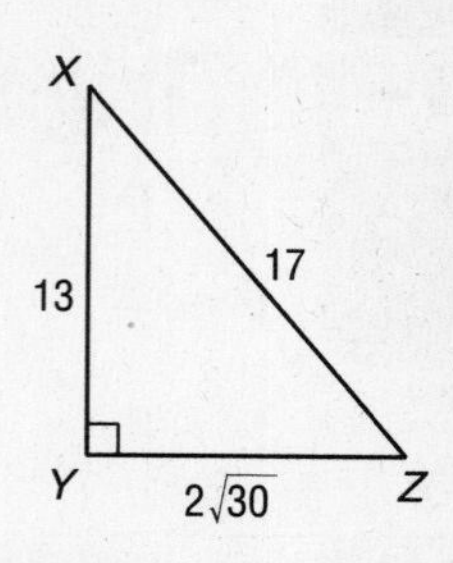

EXAMPLE Find the Sine of an Angle

2 **Find cos 65° to the nearest ten-thousandth.**

Keystrokes COS 65 ENTER .4226182617

Rounded to the nearest ten-thousandth,
cos 65° ≈ ______.

Your Turn Find tan 32° to the nearest ten-thousandth.

EXAMPLE Find the Measure of an Angle

3 **Find the measure of $\angle B$ to the nearest degree.**

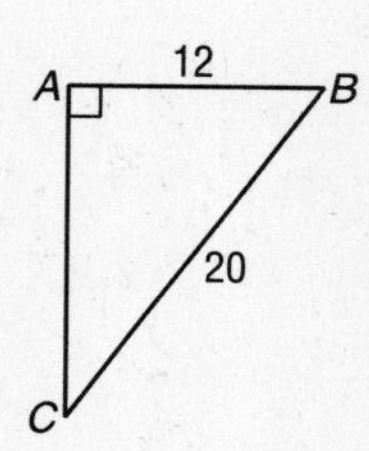

Since the lengths of the adjacent leg and the hypotenuse are known, use the cosine ratio.

$\cos B = \frac{\text{adjacent leg}}{\text{hypotenuse}}$

= ______ $\quad AB = 12$ and $BC = 20$

Angle B has a cosine ratio of $\frac{12}{20}$.

By using COS⁻¹ on a calculator, you find that the measure of $\angle B$ is about 53°.

Your Turn Find the measure of $\angle Q$ to the nearest degree.

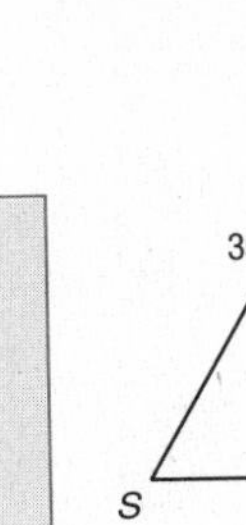

BUILD YOUR VOCABULARY (page 253)

Finding all of the measures of the ______ and the ______ in a right triangle is called **solving a triangle.**

WRITE IT

Explain how you can use the measures of the sides of a triangle to prove whether or not it is a right triangle.

______

______

______

______

EXAMPLE Solve a Triangle

4 Find all of the missing measures in $\triangle DEF$.

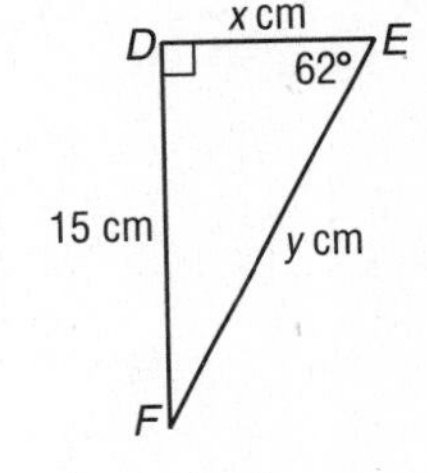

Step 1 Find the measure of $\angle F$. The sum of the measures of the angles in a triangle is 180.

______

The measure of $\angle F$ is ______.

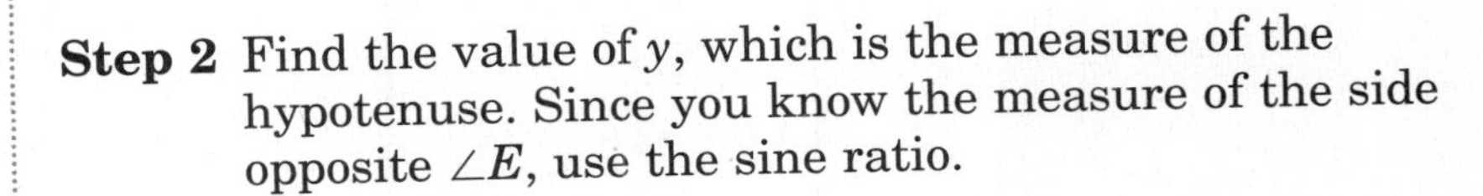

Step 2 Find the value of y, which is the measure of the hypotenuse. Since you know the measure of the side opposite $\angle E$, use the sine ratio.

$\sin 62° =$ ______	Definition of sine
______ $\approx \frac{15}{y}$	Evaluate sin 62°.
$0.8829y \approx$ ______	Multiply each side by ______.
$y \approx$ ______	Divide each side by 0.8829.

$\overline{EF}$ is about ______ centimeters long.

Step 3 Find the value of x, the measure of the side adjacent to $\angle E$. Use the tangent ratio.

$\tan 62° = \square$	Definition of tangent
$\square \approx \frac{15}{x}$	Evaluate tan 62°.
$1.8807x \approx 15$	Multiply each side by x.
$x \approx \square$	Divide each side by $\square$.

$\overline{DE}$ is about ☐ centimeters long.

The missing measures are ☐, ☐ cm, and ☐ cm.

Your Turn Find all of the missing measures in $\triangle PQR$.

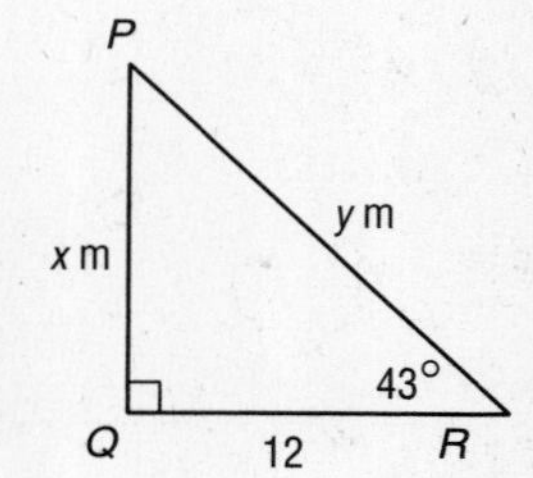

HOMEWORK ASSIGNMENT

Page(s):

Exercises:

CHAPTER 11

BRINGING IT ALL TOGETHER

STUDY GUIDE

FOLDABLES™	VOCABULARY PUZZLEMAKER	BUILD YOUR VOCABULARY
Use your **Chapter 11 Foldable** to help you study for your chapter test.	To make a crossword puzzle, word search, or jumble puzzle of the vocabulary words in Chapter 11, go to www.glencoe.com/sec/math/t_resources/free/index.php	You can use your completed **Vocabulary Builder** (pages 252–253) to help you solve the puzzle.

11–1 Simplifying Radical Expressions

Simplify.

1. $\sqrt{28x^2y^4}$ ______ **2.** $\sqrt{\frac{5}{32}}$ ______ **3.** $\frac{8}{3+\sqrt{3}}$ ______

4. What should you remember to check for when you want to determine if a radical expression is in simplest form?

Check radicands for ______ and ______,

and check fractions for ______ in the ______.

11–2 Operations with Radical Expressions

Simplify each expression.

5. $6\sqrt{3} - \sqrt{12}$ ______ **6.** $2\sqrt{12} - 7\sqrt{3}$ ______ **7.** $3\sqrt{2}\left(\sqrt{8} + \sqrt{24}\right)$ ______

8. $\left(2\sqrt{5} - 2\sqrt{3}\right)\left(\sqrt{10} + \sqrt{6}\right)$

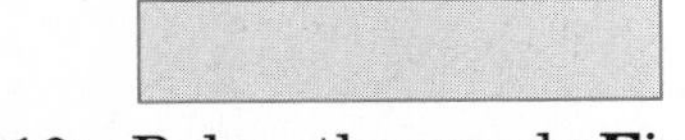

9. $\sqrt{27} + \sqrt{18} + \sqrt{300}$ ______

10. Below the words **F**irst terms, **O**uter terms, **I**nner terms, and **L**ast terms, write the products you would use to simplify the expression $\left(2\sqrt{15} + 3\sqrt{15}\right)\left(6\sqrt{3} - 5\sqrt{2}\right)$.

First terms		Outer terms		Inner terms		Last terms
______	+	______	+	______	+	______

11–3

Radical Equations

11. To solve a radical equation, you first isolate the radical on one side of the equation. Why do you then square each side of the equation?

12. Provide the reason for each step in the solution of the given radical equation.

Step	Reason
$\sqrt{5x - 1} - 4 = x - 3$	Original equation
$\sqrt{5x - 1} = x + 1$	
$(\sqrt{5x - 1})^2 = (x + 1)^2$	
$5x - 1 = x^2 + 2x + 1$	
$0 = x^2 - 3x + 2$	
$0 = (x - 1)(x - 2)$	
$x - 1 = 0$ or $x - 2 = 0$	
$x = 1$ $\quad$ $x = 2$	

13. To be sure that 1 and 2 are the correct solutions, into which equation should you substitute to check?

14. A computer screen measures 12 inches high and 17 inches wide. What is the length of the screen's diagonal? Round your answer to the nearest whole number.

11–4

The Pythagorean Theorem

Write an equation that you could solve to find the missing side length of each right triangle. Then solve.

15.

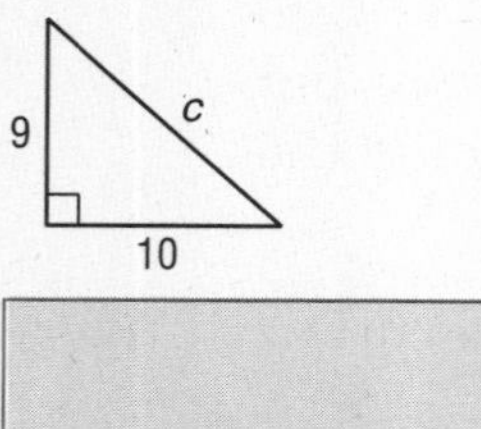

16.

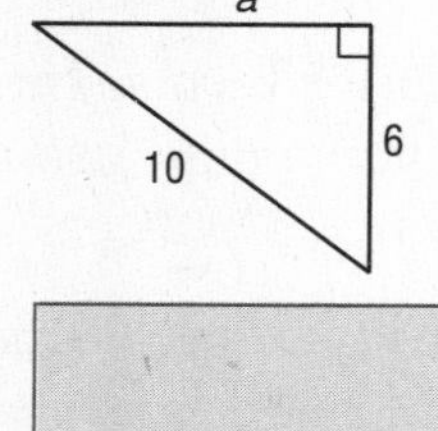

17.

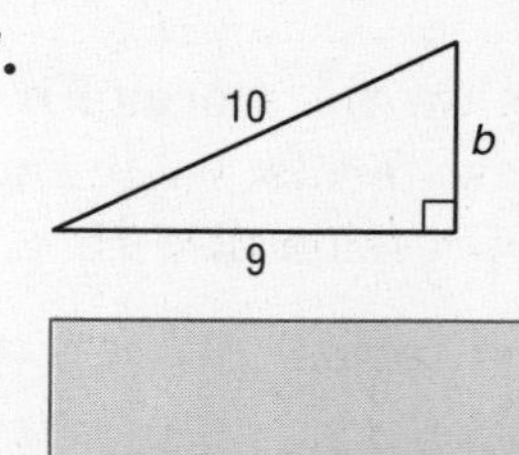

11–5 The Distance Formula

Find the distance between each pair of points whose coordinates are given. Express answers in simplest radical form and as decimal approximations rounded to the nearest hundredth if necessary.

18. (6, 4), (2, 1)

19. (3, 7), (9, −2)

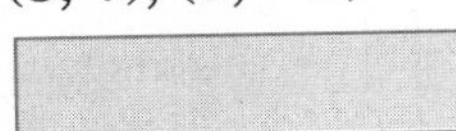

11–6 Similar Triangles

Determine whether each pair of triangles is similar. Explain how you would know that your answer is correct.

20.

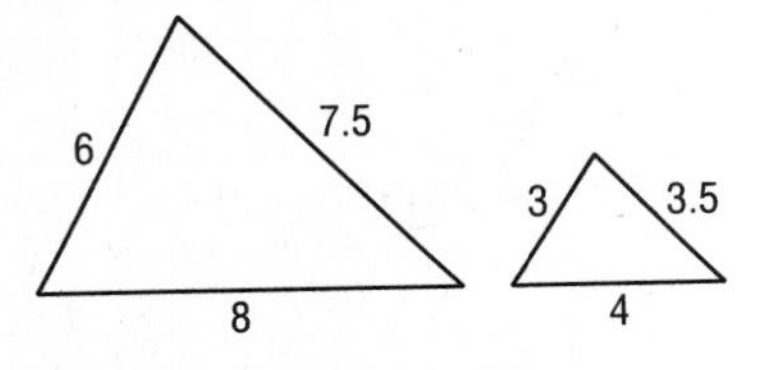

21.

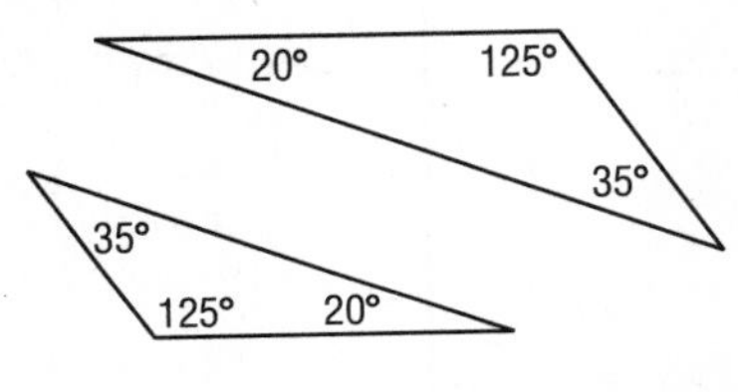

11–7 Trigonometric Ratios

Complete each sentence.

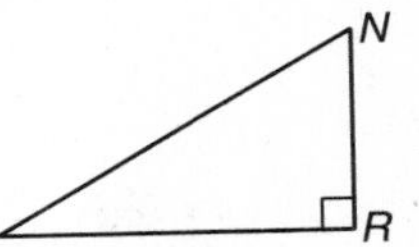

22. The legs of triangle MNR are segments  and .

23. The hypotenuse of triangle MNR is ☐.

24. The leg opposite $\angle N$ is 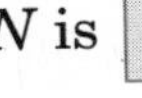 ☐, and the leg adjacent to $\angle N$ is ☐.

Write K or S to complete each equation.

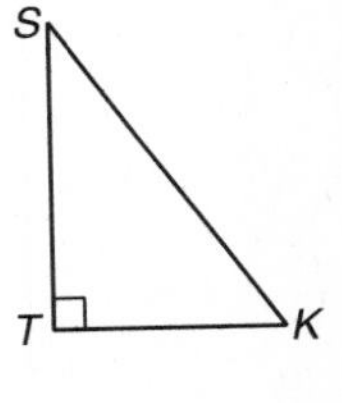

25. tan ☐ $= \dfrac{KT}{TS}$

26. sin ☐ $= \dfrac{KT}{KS}$

27. sin ☐ $= \dfrac{ST}{KS}$

28. cos ☐ $= \dfrac{ST}{KS}$

29. cos ☐ $= \dfrac{KT}{KS}$

30. tan ☐ $= \dfrac{ST}{KT}$

ARE YOU READY FOR THE CHAPTER TEST?

Visit **www.algebra1.com** to access your textbook, more examples, self-check quizzes, and practice tests to help you study the concepts in Chapter 11.

Check the one that applies. Suggestions to help you study are given with each item.

☐ **I completed the review of all or most lessons without using my notes or asking for help.**

- You are probably ready for the Chapter Test.
- You may want to take the Chapter 11 Practice Test on page 637 of your textbook as a final check.

☐ **I used my Foldable or Study Notebook to complete the review of all or most lessons.**

- You should complete the Chapter 11 Study Guide and Review on pages 632–636 of your textbook.
- If you are unsure of any concepts or skills, refer back to the specific lesson(s).
- You may also want to take the Chapter 11 Practice Test on page 637.

☐ **I asked for help from someone else to complete the review of all or most lessons.**

- You should review the examples and concepts in your Study Notebook and Chapter 11 Foldable.
- Then complete the Chapter 11 Study Guide and Review on pages 632–636 of your textbook.
- If you are unsure of any concepts or skills, refer back to the specific lesson(s).
- You may also want to take the Chapter 11 Practice Test on page 637.

Student Signature

Parent/Guardian Signature

Teacher Signature

Rational Expressions and Equations

Use the instructions below to make a Foldable to help you organize your notes as you study the chapter. You will see Foldable reminders in the margin of this Interactive Study Notebook to help you in taking notes.

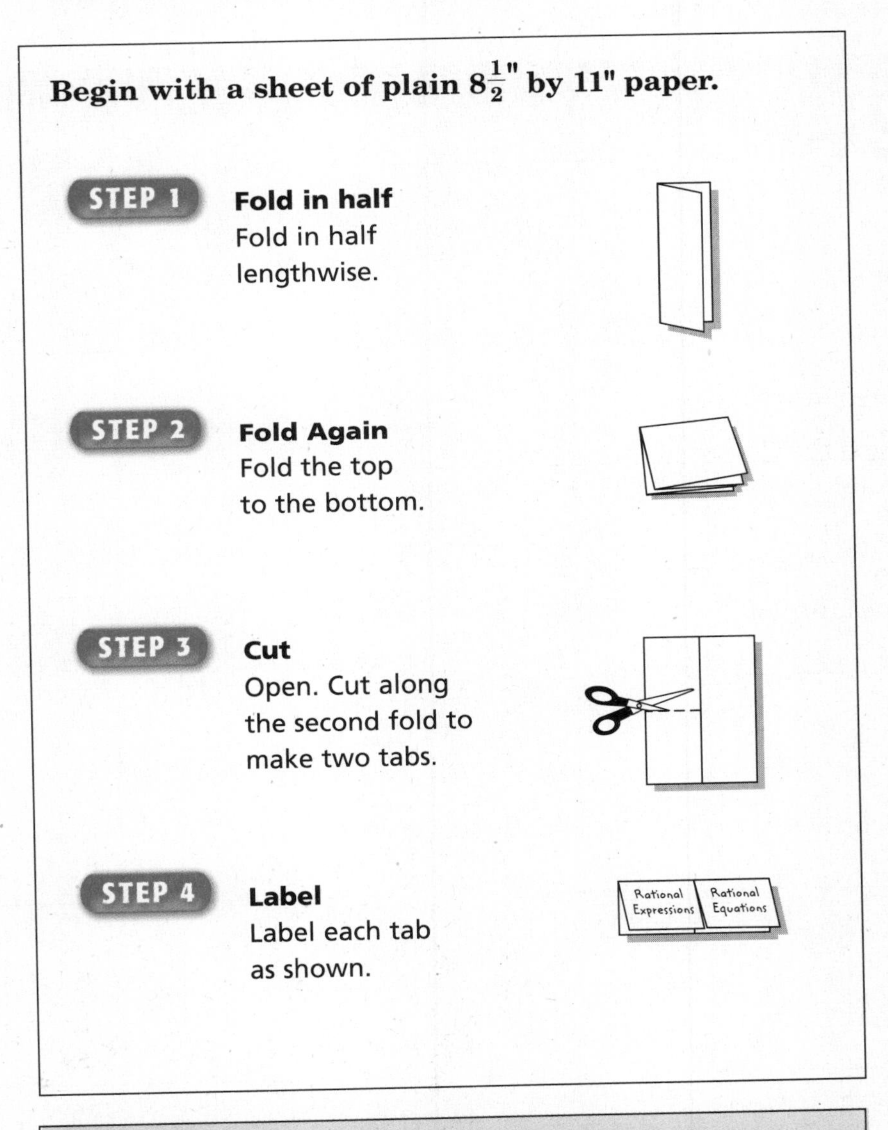

NOTE-TAKING TIP: When you take notes, it may be helpful to sit as close as possible to the front of the class. There are fewer distractions and it is easier to hear.

BUILD YOUR VOCABULARY

This is an alphabetical list of new vocabulary terms you will learn in Chapter 12. As you complete the study notes for the chapter, you will see Build Your Vocabulary reminders to complete each term's definition or description on these pages. Remember to add the textbook page number in the second column for reference when you study.

Vocabulary Term	Found on Page	Definition	Description or Example
complex fraction			
excluded values			
extraneous solutions [ehk STRAY nee uhs]			
inverse variation [ihn VUHRS]			
least common multiple			
least common denominator			
mixed expression			
product rule			
rate problem			
rational equation			
rational expression			
work problem			

12–1 Inverse Variation

WHAT YOU'LL LEARN

- Graph inverse variations.
- Solve problems involving inverse variations.

BUILD YOUR VOCABULARY (page 278)

When the product of two values remains ______ the relationship forms an **inverse variation**.

KEY CONCEPT

Inverse Variation
y varies inversely as x if there is some nonzero constant k such that $xy = k$.

EXAMPLE Graph an Inverse Variation

1 MANUFACTURING **The time t in hours that it takes to build a particular model of computer varies inversely with the number of people p working on the computer. The equation $pt = 12$ can be used to represent the people building a computer. Draw a graph of the relation.**

Solve for $p = 2$.

$pt = 12$	Original equation
___ $t = 12$	Replace p with ___.
$t = \frac{12}{\_\_\_}$	Divide each side by ___.
$t =$ ___	Simplify.

Solve the equation for the other values of p.

p	2	4	6	8	10	12
t						

Graph the ordered pairs. As the number of people p increases, the time t it takes to build a computer decreases.

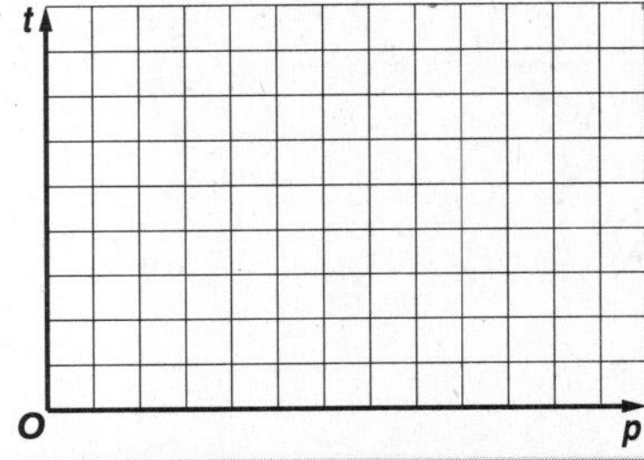

Your Turn The time t in hours that it takes to prepare packages for delivery varies inversely with the number of people p that are preparing them. The equation $pt = 36$ can be used to represent the people preparing the packages. Draw a graph of the relation.

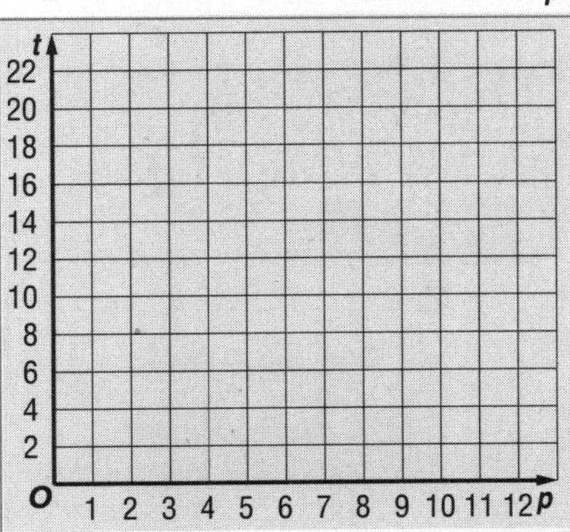

BUILD YOUR VOCABULARY (page 278)

The equation $x_1y_1 = x_2y_2$ is called the **product rule** for ______ variations.

FOLDABLES™

ORGANIZE IT

Under the tab for Rational Expressions, write the general form for inverse variation. Then give an example of an inverse variation equation.

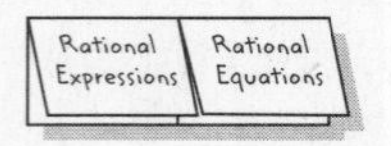

EXAMPLE Solve for *x*

2 If y varies inversely as x and $y = 5$ when $x = 12$, find x when $y = 15$.

Let $x_1 = 12$, $y_1 = 5$, and $y_2 = 15$. Solve for x_2.

Method 1 Use the product rule.

$x_1y_1 = x_2y_2$	Product rule for inverse variations
______ · ______ $= x_2$ · ______	$x_1 = 12$, $y_1 = 5$, $y_2 = 15$
______ $= x_2$	Divide each side by ______.
______ $= x_2$	Simplify.

Method 2 Use a proportion.

$\frac{x_1}{x_2} = \frac{y_2}{y_1}$ Proportion rule for inverse variations

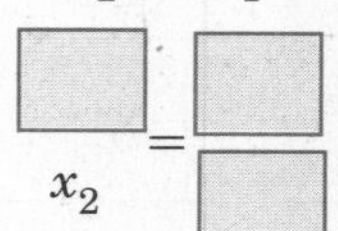

$x_1 = 12$, $y_1 = 5$, $y_2 = 15$

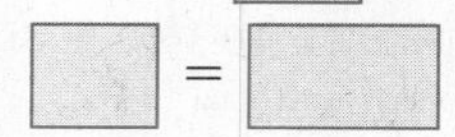

Cross multiply.

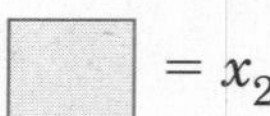

Divide each side by 15.

Both methods show that $x =$ ______ when $y =$ ______.

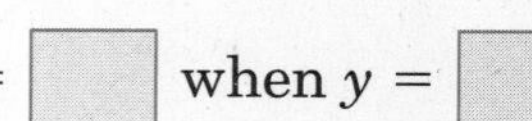

Your Turn

a. If y varies inversely as x and $y = 6$ when $x = 40$, find x when $y = 30$.

b. If y varies inversely as x and $y = -5$ when $x = 15$, find y when $x = 3$.

HOMEWORK ASSIGNMENT

Page(s):

Exercises:

12–2 Rational Expressions

WHAT YOU'LL LEARN

- Identify values excluded from the domain of a rational expression.
- Simplify rational expressions.

BUILD YOUR VOCABULARY (page 278)

A **rational expression** is an algebraic fraction whose ______ and ______ are polynomials.

Any values of a variable that result in a denominator of ______ must be excluded from the ______ of the variable and are called **excluded values** of the rational expression.

EXAMPLE One Excluded Value

1 State the excluded value of $\frac{3b-2}{b+7}$.

Exclude the values for which $b + 7 = 0$.

$b + 7 = 0$ — The denominator cannot equal ______.

$b =$ ______ — Subtract ______ from each side.

So, b cannot equal ______.

EXAMPLE Multiple Excluded Values

2 State the excluded value of $\frac{5a^2+2}{a^2-a-12}$

Exclude the values for which $a^2 - a - 12 =$ ______.

$a^2 - a - 12 = 0$ — The denominator cannot equal ______.

$($ ______ $)($ ______ $) = 0$ — Factor.

Use the Zero Product Property to solve for a.

______ or ______

$a =$ ______ $a =$ ______

So, a cannot equal ______ or ______.

Your Turn **State the excluded value for each rational expression.**

a. $\frac{2y - 7}{y + 3}$

b. $\frac{x^2 + 1}{x^2 - 5x + 6}$

EXAMPLE Expressions Involving Monomials

FOLDABLES™

ORGANIZE IT

Take notes on how to simplify a rational expression.

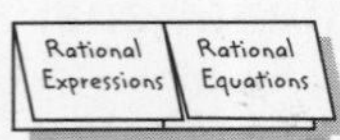

3 **Simplify $\frac{32x^5y^2}{4xy^7}$.**

$\frac{32x^5y^2}{4xy^7} = \frac{(4xy^2)(8x^4)}{(4xy^2)(y^5)}$ The GCF of the numerator and denominator is ____.

$= \frac{\overset{1}{\cancel{(4xy^2)}}(8x^4)}{\underset{1}{\cancel{(4xy^2)}}(y^5)}$ Divide the numerator and denominator by ____.

$=$ ____ Simplify.

EXAMPLE Expression Involving Polynomials

WRITE IT

How do you know that a rational expression is in simplest form?

______

______

______

______

4 **Simplify $\frac{x^2 - 9x + 14}{x^2 - 2x - 35}$.**

$\frac{x^2 - 9x + 14}{x^2 - 2x - 35} = \frac{\quad}{\quad}$ Factor.

$= \frac{\overset{1}{\cancel{(x - 7)}}(x - 2)}{\underset{1}{\cancel{(x - 7)}}(x + 5)}$ Divide the numerator and denominator by the GCF, ____.

$=$ ____ Simplify.

Your Turn **Simplify.**

a. $\frac{60a^3b^2}{18ab^3}$

b. $\frac{c^2 + 4c - 60}{c^2 - 5c - 6}$

HOMEWORK ASSIGNMENT

Page(s):

Exercises:

12–3 Multiplying Rational Expressions

WHAT YOU'LL LEARN

- Multiply rational expressions.
- Use dimensional analysis with multiplication.

EXAMPLE Expressions Involving Monomials

1 **Find $\frac{7x^2y}{12z^3} \cdot \frac{14z}{49xy^4}$.**

Method 1

Divide by the greatest common factor after multiplying.

$$\frac{7x^2y}{12z^3} \cdot \frac{14z}{49xy^4} = \frac{\square}{\square}$$ ← Multiplying the numerators. ← Multiplying the denominators.

$$= \frac{\cancel{98xyz}(\square)}{\cancel{98xyz}(\square)}$$ The GCF is $\square$.

$$= \square$$ Simplify.

Method 2

Divide the common factors before multiplying.

$$\frac{7x^2y}{12z^3} \cdot \frac{14z}{49xy^4} = \frac{\overset{1}{\cancel{7}}\,\overset{x}{\cancel{x^2}}\,\overset{1}{\cancel{y}}}{\underset{6}{\cancel{12}}\,\underset{z^2}{\cancel{z^3}}} \cdot \frac{\overset{1}{\cancel{14}}\,\overset{1}{\cancel{z}}}{\underset{1}{\cancel{49}}\,\underset{1}{\cancel{x}}\,\underset{y^3}{\cancel{y^4}}}$$ Divide by common factors $\square$ and $\square$.

$$= \square$$ Multiply.

EXAMPLE Expressions Involving Polynomials

2 **Find $\frac{x}{x+4} \cdot \frac{x^2-4x-32}{x^3}$.**

$$\frac{x}{x+4} \cdot \frac{x^2-4x-32}{x^3} = \frac{x}{x+4} \cdot \frac{(\square)(\square)}{x^3}$$ Factor the numerator.

$$= \frac{\overset{1}{\cancel{x}}}{\underset{1}{\cancel{x+4}}} \cdot \frac{(x-8)\overset{1}{\cancel{(x+4)}}}{\underset{x^2}{\cancel{x^3}}}$$ The GCF is $\square$.

$$= \square$$ Simplify.

Your Turn **Find each product.**

a. $\frac{6mn^2}{11m^3p^4} \cdot \frac{22p^3}{3n}$

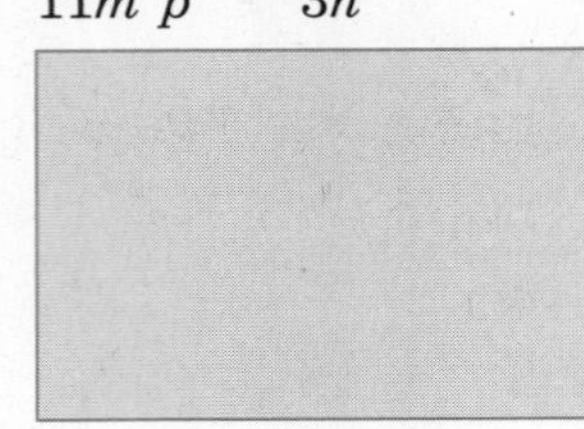

b. $\frac{14k^3w^2}{5s^2t^2} \cdot \frac{10s^4t}{21kw^5}$

c. $\frac{y + 1}{y} \cdot \frac{y^2}{y^2 + 8y + 7}$

d. $\frac{13c - 39}{c - 4} \cdot \frac{c^2 - 16}{c^2 + 3c - 18}$

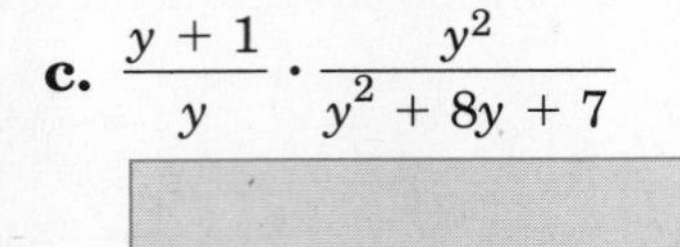

REVIEW IT

When do you need to use dimensional analysis in a word problem? *(Lesson 3–8)*

______

______

______

______

______

EXAMPLE Dimensional analysis

3 SPACE **The velocity that a spacecraft must have in order to escape Earth's gravitational pull is called the escape velocity. The escape velocity for a spacecraft leaving Earth is about 40,320 kilometers per hour. What is this speed in meters per second?**

$$\frac{40{,}320 \text{ kilometers}}{\text{hour}} \cdot \frac{1000\text{meters}}{1 \text{ kilometer}} \cdot \frac{1 \text{ hour}}{\square} \cdot \frac{\square \text{ minute}}{\square \text{ seconds}}$$

$$= \frac{40{,}320 \text{ kilometers}}{\text{hour}} \cdot \frac{1000\text{meters}}{1 \text{ kilometer}} \cdot \frac{1 \text{ hour}}{60 \text{ minutes}} \cdot \frac{1 \text{ minute}}{60 \text{ seconds}}$$

$$= \frac{\overset{1120}{40{,}320} \cdot \overset{10}{1000} \cdot 1 \cdot 1 \text{ meters}}{1 \cdot 1 \cdot \underset{1}{60} \cdot \underset{1}{60} \text{ seconds}}$$

= ______ Simplify.

= ______ Multiply.

The escape velocity is ______ meters per second.

Your Turn The speed of sound, or Mach 1, is approximately 330 meters per second at sea level. What is the speed of sound in kilometers per hour?

HOMEWORK ASSIGNMENT

Page(s):

Exercises:

12–4 Dividing Rational Expressions

What You'll Learn

- Divide rational expressions.
- Use dimensional analysis with division.

EXAMPLE 1 Expression Involving Monomials

Find $\frac{6x^4}{5} \div \frac{24x}{75}$.

$\frac{6x^4}{5} \div \frac{24x}{75} = \frac{6x^4}{5} \cdot$ ____ Multiply by ____, the reciprocal of $\frac{24x}{75}$.

$= \frac{\overset{1\ x^3}{\cancel{6}\cancel{x^4}}}{\underset{1}{\cancel{5}}} \cdot \frac{\overset{15}{\cancel{75}}}{\underset{4}{\cancel{24}}\ \underset{1}{\cancel{x}}}$ Divide by common factors ____, ____, and ____.

$=$ ____ Simplify.

EXAMPLE 2 Expression Involving Binomials

Find $\frac{3m + 12}{m + 5} \div \frac{m + 4}{m - 2}$.

$\frac{3m + 12}{m + 5} \div \frac{m + 4}{m - 2}$

$= \frac{3m + 12}{m + 5} \cdot \frac{m - 2}{m + 4}$ Multiply by $\frac{m - 2}{m + 4}$, the reciprocal of ____.

$= \frac{3(m + 4)}{m + 5} \cdot \frac{m - 2}{m + 4}$ Factor ____.

$= \frac{3\overset{1}{\cancel{(m + 4)}}}{m + 5} \cdot \frac{m - 2}{\underset{1}{\cancel{m + 4}}}$ The GCF is ____.

$=$ ____ or ____ Simplify.

Your Turn **Find each quotient.**

a. $\frac{3a}{7} \div \frac{9a^5}{14}$

b. $\frac{n - 8}{n - 12} \div \frac{6n - 48}{n + 3}$

EXAMPLE Divide by a Binomial

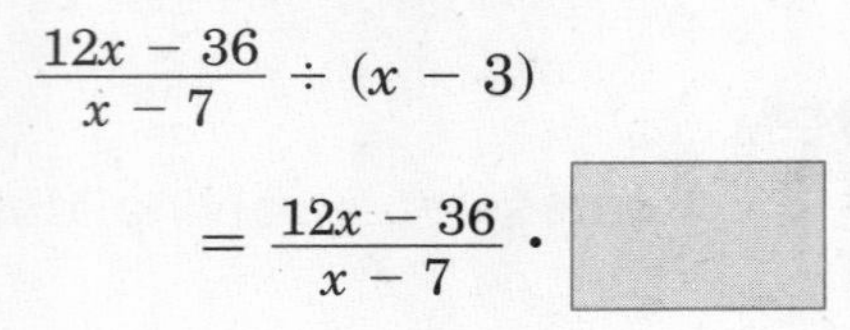

3 **Find $\frac{12x - 36}{x - 7} \div (x - 3)$.**

> **REMEMBER IT**
> 
> When you are dividing rational expressions, always multiply by the reciprocal.

$\frac{12x - 36}{x - 7} \div (x - 3)$

$= \frac{12x - 36}{x - 7} \cdot$ ______ Multiply by , the reciprocal of $(x - 3)$.

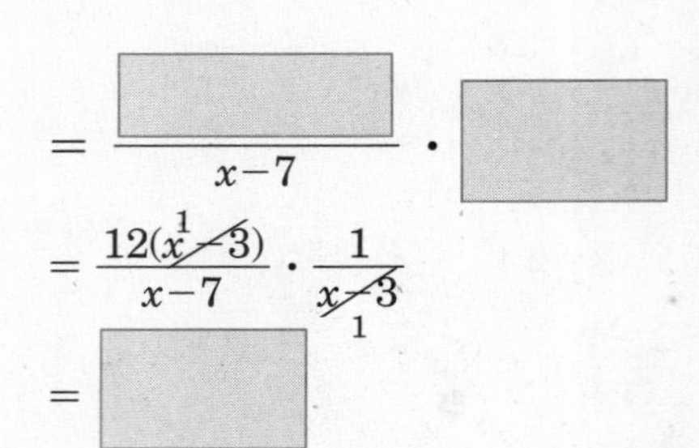

$= \frac{\text{\_\_\_\_\_\_}}{x - 7} \cdot$ ______ Factor $12x - 36$.

$= \frac{12(\cancel{x - 3})^{1}}{x - 7} \cdot \frac{1}{\cancel{x - 3}_{1}}$ The GCF is ______.

$=$ ______ Simplify.

EXAMPLE Expression Involving Polynomials

4 **Find $\frac{q^2 - 11q - 26}{7} \div \frac{q - 13}{q + 7}$.**

> **FOLDABLES™**
> **ORGANIZE IT**
> Under the tab for Rational Expressions, write the question and answer to Example 4. Then label the quotient, dividend, and divisor.
>
> Rational Expressions | Rational Equations

$\frac{q^2 - 11q - 26}{7} \div \frac{q - 13}{q + 7}$

$= \frac{q^2 - 11q - 26}{7} \cdot$ ______ Multiply by the reciprocal, ______.

$= \frac{(\text{\_\_\_\_\_\_})(\text{\_\_\_\_\_\_})}{7} \cdot \frac{q + 7}{q - 13}$ Factor $q^2 - 11q - 26$.

$= \frac{(\cancel{q - 13})^{1}(q + 2)}{7} \cdot \frac{q + 7}{\cancel{q - 13}_{1}}$ The GCF is .

$=$  or ______ Simplify.

Your Turn **Find each quotient.**

a. $\frac{3b + 15}{b - 6} \div (b + 5)$

b. $\frac{k^2 - 13k + 30}{10} \div \frac{k - 3}{k - 2}$

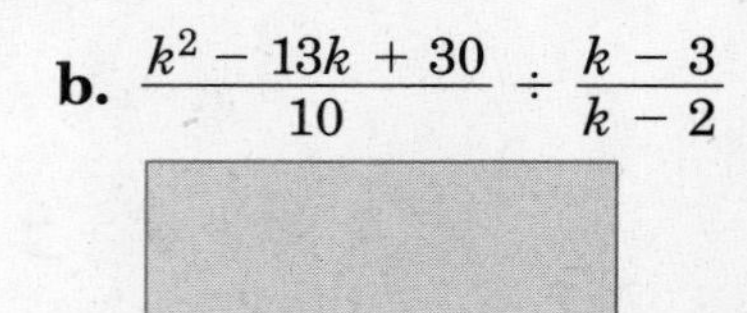

EXAMPLE Dimensional Analysis

5 AVIATION **In 1986, an experimental aircraft named Voyager was piloted by Jenna Yeager and Dick Rutan around the world non-stop, without refueling. The trip took exactly 9 days and covered a distance of 25,012 miles. What was the speed of the aircraft in miles per hour? Round to the nearest miles per hour.**

Use the formula for rate, time, and distance.

$rt = d$ — rate · time = distance

$r \cdot 9 \text{ days} = 25{,}012 \text{ mi}$ — $t =$ ____ days, $d =$ ____ mi

$r = \frac{25{,}012 \text{ mi}}{\square}$ — Divide each side by ____.

$= \frac{25{,}012 \text{ mi}}{9 \text{ days}} \cdot \square$ — Convert days to hours.

$= \frac{25{,}012 \text{ mi}}{\square}$ or about $\frac{\square}{1 \text{ hour}}$

The speed of the aircraft was about ____ miles per hour.

Your Turn Suppose that Jenna Yeager and Dick Rutan wanted to complete the trip in exactly 7 days. What would be their average speed in miles per hour for the 25,012 mile trip?

FOLDABLES™

ORGANIZE IT

Write how to divide rational expressions in your Foldable.

Rational Expressions | Rational Equations

HOMEWORK ASSIGNMENT

Page(s):

Exercises:

12–5 Dividing Polynomials

What You'll Learn

- Divide a polynomial by a monomial.
- Divide a polynomial by a binomial.

EXAMPLE Divide a Binomial by a Monomial

1 **Find $(4x^2 - 18x) \div 2x$.**

$(4x^2 - 18x) \div 2x = \frac{4x^2 - 18x}{2x}$ Write as a rational expression.

$= \frac{4x^2}{\square} - \frac{18x}{\square}$ Divide each term by $\square$.

$= \frac{\overset{2x}{\cancel{4x^2}}}{\underset{1}{\cancel{2x}}} - \frac{\overset{9}{\cancel{18x}}}{\underset{1}{\cancel{2x}}}$ Simplify each term.

$= \square$ Simplify.

EXAMPLE Divide a Polynomial by a Monomial

2 **Find $(2y^2 - 3y - 9) \div 3y$.**

$(2y^2 - 3y - 9) \div 3y = \square$ Write as a rational expression.

$= \frac{2y^2}{3y} - \frac{3y}{3y} - \frac{9}{3y}$ Divide each term by $\square$.

$= \frac{\overset{2y}{\cancel{2y^2}}}{\underset{3}{\cancel{3y}}} - \frac{\overset{1}{\cancel{3y}}}{\underset{1}{\cancel{3y}}} - \frac{\overset{3}{\cancel{9}}}{\underset{y}{\cancel{3y}}}$ Simplify each term.

$= \square$ Simplify.

EXAMPLE Divide a Polynomial by a Binomial

3 **Find $(2r^2 + 5r - 3) \div (r + 3)$.**

$(2r^2 + 5r - 3) \div (r + 3) = \square$ Write as a rational expression.

$= \frac{(\square)(\square)}{r + 3}$ Factor the numerator.

$= \frac{(2r - 1)\overset{1}{\cancel{(r + 3)}}}{\underset{1}{\cancel{r + 3}}}$ Divide by the GCF.

$= \square$ Simplify.

EXAMPLE Long Division

4 **Find $(x^2 + 7x - 15) \div (x - 2)$.**

Step 1 Divide the first term of the dividend, x^2, by the first term of the divisor, x.

$$x - 2\overline{)x^2 + 7x - 15}\quad \text{quotient: } x$$

$x^2 \div x = x$.

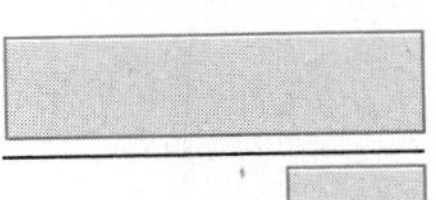

Multiply x and $x - 2$.

Subtract.

Step 2 Divide the first term of the partial dividend, $9x - 15$, by the first term of the divisor, x.

$$x - 2\overline{)x^2 + 7x - 15}\quad \text{quotient: } x + 9$$

$9x \div x = 9$

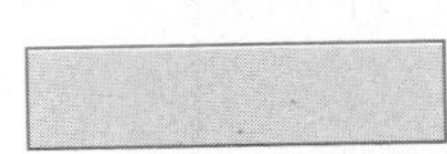

Subtract and bring down ____.

$9x - 15$

Multiply 9 and $x - 2$.

Subtract.

The quotient of $(x^2 + 7x - 15) \div (x - 2)$ is ____ with a remainder of ____, which can be written as

____

WRITE IT

Explain how dividing polynomials is similar to dividing whole numbers.

____
____
____
____

Your Turn **Find each quotient.**

a. $(48z^2 + 18z) \div 6z$

b. $(-8x^2 + 6x - 28) \div 4x$

c. $(2c^2 - 3c - 9) \div (c - 3)$

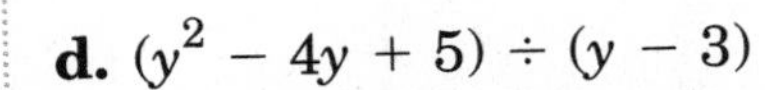

d. $(y^2 - 4y + 5) \div (y - 3)$

HOMEWORK ASSIGNMENT

Page(s):

Exercises:

12–6 Rational Expression with Like Denominators

What You'll Learn

- Add rational expressions with like denominators.
- Subtract rational expressions with like denominators.

EXAMPLE Numbers in Denominator

1 Find $\frac{4b}{15} + \frac{16b}{15}$.

$\frac{4b}{15} + \frac{16b}{15} = $ ______ The common denominator is ______.

$= \frac{\text{\_\_\_\_\_\_}}{\text{\_\_\_\_\_\_}}$ Add the numerators.

$= \frac{\overset{4b}{\cancel{20b}}}{\underset{3}{\cancel{15}}}$ or $\frac{4b}{3}$ Divide by the common factor ______ and simplify.

Remember It You must have like denominators before adding or subtracting rational expressions.

EXAMPLE Binomials in Denominator

2 Find $\frac{6c}{c + 2} + \frac{12}{c + 2}$.

$\frac{6c}{c + 2} + \frac{12}{c + 2} = \frac{6c + 12}{c + 2}$ The common denominator is .

$= \frac{6(\text{\_\_\_\_\_\_})}{c+2}$ Factor the numerator.

$= \frac{6\overset{1}{\cancel{(c+2)}}}{\underset{1}{\cancel{c+2}}}$ or 6 Divide by the common factor, $c + 2$ and simplify.

EXAMPLE Subtract Rational Expressions

3 Find $\frac{7x + 9}{x - 3} - \frac{x - 5}{x - 3}$.

$\frac{7x + 9}{x - 3} - \frac{x - 5}{x - 3}$

$=$ ______ The common denominator is .

$= \frac{(7x + 9) + [-(x - 5)]}{x - 3}$ The additive inverse of $(x - 5)$ is ______.

$=$ ______ Distributive Property

$=$ ______ or ______ Simplify.

EXAMPLE Inverse Denominators

4 **Find $\frac{3s}{11 - s} + \frac{-5s}{s - 11}$.**

Rewrite the second expression so that it has the same denominator as the first.

$\frac{3s}{11 - s} + \frac{-5s}{s - 11}$

$=$ ______ $\quad s - 11 =$ ______

$=$ ______ $\quad$ Rewrite using common denominators.

$= \frac{3s + 5s}{11 - s}$ $\quad$ The common denominator is .

$=$ ______ $\quad$ Simplify.

Your Turn **Find each sum or difference.**

a. $\frac{7k}{9} + \frac{17k}{9}$

b. $\frac{5y}{y + 4} + \frac{20}{y + 4}$

c. $\frac{11y - 3}{y + 1} - \frac{5y + 6}{y + 1}$

d. $\frac{8n}{n - 4} + \frac{n}{4 - n}$

HOMEWORK ASSIGNMENT

Page(s):

Exercises:

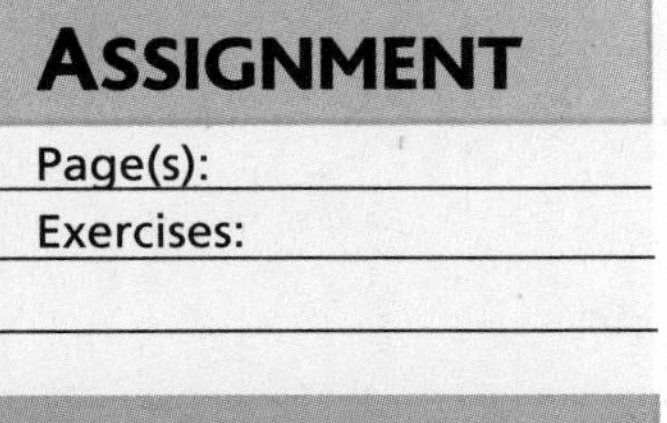

12–7 Rational Expression with Unlike Denominators

BUILD YOUR VOCABULARY (page 278)

The **least common multiple** is the ______ number that is a ______ multiple of two or more numbers.

The least common multiple of the ______ of two or more ______ is known as the **least common denominator**.

WHAT YOU'LL LEARN

- Add rational expressions with unlike denominators.
- Subtract rational expressions with unlike denominators.

KEY CONCEPT

Add Rational Expressions Use the following steps to add rational expressions with unlike denominators.

Step 1 Find the LCD.

Step 2 Change each rational expression into an equivalent expression with the LCD as the denominator.

Step 3 Add just as with rational expressions with like denominators.

Step 4 Simplify if necessary.

EXAMPLE LCM of Monomials

1 Find the LCM of $12b^4c^5$ and $32bc^2$.

Find the prime factors of each coefficient and variable expression.

$12b^4c^5 =$ ______

$32bc^2 =$ ______

Use each prime factor the greatest number of times it appears in any of the factorizations.

LCM = ______

EXAMPLE LCM of Polynomials

2 Find the LCM of $x^2 - 3x - 28$ and $x^2 - 8x + 7$.

Express each polynomial in factored form.

$x^2 - 3x - 28 =$ (______)(______)

$x^2 - 8x + 7 =$ (______)(______)

Use each factor the greatest number of times it appears.

LCM = ______.

Your Turn **Find the LCM for each pair of expressions.**

a. $21a^2b^4$ and $35a^3b^2$ **b.** $y^2 + 12y + 36$ and $y^2 + 2y - 24$

EXAMPLE Monomial Denominators

3 **Find** $\frac{z+2}{5z} + \frac{z-6}{z}$.

Factor each denominator and find the LCD. The LCD is ____.

$\frac{z+2}{5z} + \frac{z-6}{z}$

$= \frac{z+2}{5z} + \text{\_\_\_\_}$ Rename $\frac{z-6}{z}$.

$= \frac{z+2}{5z} + \frac{\text{\_\_\_\_}}{5z}$ Distributive Property

$= \frac{\text{\_\_\_\_}}{5z}$ Add the numerators.

$=$ ____ or ____ Simplify.

EXAMPLE Polynomial Denominators

4 **Find** $\frac{x+7}{x^2 - 6x + 9} + \frac{x+3}{x-3}$.

$\frac{x+7}{x^2 - 6x + 9} + \frac{x+3}{x-3}$

$= \frac{x+7}{(\text{\_\_\_\_})^2} + \frac{x+3}{x-3}$ Factor the denominators.

$= \frac{x+7}{(x-3)^2} + \frac{x+3}{x-3} \cdot \text{\_\_\_\_}$ The LCD is ____.

$= \frac{x+7}{(x-3)^2} + \frac{x^2 - 9}{(x-3)^2}$ $(x + 3)(x - 3) =$ ____.

$= \frac{\text{\_\_\_\_}}{(x-3)^2}$ Add the numerators.

$=$ ____ or ____ Simplify.

FOLDABLES™

ORGANIZE IT

Under the tab for Rational Equations, write each new Vocabulary Builder word. Then give an example of each word.

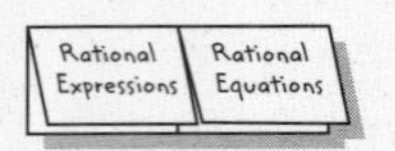

EXAMPLE Binomials in Denominators

5 **Find $\frac{c}{20 + 4c} + \frac{6}{5 - c}$.**

$\frac{c}{20 + 4c} + \frac{6}{5 - c}$

$= \frac{c}{4(\quad)} + \frac{6}{(5 - c)}$ Factor.

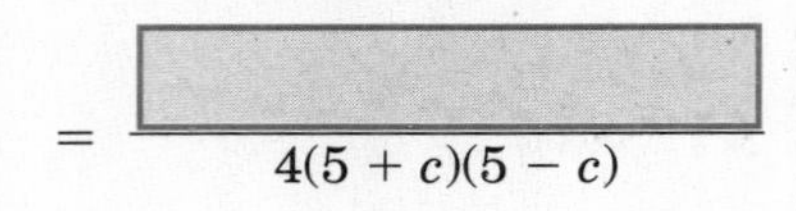

$= \frac{\quad}{4(5 + c)(5 - c)} + \frac{\quad}{4(5 - c)(5 + c)}$ The LCD is ______.

= ______ Add the numerators.

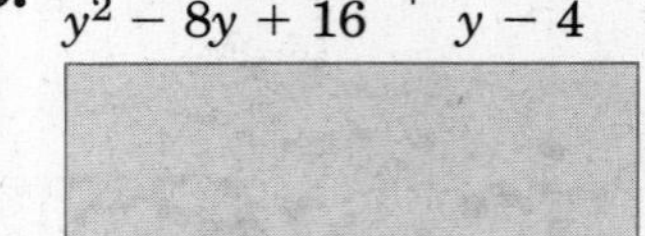

$= \frac{\quad}{4(5 + c)(5 - c)}$ Multiply.

$= \frac{-c^2 + 29c + 120}{4(5 + c)(5 - c)}$ or ______ Simplify.

Your Turn **Find each sum or difference.**

a. $\frac{b - 2}{4b} + \frac{b - 7}{b}$

b. $\frac{y - 14}{y^2 - 8y + 16} + \frac{y + 4}{y - 4}$

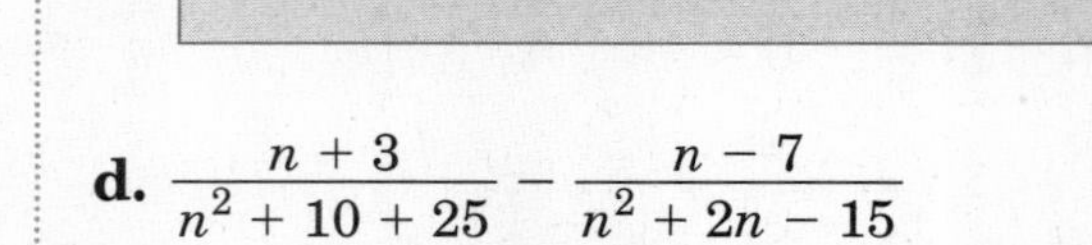

c. $\frac{3}{b + 1} - \frac{b}{4b - 4}$

d. $\frac{n + 3}{n^2 + 10 + 25} - \frac{n - 7}{n^2 + 2n - 15}$

HOMEWORK ASSIGNMENT

Page(s):

Exercises:

12–8 Mixed Expression and Complex Fractions

What You'll Learn

- Simplify mixed expressions.
- Simplify complex fractions.

BUILD YOUR VOCABULARY (page 278)

An expression that contains the sum of a  and a rational expression is known as a **mixed expression**.

If a fraction has  or more fractions in the numerator or denominator, it is called a **complex fraction**.

EXAMPLE Mixed Expression to Rational Expression

1 **Simplify $3 + \frac{7}{x-2}$.**

$3 + \frac{7}{x-2} = \frac{3(x-2)}{x-2} + \frac{7}{x-2}$ — The LCD is $x - 2$.

$= \frac{3(x-2)+7}{x-2}$ — Add the numerators.

$=$ ______ — Distributive Property

$=$ ______ — Simplify.

EXAMPLE Complex Fraction Involving Monomials

2 **Simplify $\dfrac{\frac{a^5b}{c^2}}{\frac{ab^4}{c^4}}$.**

$\dfrac{\frac{a^5b}{c^2}}{\frac{ab^4}{c^4}} = \frac{a^5b}{c^2} \div \frac{ab^4}{c^4}$ — Rewrite as a ______ sentence.

$= \frac{a^5b}{c^2} \cdot$ ______ — Rewrite as multiplication by the reciprocal.

$= \frac{\overset{a^4}{\cancel{a^5}}\ \overset{1}{\cancel{b}}}{\underset{1}{\cancel{c^2}}} \cdot \frac{\overset{c^2}{\cancel{c^4}}}{\underset{1}{\cancel{a}}\ \underset{b^3}{\cancel{b^4}}}$ — Divide by common factors.

$=$ ______ — Simplify.

KEY CONCEPT

Simplifying a Complex Fraction Any complex fraction $\dfrac{\frac{a}{b}}{\frac{c}{d}}$, where $b \neq 0$, $c \neq 0$, and $d \neq 0$, can be expressed as $\frac{ad}{bc}$.

EXAMPLE Complex Fraction Involving Polynomials

3 **Simplify** 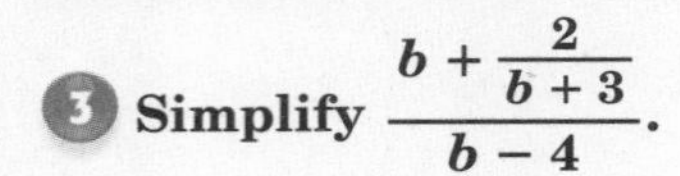 $\dfrac{b + \dfrac{2}{b+3}}{b-4}$.

The numerator contains a mixed expression. Rewrite it as a rational expression first.

$$\frac{b + \frac{2}{b+3}}{b-4} = \frac{\frac{b(b+3)}{b+3} + \frac{2}{b+3}}{b-4}$$

The LCD of the fractions in the numerator is ______.

$= \dfrac{\square}{b-4}$ 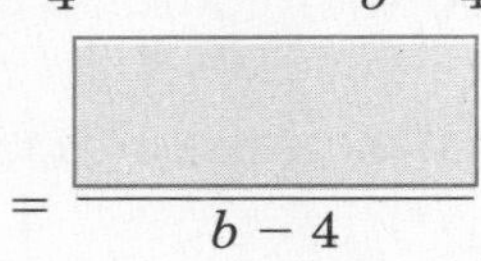

Simplify the numerator.

$= \dfrac{\dfrac{\square\ \square}{b+3}}{b-4}$ 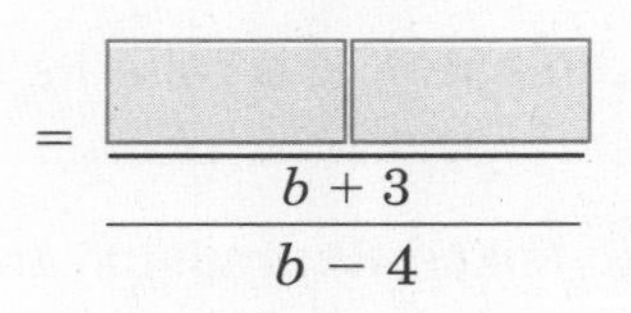

Factor.

$$= \frac{(b+1)(b+2)}{b+3} \div (b-4)$$

Rewrite as a division sentence.

$= \dfrac{(b+1)(b+2)}{b+3} \cdot$ ______

Multiply by the reciprocal of $b - 4$.

$=$ 

Simplify.

Your Turn **Simplify each expression.**

a. 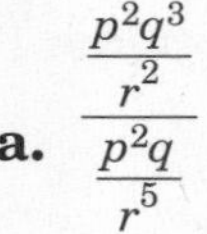 $\dfrac{\dfrac{p^2q^3}{r^2}}{\dfrac{p^2q}{r^5}}$

b. 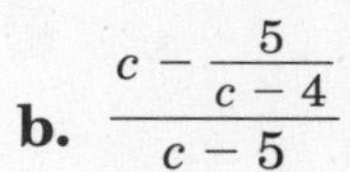 $\dfrac{c - \dfrac{5}{c-4}}{c-5}$

c. $5 + \dfrac{2}{y-4}$

FOLDABLES™

ORGANIZE IT

Under the tab for Rational Equations, write why the fraction bar in a complex fraction is considered a grouping symbol.

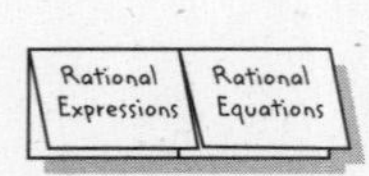

HOMEWORK ASSIGNMENT

Page(s):

Exercises:

12–9 Solving Rational Equations

What You'll Learn

- Solve rational equations.
- Eliminate extraneous solutions.

BUILD YOUR VOCABULARY (page 278)

A **rational equation** is an equation that contains rational expressions.

EXAMPLE Use Cross Products

1 **Solve $\frac{8}{x+3} = \frac{2}{x-6}$.**

$\frac{8}{x+3} = \frac{2}{x-6}$	Original equation
$8(\_\_\_\_) = 2(\_\_\_\_)$	Cross multiply.
$8x - 48 = 2x + 6$	Distributive Property
$6x = \_\_\_\_$	Add ____ and ____ to each side.
$x = \_\_\_\_$	Divide each side by 6.

EXAMPLE Use the LCD

2 **Solve $\frac{5}{x+1} + \frac{1}{x} = \frac{2}{x^2+x}$.**

$\frac{5}{x+1} + \frac{1}{x} = \frac{2}{x^2+x}$	Original equation
$\_\_\_\_\left(\frac{5}{x+1} + \frac{1}{x}\right) = \_\_\_\_\left(\frac{2}{x^2+x}\right)$	The LCD is ____.
$\left(\frac{x(x+1)}{1} \cdot \frac{5}{x+1}\right) + \left(\frac{x(x+1)}{1} \cdot \frac{1}{x}\right) = \frac{x(x+1)}{1} \cdot \frac{2}{x^2+x}$	Distributive Property
$\_\_\_\_ = 2$	Simplify.
$\_\_\_\_ = 2$	Add.
$6x = 1$	Subtract.
$x = \frac{1}{6}$	Divide.

Your Turn **Solve each equation.**

a. $\frac{2}{x-6} = \frac{4}{5x-3}$

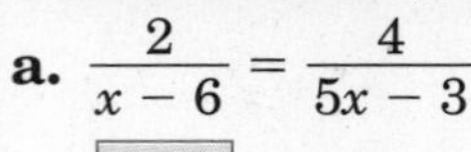

b. $\frac{q+1}{q} = \frac{q+4}{q+2} - \frac{3}{5q}$

EXAMPLE Multiple Solutions

3 **Solve** $a + \frac{a \ \ - 5}{a \ \ - 1} = \frac{a \ \ + a + 2}{a + 1}$.

$$a + \frac{a^2 - 5}{a^2 - 1} = \frac{a^2 + a + 2}{a + 1}$$ Original equation

$$\_\_\_\_\left(a + \frac{a^2 - 5}{a^2 - 1}\right) = \_\_\_\_\left(\frac{a^2 + a + 2}{a + 1}\right)$$

$$(a^2 - 1)a + \left(\frac{\cancel{a^2 - 1}^{1}}{1} \cdot \frac{a^2 - 5}{\cancel{a^2 - 1}_{1}}\right) = (\cancel{a^2 - 1}^{a-1})\left(\frac{a^2 + a + 2}{\cancel{a + 1}_{1}}\right)$$ Distributive Property

$$\_\_\_\_ = a^3 + a - 2$$ Simplify.

$$a^2 - 2a - 3 = 0$$ Set equal to .

$$\_\_\_\_ \ \_\_\_\_ = 0$$ Factor.

$\_\_\_\_ = 0$ or $\_\_\_\_ = 0$

$a = \_\_\_\_$ $\quad a = \_\_\_\_$

The number ____ is an ____ value for x.

Thus, the solution is ____.

Your Turn Solve $\frac{3}{m-2} = \frac{m+1}{m} - \frac{1}{m}$.

REVIEW IT

When checking solutions to equations, why do you check both solutions in the original equation? *(Lesson 11–3)*

EXAMPLE No Solution

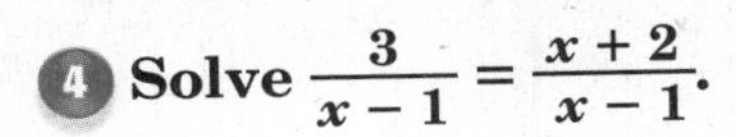

4 **Solve** $\dfrac{3}{x-1} = \dfrac{x+2}{x-1}$.

$\dfrac{3}{x-1} = \dfrac{x+2}{x-1}$	Original equation
$\square\left(\dfrac{3}{x-1}\right) = \square\left(\dfrac{x+2}{x-1}\right)$	The LCD is $\square$.
$(\cancel{x-1})^{1}\left(\dfrac{3}{\cancel{x-1}_{1}}\right) = (\cancel{x-1})^{1}\left(\dfrac{x+2}{\cancel{x-1}_{1}}\right)$	Distributive Property
$\square = \square$	Simplify.
$1 = x$	Subtract 2 from each side.

Since 1 is an excluded value for x, the number 1 is an extraneous solution. Thus, the equation has no solution.

FOLDABLES™

ORGANIZE IT

Under the tab for Rational Equations, write the definition of an extraneous solution in your own words.

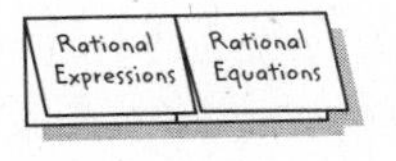

EXAMPLE Extraneous Solution

5 **Solve** $\dfrac{x^2}{x-2} = \dfrac{4}{x-2}$.

$\dfrac{x^2}{x-2} = \dfrac{4}{x-2}$	Original equation
$\square\left(\dfrac{x^2}{x-2}\right) = \square\left(\dfrac{4}{x-2}\right)$	The LCD is $\square$.
$(\cancel{x-2})^{1}\left(\dfrac{x^2}{\cancel{x-2}_{1}}\right) = (\cancel{x-2})^{1}\left(\dfrac{4}{\cancel{x-2}_{1}}\right)$	Distributive Property
$x^2 = 4$	Simplify.
$x^2 - \square = 0$	Subtract.
$(x-2)(x+2) = 0$	Factor.
$x - 2 = 0$ or $x + 2 = 0$	Zero Product Property
$x = \square \qquad x = \square$	

The number 2 is an extraneous solution, since 2 is an excluded value for x. Thus, -2 is the solution of the equation.

Your Turn **Solve each equation. State any extraneous solutions.**

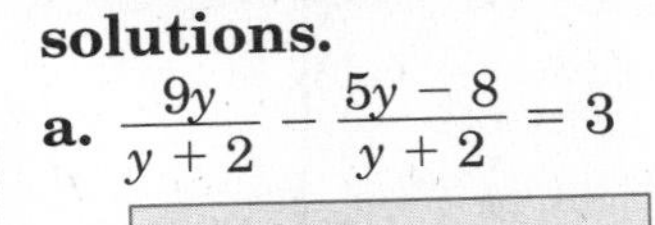

a. $\dfrac{9y}{y+2} - \dfrac{5y-8}{y+2} = 3$

b. $\dfrac{3w}{w-2} - 2 = \dfrac{5w+14}{w^2-4}$

HOMEWORK ASSIGNMENT

Page(s): ____

Exercises: ____

CHAPTER 12

BRINGING IT ALL TOGETHER

STUDY GUIDE

FOLDABLES™	VOCABULARY PUZZLEMAKER	BUILD YOUR VOCABULARY
Use your **Chapter 12 Foldable** to help you study for your chapter test.	To make a crossword puzzle, word search, or jumble puzzle of the vocabulary words in Chapter 12, go to: www.glencoe.com/sec/math/t_resources/free/index.php	You can use your completed **Vocabulary Builder** (page 278) to help you solve the puzzle.

12–1

Inverse Variation

Write *direct variation*, *inverse variation*, or *neither* to describe the relationship between x and y described by each equation.

1. $y = 3x$

2. $xy = 5$

3. $y = -8x$

4. $y = \frac{2}{x}$

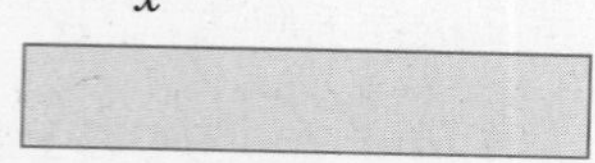

5. $x = \frac{10}{y}$

6. $y = 7x - 1$

For each problem, assume that y varies inversely as x. Use the Product Rule to write an equation you could use to solve the problem. Then write a proportion and solve the problem.

	Problem	Product Rule	Proportion	Solve
7.	If $y = 8$ when $x = 12$, find y when $x = 4$.			
8.	If $x = 50$ when $y = 6$, find x when $y = 30$			

12–2 Rational Expressions

Simplify each expression. State the excluded values of the variables.

9. $\frac{21bc}{28bc^2}$

10. $\frac{2x + 10}{x^2 - 2x - 35}$

11. $\frac{2y^2 + 9y + 4}{4y^2 - 4y - 3}$

12–3 Multiplying Rational Expressions

Find each product.

12. $\frac{18a^2}{10b^2} \cdot \frac{15b^2}{24a}$

13. $\frac{x - 4}{x^2 - x - 12} \cdot \frac{x + 3}{x - 6}$

14. $\frac{y^2 + 5y + 4}{y^2 - 36} \cdot \frac{y^2 + 5y - 6}{y^2 + 2y - 8}$

15. The number of calories used to play basketball depends on your weight and how long you play. Playing basketball expends about 3.8 calories per hour per pound of weight. If you weigh 140 pounds, how many calories do you lose in 1.25 hours?

12–4 Dividing Rational Equations

State the reciprocal of the divisor in each of the following.

16. $\frac{3b + 15}{b + 1} \div (b - 2)$

17. $\frac{2c^2}{d} \div \frac{c}{3d}$

18. Supply the reason for the steps below.

$\frac{y + 1}{y^2 + 5y + 6} \div \frac{1}{y + 3}$ Original Expression

$= \frac{y + 1}{y^2 + 5y + 6} \cdot \frac{y + 3}{1}$ Multply by the ______.

$= \frac{y + 1}{(y + 2)(y + 3)} \cdot \frac{y + 3}{1}$ ______ $y^2 + 5y + 6$.

$= \frac{y + 1}{(y + 2)(y + 3)} \cdot \frac{y + 3}{1}$ Divide by the ______.

$= \frac{y + 1}{y + 2}$ ______.

12–5

Dividing Polynomials

Find each quotient.

19. $(20y^2 + 12y) \div 4y$

20. $\dfrac{2x^2 - 5x - 3}{2x + 1}$

21. $\dfrac{6a^3 + a^2 - 2a + 17}{2a + 3}$

12–6

Rational Expressions with Like Denominators

For each addition or subtraction problem, write the needed expression in each box on the right side of the equation.

22. $\dfrac{5n}{7} + \dfrac{8}{7} = \dfrac{5n + \square}{7}$

23. $\dfrac{d - c}{c + 2d} - \dfrac{c - d}{c + 2d} = \dfrac{\square - (c - d)}{c + 2d}$

24. $\dfrac{8}{6x - 1} + \dfrac{9}{1 - 6x} = \dfrac{8 + (\square)}{6x - 1}$

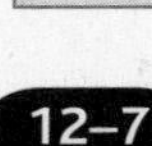

12–7

Rational Expressions with Unlike Denominators

25. What is the LCM of $49k^2n^2$ and $21kn^5$?

Find each sum or difference.

26. $\dfrac{3}{y} + \dfrac{4}{y^2}$

27. $\dfrac{5x}{3y^2} - \dfrac{2x}{9y}$

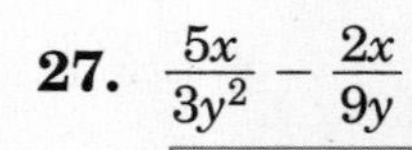

28. $\dfrac{a}{a - 5} + \dfrac{a - 1}{a + 5}$

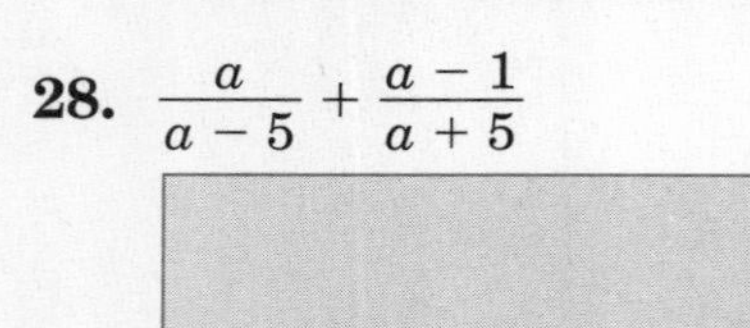

29. $\dfrac{a + 3}{a^2 - 3a - 10} - \dfrac{4a - 8}{a^2 - 10a + 25}$

12–8 Mixed Expressions and Complex Fractions

Tell whether each expression is a mixed expression or complex fraction. Write M for mixed expression and C for complex fraction.

30. $7x + \frac{x+2}{x-5}$

31. $\frac{5 + \frac{2}{s-1}}{s^2}$

32. $(b - 6) + \frac{b+3}{b+2}$

33. Simplify $\frac{\frac{x+4}{x}}{\frac{x^2-16}{x}}$.

12–9 Solving Rational Equations

34. Is $\frac{\sqrt{x-3}}{4} = \frac{3}{x}$ a rational equation? Explain.

Solve each equation. State any extraneous solutions.

35. $\frac{5}{x+2} = \frac{7}{x+6}$

36. $\frac{-2}{w+1} + \frac{2}{w} = 1$

37. $\frac{3}{2t} + \frac{2t}{t-3} = 2$

ARE YOU READY FOR THE CHAPTER TEST?

Math Online

Visit **algebra1.com** to access your textbook, more examples, self-check quizzes, and practice tests to help you study the concepts in Chapter 12.

Check the one that applies. Suggestions to help you study are given with each item.

☐ **I completed the review of all or most lessons without using my notes or asking for help.**

- You are probably ready for the Chapter Test.
- You may want take the Chapter 12 Practice Test on page 701 of your textbook as a final check.

☐ **I used my Foldable or Study Notebook to complete the review of all or most lessons.**

- You should complete the Chapter 12 Study Guide and Review on pages 696–700 of your textbook.
- If you are unsure of any concepts or skills, refer to the specific lesson(s).
- You may also want to take the Chapter 12 Practice Test on page 701.

☐ **I asked for help from someone else to complete the review of all or most lessons.**

- You should review the examples and concepts in your Study Notebook and Chapter 12 Foldable.
- Then complete the Chapter 12 Study Guide and Review on pages 696–700 of your textbook.
- If you are unsure of any concepts or skills, refer to the specific lesson(s).
- You may also want to take the Chapter 12 Practice Test on page 701.

Student Signature

Parent/Guardian Signature

Teacher Signature

Statistics

Use the instructions below to make a Foldable to help you organize your notes as you study the chapter. You will see Foldable reminders in the margin of this Interactive Study Notebook to help you in taking notes.

Begin with three sheets of plain $8\frac{1}{2}$" by 11" paper.

STEP 1 **Stack Pages**
Stack sheets of paper with edges $\frac{3}{4}$ inch apart.

STEP 2 **Fold Up Bottom Edges**
All tabs should be the same size.

STEP 3 **Crease and Staple**
Staple along fold.

STEP 4 **Turn and Label**
Label the tabs with topics from the chapter.

NOTE-TAKING TIP: When you take notes, define terms, record concepts, and write examples as concise and legible as possible.

BUILD YOUR VOCABULARY

This is an alphabetical list of new vocabulary terms you will learn in Chapter 13. As you complete the study notes for the chapter, you will see Build Your Vocabulary reminders to complete each term's definition or description on these pages. Remember to add the textbook page number in the second column for reference when you study.

Vocabulary Term	Found on Page	Definition	Description or Example
biased sample			
box-and-whisker plot			
census			
convenience sample [kuhn-VEEN-yuhn(t)s]			
dimensions			
frequency table			
histogram			
interquartile range [ihn-tuhr-KWAR-TYL]			
matrix [MAT-triks]			

Vocabulary Term	Found on Page	Definition	Description or Example
measurement classes			
measures of variation			
outlier [OWT-LY-uhr]			
population			
quartiles			
sample			
scalar multiplication [SKAY-luhr]			
simple random sample			
stratified random sample			
systematic random sample [SIHS-tuh-MA-tihk]			
voluntary response sample			

13–1 Sampling and Bias

BUILD YOUR VOCABULARY (pages 306–307)

A **sample** is some portion of a ______ group, called the **population**, selected to represent that group. If all of the ______ within a population are included, it is called a **census**.

In a **biased sample**, one or more parts of a population are ______ over others.

WHAT YOU'LL LEARN

- Identify various sampling techniques.
- Recognize a biased sample.

KEY CONCEPTS

Simple Random Sample: A sample that is as likely to be chosen as any other from the population.

Stratified Random Sample In a stratified random sample, the population is first divided into similar, nonoverlapping groups. A simple random sample is then selected from each group.

Systematic Random Sample In a systematic random sample, the items are selected according to a specified time or item interval.

EXAMPLE Classify a Random Sample

1 a. RETAIL **Each day, a department store chain selects one male and one female shopper randomly from each of their 57 stores, and asks them survey questions about their shopping habits.**

Identify the sample and suggest a population from which it was selected.

The sample is ______ male and ______ female shoppers.

The population is ______.

b. Classify the sample as *simple, stratified,* or *systematic*.

The population is divided into similar, nonoverlapping groups. This is a ______ sample.

Your Turn **At an automobile factory, every tenth item is checked for quality controls.**

a. Identify the sample and suggest a population from which it was selected.

______

b. Classify the sample as *simple*, *stratified*, or *systematic*.

______

FOLDABLES

ORGANIZE IT

On the tab for Lesson 13-1, write your own example of a biased sample.

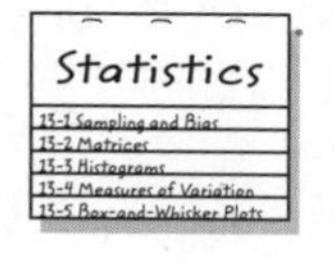

EXAMPLE Identify Sample as Biased or Unbiased

2 STUDENT COUNCIL **The student council surveys the students in one classroom to decide the theme for the spring dance. Identify the sample as *biased* or *unbiased*. Explain your reasoning.**

The sample includes only students in one classroom.

The sample is ______.

Your Turn Identify the sample as *biased* or *unbiased*. Explain your reasoning.

A local news station interviews one person on every street in Los Angeles to give their opinion on their mayor.

______

EXAMPLE Identify and Classify a Biased Sample

3 **a.** COMMUNITY **The residents of a neighborhood are to be surveyed to find out when to hold a neighborhood clean up day. The neighborhood chairperson decides to ask her immediate neighbors and the neighbors in the houses directly across the street from her house.**

Identify the sample, and suggest a population from which it was selected.

The sample is the ______ ______ and the neighbors across the street. The ______ is the residents of the neighborhood.

b. Classify the sample as a *convenience sample*, or a *voluntary response sample*.

This is a ______ sample because the chairperson asked only her closest neighbors.

Your Turn Mark wanted to find out what the average student in the United States does on the weekend. He decides to interview people in his dorm. Identify the sample, and suggest a population from which it was selected. Then classify the sample as a *convenience sample*, or a *voluntary response sample*.

______

KEY CONCEPTS

Biased Samples
A **convenience sample** includes members of a population that are easily accessed.
A **voluntary response sample** involves only those who want to participate in the sampling.

HOMEWORK ASSIGNMENT

Page(s):

Exercises:

13–2 Introduction to Matrices

BUILD YOUR VOCABULARY (pages 306–307)

A **matrix** is a rectangular arrangement of numbers in

A matrix is usually described by its **dimensions** or the number of ______ and ______.

WHAT YOU'LL LEARN

- Organize data into matrices.
- Solve problems by adding or subtracting matrices or by multiplying by a scalar.

EXAMPLE Name Dimensions of Matrices

1 State the dimensions of the matrix. Then identify the position of the circled element in the matrix.

$$\begin{bmatrix} 12 \\ \textcircled{7} \\ 5 \end{bmatrix}$$

The matrix has ______ rows and ______ column.

It is a ______ matrix. The circled element is the second

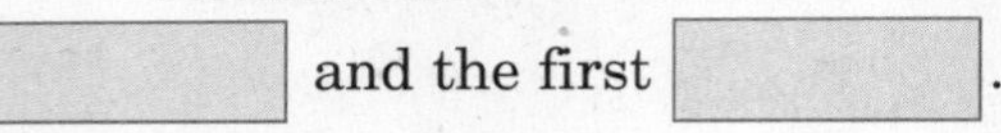

REMEMBER IT When giving the dimensions of a matrix, the first number tells the number of rows, while the second number tells the number of columns.

Your Turn State the dimensions of the matrix. Then identify the position of the circled element.

$$\begin{bmatrix} -2 & 1 & 5 & -7 \\ -3 & 12 & -\textcircled{4} & 2 \end{bmatrix}$$

EXAMPLE Add Matrices

2 Find the sum. If the sum does not exist, write *impossible*.

$$\begin{bmatrix} 2 & 6 \\ -8 & 11 \\ 10 & -5 \end{bmatrix} + \begin{bmatrix} 13 & 7 \\ 9 & 4 \\ -7 & -12 \end{bmatrix} = \begin{bmatrix} \square & \square \\ \square & \square \\ 10 + (-7) & -5 + (-12) \end{bmatrix}$$

$$= \begin{bmatrix} 15 & 13 \\ \square & \square \\ 3 & -17 \end{bmatrix}$$

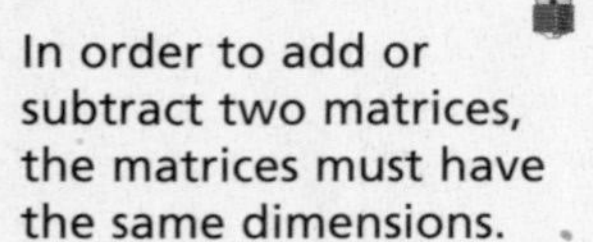

REMEMBER IT In order to add or subtract two matrices, the matrices must have the same dimensions.

FOLDABLES

ORGANIZE IT

On the tab for Lesson 13–2, write two matrices and then find their sum.

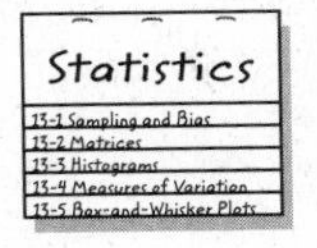

WRITE IT

What must be the same in order to add or subtract two matrices?

____

____

____

____

Your Turn **Find the sum. If the sum does not exist, write *impossible*.**

a. $\begin{bmatrix} -2 & 1 \\ 13 & -4 \\ -5 & 7 \end{bmatrix} + \begin{bmatrix} 9 & -5 & 3 \\ -4 & 2 & -3 \end{bmatrix}$

b. $\begin{bmatrix} -4 & 7 & -8 \\ -12 & 4 & -6 \\ 15 & -1 & 0 \end{bmatrix} + \begin{bmatrix} -3 & -5 & 11 \\ -1 & -2 & 6 \\ 19 & 5 & -9 \end{bmatrix}$

EXAMPLE Subtract Matrices

3 COLLEGE FOOTBALL **The Division 1-A current football coaches with the five best overall records as of 2000 are listed below. Use subtraction of matrices to determine the regular season records of these coaches.**

Overall Record			
Coach	**Won**	**Lost**	**Tied**
Joe Paterno	322	90	3
Bobby Bowden	315	87	4
Lou Holtz	224	110	7
Jackie Sherrill	172	93	4
Ken Hatfield	147	104	4

Bowl Record			
Coach	**Won**	**Lost**	**Tied**
Joe Paterno	20	9	1
Bobby Bowden	17	6	1
Lou Holtz	11	8	2
Jackie Sherrill	8	6	0
Ken Hatfield	4	6	0

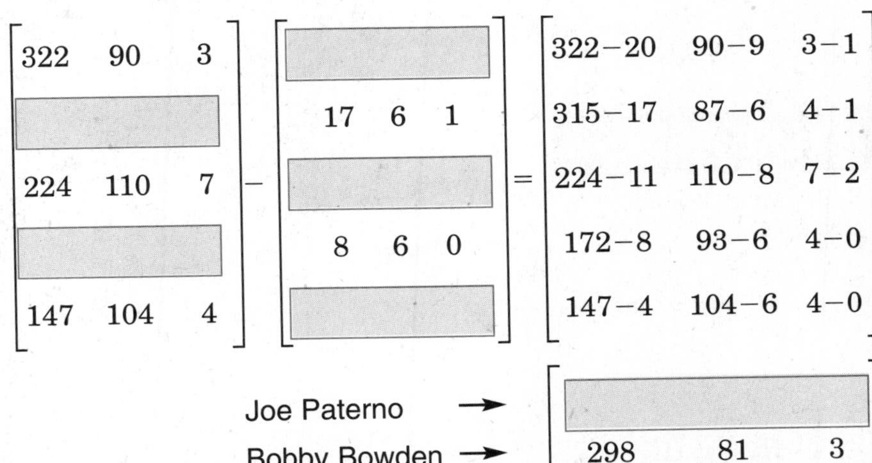

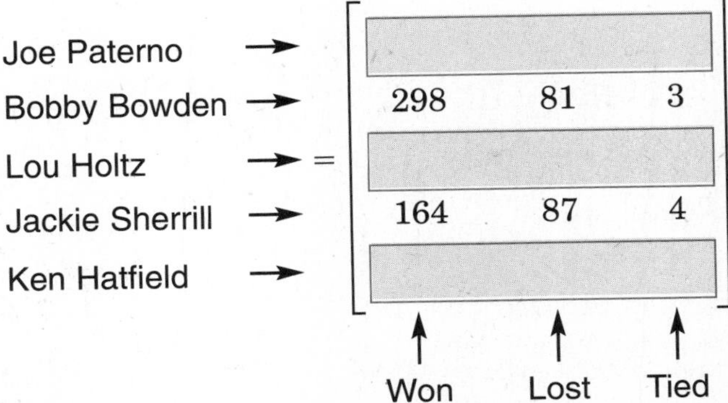

Your Turn Four performing theaters have seating capacities listed below. Use subtraction of matrices to find the seating capacities on the main floors of the theaters

Overall Seating Capacity			
Theater	**Left Section**	**Middle Section**	**Right Section**
King	1065	640	1065
Falcon	840	620	840
Emerald	615	550	615
Daisy	530	410	530

Number of Seats in Balcony			
Theater	**Left Section**	**Middle Section**	**Right Section**
King	385	180	385
Falcon	210	190	210
Emerald	175	140	175
Daisy	160	120	160

EXAMPLE Perform Scalar Multiplication

4 **If** $R = \begin{bmatrix} -5 & 8 \\ 12 & -3 \end{bmatrix}$ **find 5*R*.**

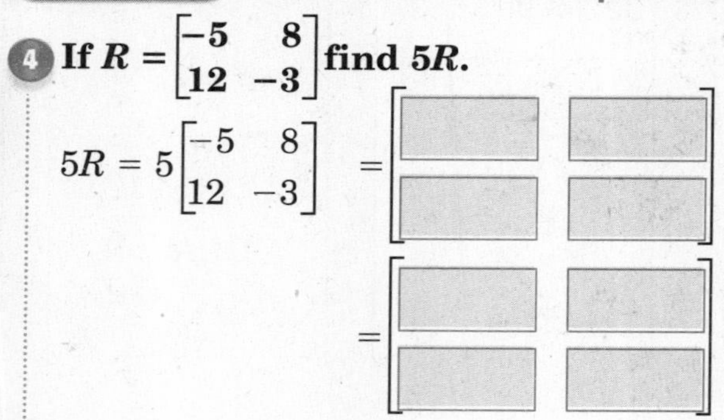

$5R = 5\begin{bmatrix} -5 & 8 \\ 12 & -3 \end{bmatrix} = \begin{bmatrix} \square & \square \\ \square & \square \end{bmatrix}$

$= \begin{bmatrix} \square & \square \\ \square & \square \end{bmatrix}$

KEY CONCEPT

Scalar Multiplication of a Matrix

$m\begin{bmatrix} a & b & c \\ d & e & f \end{bmatrix} = \begin{bmatrix} ma & mb & mc \\ md & me & mf \end{bmatrix}$

Your Turn If $A = \begin{bmatrix} 4 & -13 & -5 \\ 11 & 2 & -7 \end{bmatrix}$, find 6*A*.

HOMEWORK ASSIGNMENT

Page(s):

Exercises:

13–3 Histograms

What You'll Learn

- Interpret data displayed in histograms.
- Display data in histograms.

BUILD YOUR VOCABULARY (pages 306–307)

A **frequency table** shows the frequency of events.

A **histogram** is a bar graph in which the data are organized into equal ________.

In a histogram, the ________ axis shows the ________ of data values separated into measurement classes.

EXAMPLE Determine Information from a Histogram

1 STUDENT POPULATION **In what measurement class does the median occur? Use the histogram shown.**

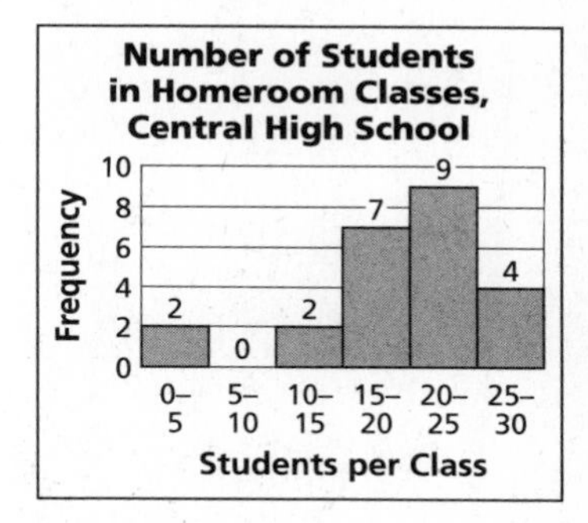

Add up the frequencies to determine the number of students in homeroom classes.

$2 + 0 + 2 + 7 + 9 + 4 =$ ____

There are ____ homeroom classes, so the middle data value is between the 12th and 13th data values. Both the 12th and 13th data values are located in the ________ measurement class. The median occurs in this class.

Your Turn **Refer to the histogram shown.**

Cost of Chemistry Lab Equipment

Dollars	0–10	10–20	20–30	30–40	40–50
Frequency	7	9	8	4	3

a. In what measurement class does the median occur?

____

b. Describe the distribution of the data.

____

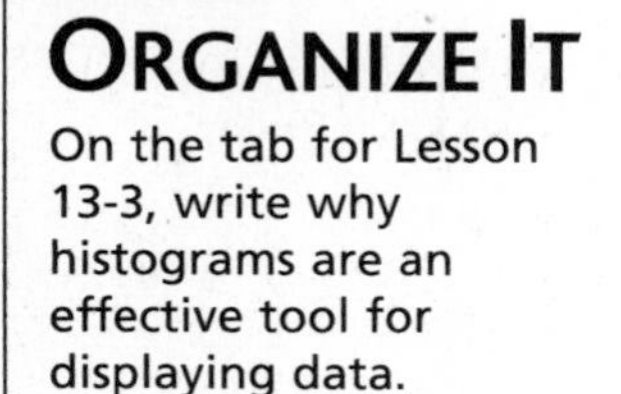

Organize It

On the tab for Lesson 13-3, write why histograms are an effective tool for displaying data.

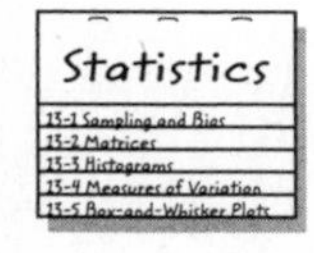

EXAMPLE Create a Histogram

2 FOOTBALL **Create a histogram to represent the following scores of top 25 winning college football teams during one week of the 2001 season.**

43, 52, 38, 36, 42, 46, 26, 38, 38, 31, 38, 37, 38, 48, 45, 27, 47, 35, 35, 26, 47, 24, 41, 21, 32

> **REMEMBER IT**
> When making a histogram, it helps to create a frequency table first.

Step 1 Identify the greatest and least values in the data set.

The scores range from ______ to ______ points.

Step 2 Create measurement classes of equal width.

For these data, use measurement classes from 20 to 55 with a ______ interval for each class.

Step 3 Create a frequency table using the measurement classes.

Score	Tally	Frequency
$20 \le s < 25$	II	2
$25 \le s < 30$	III	3
$30 \le s < 35$	II	2
$35 \le s < 40$	𝍸 IIII	9
$40 \le s < 45$	III	3
$45 \le s < 50$	𝍸	5
$50 \le s < 55$	I	1

Step 4 Draw the histogram.

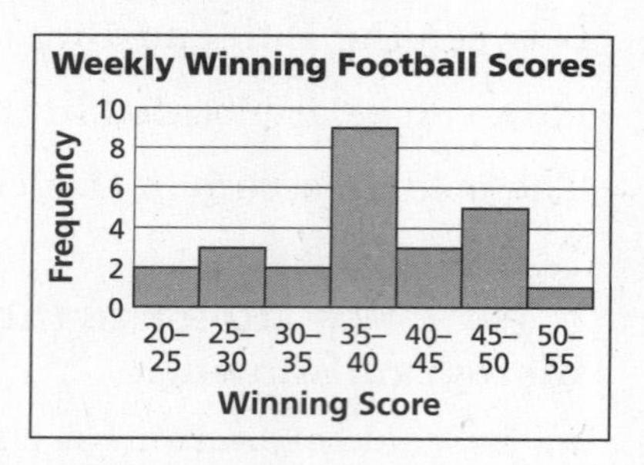

Your Turn **Create a histogram to represent the attendance at the weekly meeting of the Math Club.**

75, 58, 71, 67, 73, 58, 67, 78, 65, 77, 72, 68, 76, 64, 72, 57, 71, 75, 64, 74, 60, 54, 66, 74

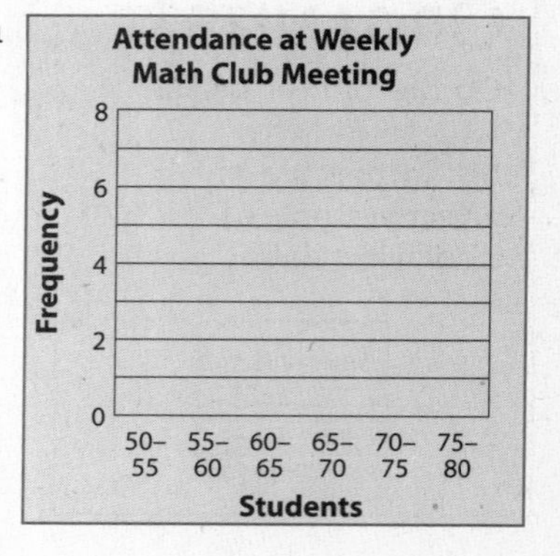

HOMEWORK ASSIGNMENT

Page(s):

Exercises:

13–4 Measures of Variation

WHAT YOU'LL LEARN

- Find the range of a set of data.
- Find the quartiles and interquartile range of a set of data.

BUILD YOUR VOCABULARY (pages 306–307)

Measures that describe the ______ of the values in a set of ______ are called **measures of variation.**

In a set of data, the **quartiles** are values that separate the data into ______ equal subsets.

KEY CONCEPT

Definition of Range The **range** of a set of data is the difference between the greatest and the least values of the set.

EXAMPLE Find the Range

1 COLLEGE FOOTBALL **The teams with the top 15 offensive yardage gains for the 2000 season are listed in the table. Find the range of the data.**

Team	Yardage
Air Force	4971
Boise St.	5459
Clemson	4911
Florida St.	6588
Georgia Tech	4789
Idaho	4985
Indiana	4830
Kentucky	4900
Miami	5069
Michigan	4900
Nebraska	5059
Northwestern	5232
Purdue	5183
Texas	4825
Tulane	4989

The greatest amount of yardage gains is ______ and the least amount of yardage gains is ______. So, the range of the yardage is ______ or ______ yards.

Your Turn BASEBALL The baseball players with the top 12 all-time runs batted in (RBI) are listed in the table. Find the range of the data.

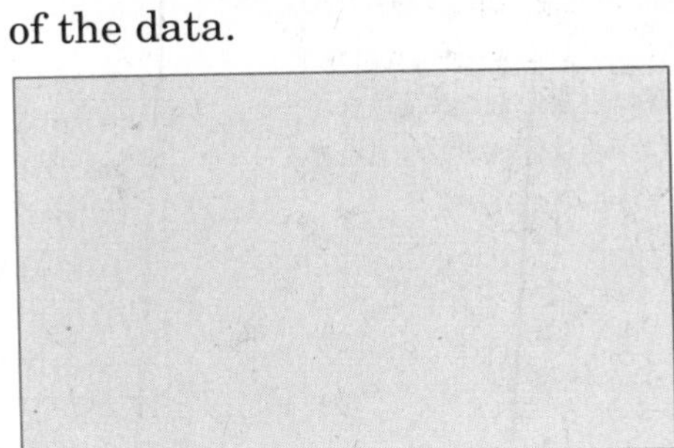

Player	RBI
Hank Aaron	2297
Cap Anson	2076
Ty Cobb	1937
Jimmie Foxx	1922
Lou Gehrig	1995
Willie Mays	1903
Eddie Murray	1917
Stan Musial	1951
Mel Ott	1860
Babe Ruth	2213
Ted Williams	1839
Carl Yastrzemski	1844

KEY CONCEPT

Definition of Interquartile Range The difference between the upper quartile and the lower quartile of a set of data is called the **interquartile range**. It represents the middle half, or 50%, of the data in the set.

FOLDABLES

On the tab for Lesson 13-4, tell how to find the quartiles and interquartile range of a set of data.

EXAMPLE Find the Quartiles and the Interquartile Range

2 GEOGRAPHY **The areas of the 5 largest states are listed in the table. Find the median, the lower quartile, the upper quartile, and the interquartile range of the areas.**

State	Area (thousand square miles)
Alaska	656
California	164
Montana	147
New Mexico	124
Texas	269

First, list the areas from least to greatest. Then find the median of the data. The median will divide the data into two sets of data. To find the upper and lower quartiles, find the median of each of these sets of data. Finally subtract the lower quartile from the upper quartile to find the interquartile range.

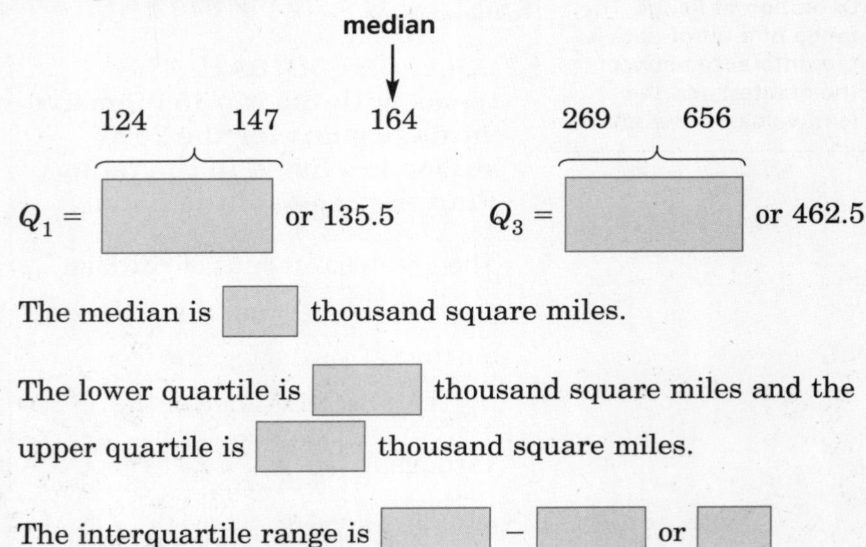

The median is ______ thousand square miles.

The lower quartile is ______ thousand square miles and the upper quartile is ______ thousand square miles.

The interquartile range is ______ – ______ or ______ thousand square miles.

Your Turn **The following table shows the areas of some major U.S. cities. Find the median, the lower quartile, the upper quartile, and the interquartile range of the areas.**

City	Area (thousand square miles)
Atlanta	136
Boston	47
Dallas	378
Indianapolis	352
Kansas City	317
Los Angeles	467
Miami	34
New York City	322
Philadelphia	136
Washington, D.C.	68

HOMEWORK ASSIGNMENT

Page(s):

Exercises:

13–5 Box-and-Whisker Plots

WHAT YOU'LL LEARN

- Organize and use data in box-and-whisker plots.
- Organize and use data in parallel box-and-whisker plots.

FOLDABLES™

ORGANIZE IT

On the tab for Lesson 13-5, write the three steps for drawing a box-and-whisker plot.

BUILD YOUR VOCABULARY (pages 306–307)

A visual way to display the ________ of a set of data is called a **box-and-whisker plot.**

EXAMPLE Draw a Box-and-Whisker Plot

1 ECOLOGY **The average water level in Lake Travis in central Texas during August is a good indicator of whether the region has had normal rainfall, or is suffering from a drought. The following is a list of the water level in feet above sea level during August for the years 1990 to 2000. Draw a box-and-whisker plot for these data.**

674, 673, 678, 673, 670, 677, 653, 679, 664, 672, 645

Step 1 Determine the quartiles and any outliers.

Order the data from least to greatest. Use this list to determine the quartiles.

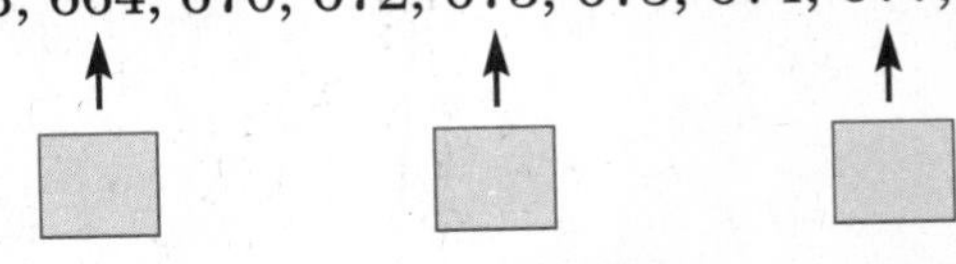

The interquartile range is ________ – ________ or 13.

664 – 1.5(13) = 644.5 677 + 1.5(13) = 696.5

Any numbers less than 644.5 or greater than 696.5 are outliers. There are none.

Step 2 Draw a number line.

Assign a scale to the number line that includes the extreme values. Above the number line, place bullets to represent the three quartile points, any outliers, the least number that is *not* an outlier, and the greatest number that is *not* an outlier.

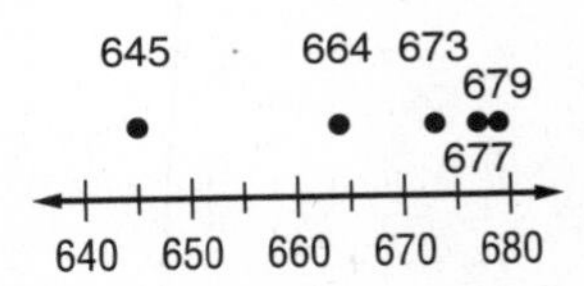

Step 3 Complete the box-and-whisker plot.

Draw a box to designate the data between the upper and lower quartiles. Draw a vertical line through the point representing the median. Draw a line from the lower quartile to the least value that is *not* an outlier. Draw a line from the upper quartile to the greatest value that is *not* an outlier.

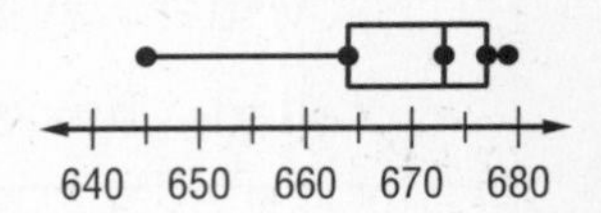

Your Turn ECOLOGY The average yearly rainfall is listed for 11 major cities.
Draw a box-and-whisker plot for these data.

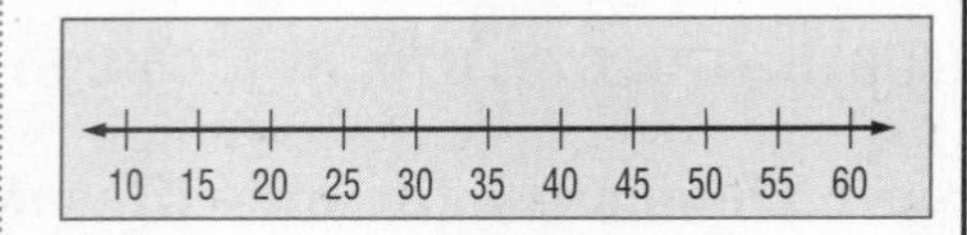

Average Rainfall for Some U.S. Cities	
City	**Rainfall (in.)**
Chicago	33.34
Cleveland	35.40
Helena	11.37
Louisville	43.56
Milwaukee	30.94
Philadelphia	41.42
Portland	37.39
St. Louis	33.91
Savannah	49.70
Seattle	38.60
Tampa	46.73

EXAMPLE Draw Parallel Box-and-Whisker Plots

WRITE IT

If the box portion of a box-and-whisker plot is short, what does this tell you about the middle part of the data?

______

______

______

______

2 CLIMATE **Pilar, who grew up on the island of Hawaii, is going to go to college in either Dallas or Nashville. She does not want to live in a place that gets too cold in the winter, so she decided to compare the average monthly low temperatures of each city.**

Draw a parallel box-and-whisker plot for the data.

Determine the quartiles and outliers for each city.

Average Monthly Low Temperatures (°F)		
Month	**Dallas**	**Nashville**
Jan.	32.7	26.5
Feb.	36.9	29.9
Mar.	45.6	39.1
Apr.	54.7	47.5
May	62.6	56.6
June	70	64.7
July	74.1	68.9
Aug.	73.6	67.7
Sept.	66.9	61.1
Oct.	55.8	48.3
Nov.	45.4	39.6
Dec.	36.3	30.9

Dallas

32.7, 36.3, 36.9, 45.4, 45.6, 54.7, 55.8, 62.6, 66.9, 70, 73.6, 74.1

Q_1 = ______ Q_2 = ______ Q_3 = ______

Nashville

26.5, 29.9, 30.9, 39.1, 39.6, 47.5, 48.3, 56.6, 61.1, 64.7, 67.7, 68.9

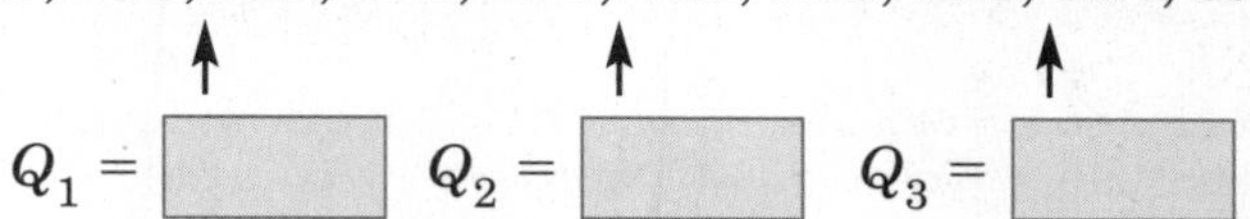

Neither city has any outliers. Draw the box-and-whisker plots using the same number line.

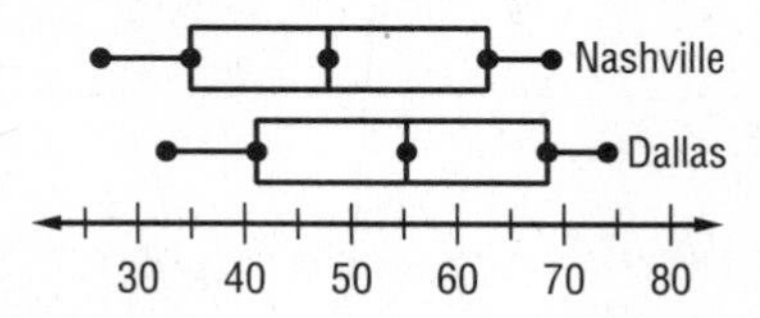

Your Turn Suppose a baseball manager has two job offers to manage the Yankees or the Rangers. He compares the teams using their average runs batted in (RBI) for their top twelve players. Draw a parallel box-and-whisker plot for the data.

RBI for Top 12 Yankee and Rangers Hitters (2001)		
Player	Yankees	Rangers
1	113	135
2	95	123
3	94	72
4	74	67
5	73	65
6	70	54
7	51	49
8	49	36
9	46	35
10	44	34
11	32	29
12	19	25

HOMEWORK ASSIGNMENT

Page(s):

Exercises:

CHAPTER 13

BRINGING IT ALL TOGETHER

STUDY GUIDE

FOLDABLES	VOCABULARY PUZZLEMAKER	BUILD YOUR VOCABULARY
Use your **Chapter 13 Foldable** to help you study for your chapter test.	To make a crossword puzzle, word search, or jumble puzzle of the vocabulary words in Chapter 13, go to www.glencoe.com/sec/math/t_resources/free/index.php	You can use your completed **Vocabulary Builder** (pages 306–307) to help you solve the puzzle.

13–1 Sampling and Bias

Suppose the principal at a school wants to use Saturdays as make-up days when school is closed due to weather. The principal selects and then polls a group of students to see if the student body supports the idea. Complete the sentences.

1. The student body is the ______ from which a ______ of students is selected to be polled. If all the students are polled, it is called a ______.

2. If all students are requested to enter school through the administration building and every twenty-fifth student is selected to be polled, then the sample is a ______ ______ sample.

13–2 Introduction to Matrices

Find each sum or difference.

3. $\begin{bmatrix} 7 & -8 \\ -12 & 6.2 \end{bmatrix} + \begin{bmatrix} -8 & 9.4 \\ -9 & 17 \end{bmatrix}$

4. $\begin{bmatrix} -3 & 4 \\ 0 & -1 \end{bmatrix} - \begin{bmatrix} -5 & 6 \\ 9 & -4 \end{bmatrix}$

5. How can you tell whether two matrices can be added or subtracted?

13–3

Histograms

Use the histogram to complete the sentence.

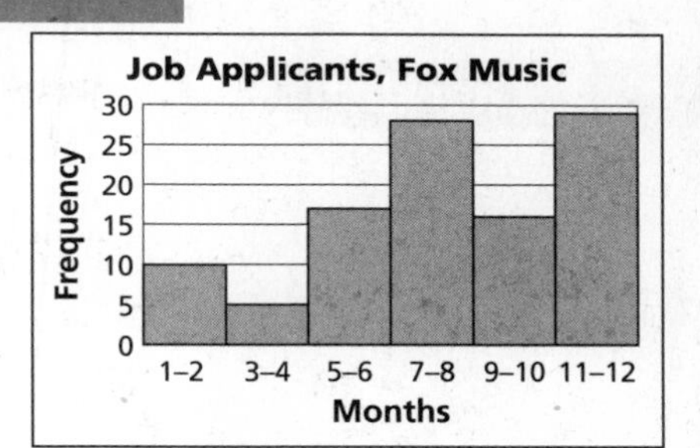

6. From the histogram, you can see there are six ________ classes, each with a width of ____ months.

7. To determine the median, add the ________ to find the number of ________. Then locate the measurement class in which the median lies. If the median in the histogram is 53, it occurs in the measurement class ____.

13–4

Measures of Variation

8. Find the range, median, lower quartile, upper quartile, and interquartile range of the data set. Identify any outliers.
200, 250, 230, 180, 160, 140, 210, 190, 170, 220

____

9. The running times in seconds for the 400-meter hurdles were 48.99, 50.73, 50.05, 49.98, 48.48, and 49.11. Find the range.

____

13–5

Box-and-Whisker Plots

Use the parallel box-and-whisker plots at the right to complete the sentences that follow.

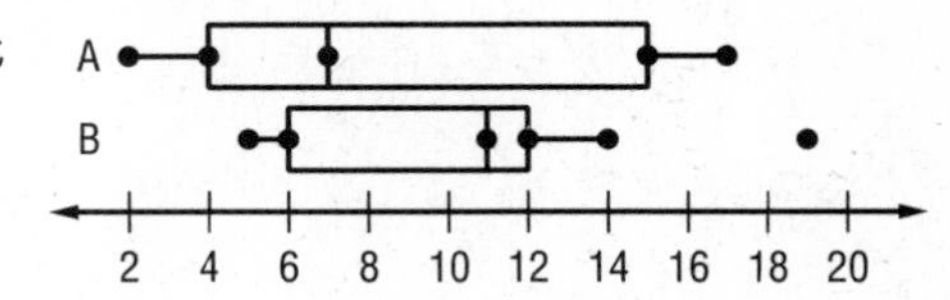

10. The bullets located at 2 and 17 in plot A and 5 and 19 in plot B represent the ________ values of the data sets.

11. The bullets at the ends of the whiskers are the ____ and ________ values that are *not* ________. Plot ____ has an outlier.

ARE YOU READY FOR THE CHAPTER TEST?

Visit **www.algebra1.com** to access your textbook, more examples, self-check quizzes, and practice tests to help you study the concepts in Chapter 13.

Check the one that applies. Suggestions to help you study are given with each item.

☐ **I completed the review of all or most lessons without using my notes or asking for help.**

- You are probably ready for the Chapter Test.
- You may want to take the Chapter 13 Practice Test on page 749 of your textbook as a final check.

☐ **I used my Foldable or Study Notebook to complete the review of all or most lessons.**

- You should complete the Chapter 13 Study Guide and Review on pages 745–748 of your textbook.
- If you are unsure of any concepts or skills, refer back to the specific lesson(s).
- You may also want to take the Chapter 13 Practice Test on page 749.

☐ **I asked for help from someone else to complete the review of all or most lessons.**

- You should review the examples and concepts in your Study Notebook and Chapter 13 Foldable.
- Then complete the Chapter 13 Study Guide and Review on pages 745–748 of your textbook.
- If you are unsure of any concepts or skills, refer back to the specific lesson(s).
- You may also want to take the Chapter 13 Practice Test on page 749.

Student Signature

Parent/Guardian Signature

Teacher Signature

Probability

Use the instructions below to make a Foldable to help you organize your notes as you study the chapter. You will see Foldable reminders in the margin of this Interactive Study Notebook to help you in taking notes.

Begin with a sheet of plain $8\frac{1}{2}$" by 11" paper.

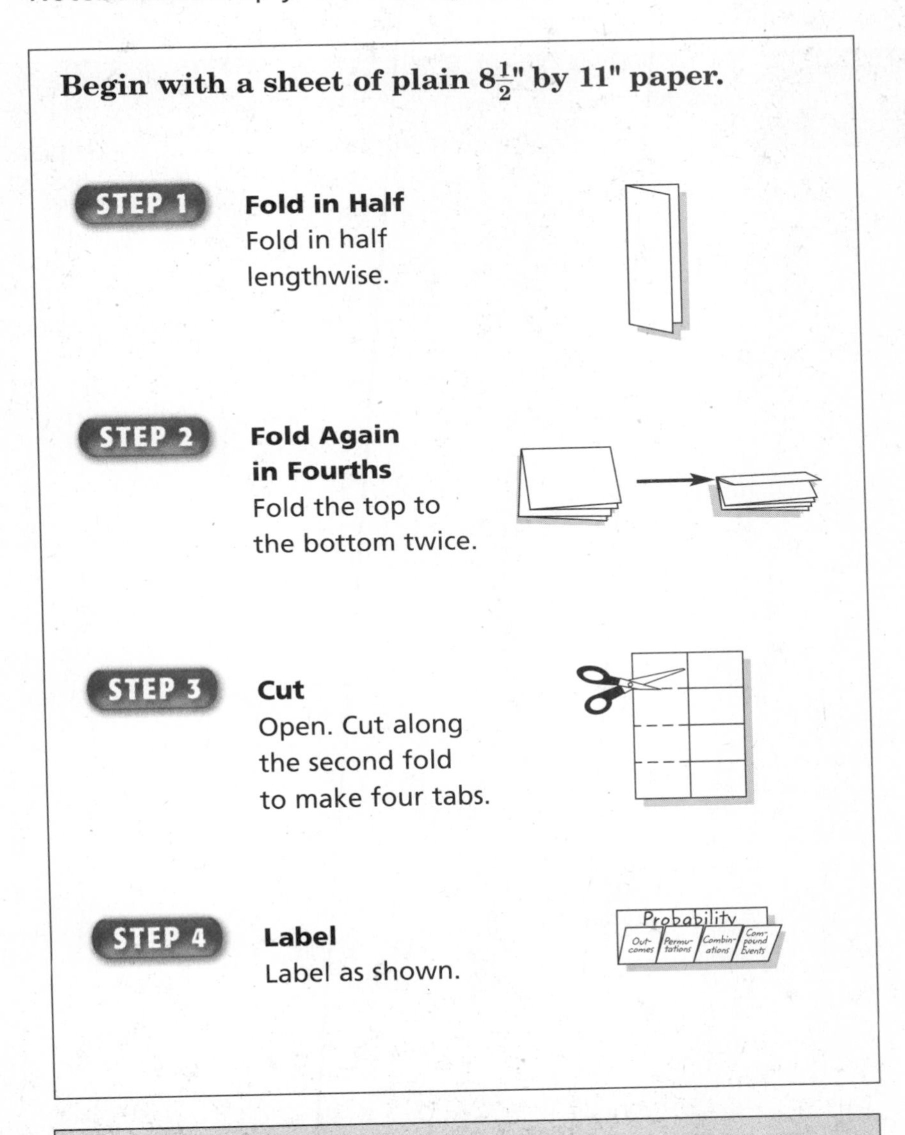

Fold in Half
Fold in half lengthwise.

Fold Again in Fourths
Fold the top to the bottom twice.

Cut
Open. Cut along the second fold to make four tabs.

Label
Label as shown.

NOTE-TAKING TIP: If your instructor points out definitions or procedures from your text, write a reference page in your notes. You can then write these referenced items in their proper place in your notes after class.

BUILD YOUR VOCABULARY

This is an alphabetical list of new vocabulary terms you will learn in Chapter 14. As you complete the study notes for the chapter, you will see Build Your Vocabulary reminders to complete each term's definition or description on these pages. Remember to add the textbook page number in the second column for reference when you study.

Vocabulary Term	Found on Page	Definition	Description or Example
combination			
complements			
compound event			
dependent events			
empirical study [ihm-PIHR-ih-kuhl]			
event			
experimental probability			
factorial [fak-TOHR-ee-uhl]			
Fundamental Counting Principle			

Vocabulary Term	Found on Page	Definition	Description or Example
inclusive			
independent events			
mutually exclusive			
permutation [PUHR-myu-TAY-shuhn]			
probability distribution			
probability histogram			
relative frequency			
sample space			
simulation [SIHM-yuh-LAY-shuhn]			
theoretical probability			
tree diagram			

14–1 Counting Outcomes

WHAT YOU'LL LEARN

- Count outcomes using a tree diagram.
- Count outcomes using the Fundamental Counting Principle.

BUILD YOUR VOCABULARY (pages 324–325)

One method used for counting the number of possible ______ is to draw a **tree diagram.**

The list of all possible ______ is called the **sample space**. An **event** is any collection of one or more outcomes in the sample space.

ORGANIZE IT

Under the tab for Outcomes, explain how to use a tree diagram to show the number of possible outcomes.

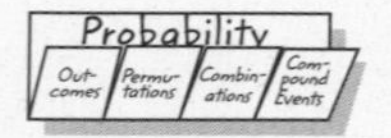

EXAMPLE Tree Diagram

1 **At football games, a concession stand sells sandwiches on either wheat or rye bread. The sandwiches come with salami, turkey, or ham, and either chips, a brownie, or fruit. Use a tree diagram to determine the number of possible sandwich combinations.**

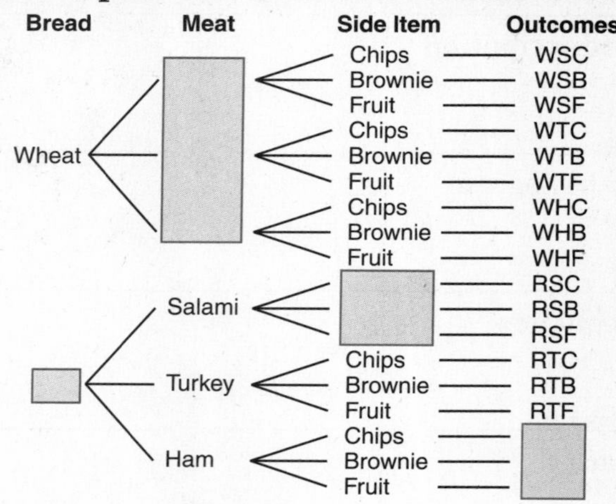

There are ______ possible combinations.

Your Turn A buffet offers a combination of a meat, a vegetable, and a drink. The choices of meat are chicken or pork; the choices of vegetable are carrots, broccoli, green beans, or potatoes; and the choices of drink are milk, lemonade, or a soft drink. Use a tree diagram to determine the number of possible combinations.

KEY CONCEPT

Fundamental Counting Principle If an event M can occur in m ways and is followed by an event N that can occur in n ways, then the event M followed by event N can occur in $m \cdot n$ ways.

EXAMPLE Fundamental Counting Principle

2 **The Too Cheap computer company sells custom made personal computers. Customers have a choice of 11 different hard drives, 6 different keyboards, 4 different mice, and 4 different monitors. How many different custom computers can you order?**

Multiply to find the number of custom computers.

hard drive choices		keyboard choices		mice choices		monitor choices		custom computers
☐	·	☐	·	☐	·	☐	=	☐

The number of different custom computers is ☐.

Your Turn A baseball team is organizing their draft. In the first five rounds, they want a pitcher, a catcher, a first baseman, a third baseman, and an outfielder. They are considering 7 pitchers, 9 catchers, 3 first baseman, 4 third baseman, and 12 outfielders. How many top picks are there to choose from?

EXAMPLE Factorial **(parallels Example 4 in text)**

KEY CONCEPT

Factorial The expression $n!$, read n factorial, where n is greater than zero, is the product of all positive integers beginning with n and counting backward to 1.

3 **Find the value of 9!.**

$9! = 9 \cdot 8 \cdot 7 \cdot 6 \cdot 5 \cdot 4 \cdot 3 \cdot 2 \cdot 1$ Definition of factorial

$=$ ☐ Simplify.

Your Turn **Find the value of each expression.**

a. 7!

b. 0!

HOMEWORK ASSIGNMENT

Page(s):

Exercises:

14–2 Permutations and Combinations

WHAT YOU'LL LEARN

- Determine probabilities using permutations.
- Determine probabilities using combinations.

BUILD YOUR VOCABULARY (pages 324–325)

An arrangement or listing in which order or placement is important is called a **permutation**.

An arrangement or listing in which order is not important is called a **combination**.

KEY CONCEPT

Permutation The number of permutations of n objects taken r at a time is the quotient of $n!$ and $(n - r)!$.

FOLDABLES Under the permutation tab, record this definition in words and in symbols

EXAMPLE Permutation and Probability

1 **Shaquille has a 5-digit code to access his e-mail account. The code is made up of the even digits 2, 4, 6, 8, and 0. Each digit can be used only once.**

a. How many different pass codes could Shaquille have?

This situation is a permutation of 5 digits taken 5 at a time.

$_nP_t = \frac{n!}{(n-r)!}$ Definition of permutation

$_5P_5 = \frac{5!}{(5-5)!}$ $n = 5, r = 5$

$_5P_5 = \frac{\text{\_\_\_\_\_\_}}{1}$ or ___ Definition of factorial

There are ___ possible pass codes.

b. What is the probability that the first two digits of his code are both greater than 5?

There are ___ digits greater than 5 and ___ digits less than 5. The number of choices for the first two digits is ___. The number of choices for the remaining digits is ___. So, the number of favorable outcomes is $2 \cdot 1 \cdot 3 \cdot 2 \cdot 1$ or ___.

$P(\text{first 2 digits} > 5) =$ ___ ← number of favorable outcomes / ← number of possible outcomes

$=$ ___ or 10% Simplify.

Your Turn Bridget and Brittany are trying to find a house, but they cannot remember the address. They can remember only that the digits used are 1, 2, 5, and 8, and that no digit is used twice. Find the number of possible addresses. Then find the probability that the first two numbers are odd.

EXAMPLE Use Combinations

2 Diane has a bag full of coins. There are 10 pennies, 6 nickels, 4 dimes, and 2 quarters in the bag. What is the probability that she will pull two pennies and two nickels out of the bag?

The number of combinations of 22 coins taken 4 at a time is

$_{22}C_4 = \frac{22!}{(22-4)!4!}$ or ______.

Using the Fundamental Counting Principle, the answer can be determined with the product of the two combinations.

$(_{10}C_2)(_6C_2) = \frac{10!}{(10-2)!2!} \cdot \frac{6!}{(6-2)!2!}$ — Definition of combination

$= \text{______} \cdot \frac{6!}{4!2!}$ — Simplify.

$= \frac{10 \cdot 9}{2!} \cdot \frac{6 \cdot 5}{2!}$ — Divide the first term by its GCF and the second term by its GCF.

$=$ ______

There are ______ ways to choose this particular combination out of 7315 possible combinations.

$P(\text{2 pennies, 2 nickels}) =$ ______ ← number of favorable outcomes / ← number of possible outcomes

$=$ ______ — Simplify.

Your Turn At a factory, there are 10 union workers, 12 engineers, and 5 foremen. The company needs 6 of these workers to attend a national conference. If the workers are chosen randomly, what is the probability that 3 union workers, 2 engineers, and 1 foreman are selected?

KEY CONCEPT

Combination The number of combinations of n objects taken r at a time is the quotient of $n!$ and $(n - r!)r!$.

FOLDABLES Under the combination tab, record this definition in words and in symbols.

HOMEWORK ASSIGNMENT

Page(s):

Exercises:

14–3 Probability of Compound Events

What You'll Learn

- Find the probability of two independent events or dependent events.
- Find the probability of two mutually exclusive events or inclusive events.

Build Your Vocabulary (pages 324–325)

A **compound event** is made up of ______ or more ______ events.

Independent events are events in which the outcome of one event does not ______ the outcome of the other.

When the outcome of one event ______ the outcome of another event, the events are dependent events.

Key Concepts

Probability of Independent Events If two events, *A* and *B*, are independent, then the probability of both events occuring is the product of the probability of *A* and the probability of *B*.

Probability of Dependent Events If two events, *A* and *B*, are dependent, then the probability of both events occuring is the product of the probability of *A* and the probability of *B* after *A* occurs.

EXAMPLE Independent Events

1 Roberta is flying from Birmingham to Chicago. She has to fly from Birmingham to Houston on the first leg of her trip. In Houston she changes planes and heads to Chicago. The airline reports that the flight from Birmingham to Houston has a 90% on time record, and the flight from Houston to Chicago has a 50% on time record. What is the probability that both flights will be on time?

$P(A \text{ and } B) = P(A) \cdot P(B)$ Definition of independent events

$P(\text{B–H on time and H–C on time})$

$= P(\text{B–H on time}) \cdot p(\text{H–C on time})$

$=$ ______ $\cdot$ ______

$=$ ______ or 45% Multiply.

Your Turn Two cities, Fairfield and Madison, lie on different faults. There is a 60% chance that Fairfield will experience an earthquake by the year 2010 and a 40% chance that Madison will experience an earthquake by 2010. Find the probability that both cities will experience an earthquake by 2010.

EXAMPLE Dependent Events

2 **At the school carnival, winners in the ring-toss game are randomly given a prize from a bag that contains 4 sunglasses, 6 hairbrushes, and 5 key chains. Three prizes are randomly drawn from the bag and not replaced. Find *P*(sunglasses, hairbrush, key chain).**

The selection of the first prize affects the selection of the next prize since there is one less prize from which to choose. So, the events are dependent.

1st prize: $P(\text{sunglasses}) = \square$ ← number of sunglasses / ← total number of prizes

2nd prize: $P(\text{hairbrush}) = \square$ or $\frac{3}{7}$ ← number of hairbrushes / ← total number of prizes

3rd prize: $P(\text{key chain}) = \square$ ← number of key chains / ← total number of prizes

$P(\text{sunglasses, hairbrush, key chain})$

$= \square \cdot \square \cdot \square$

$= \square$ or $\frac{4}{91}$

Your Turn **A gumball machine contains 16 red gumballs, 10 blue gumballs, and 18 green gumballs. Once a gumball is removed from the machine, it is not replaced. Find each probability if the gumballs are removed in the order indicated.**

a. P(red, green, blue)

b. P(green, blue, not red)

BUILD YOUR VOCABULARY (pages 324–325)

The events for drawing a marble that is green and for drawing a marble that is ______ green are called **complements**.

Events that cannot occur at the ______ time are called **mutually exclusive**.

Two events that ______ occur at the same time are called **inclusive events**.

KEY CONCEPTS

Mutually Exclusive Events If two events, A and B, are mutually exlcusive, then the probability that either A or B occurs is the sum of their probabilities.

Probability of Inclusive Events If two events, A and B, are inclusive, then the probability that either A or B occurs is the sum of their probabilities decreased by the probability of both occuring.

FOLDABLES Take notes on how to find the probability of compound events.

EXAMPLE Mutually Exclusive Events

3 Alfred is going to the Lakeshore Animal Shelter to pick a new pet. Today, the shelter has 8 dogs, 7 cats, and 5 rabbits available for adoption. If Alfred randomly picks an animal to adopt, what is the probability that the animal would be a cat or a dog?

$P(\text{cat}) = \square$ $\qquad$ $P(\text{dog}) = \square$

$P(\text{cat or dog}) = P(\text{cat}) + P(\text{dog})$ $\qquad$ Mutually exclusive events

$= \square + \square$ $\qquad$ Substitution

$= \frac{15}{20}$ or $\frac{3}{4}$ $\qquad$ Add.

EXAMPLE Inclusive Events

4 A dog has just given birth to a litter of 9 puppies. There are 3 brown females, 2 brown males, 1 mixed-color female, and 3 mixed-color males. If you choose a puppy at random from the litter, what is the probability that the puppy will be male or mixed-color?

These events are inclusive.

$P(\text{male or mixed-color})$

$= P(\text{male}) + P(\text{mixed-color}) - P(\text{male and mixed-color})$

$= \square + \square - \square$

$= \square$ or $\frac{2}{3} \approx 67\%$ $\qquad$ Simplify.

Your Turn

a. The French Club has 16 seniors, 12 juniors, 15 sophomores, and 21 freshmen as members. What is the probability that a member chosen at random is a junior or a senior?

b. In Mrs. Kline's class, 7 boys have brown eyes and 5 boys have blue eyes. Out of the girls, 6 have brown eyes and 8 have blue eyes. If a student is chosen at random from the class, what is the probability that the student will be a boy or have brown eyes?

HOMEWORK ASSIGNMENT

Page(s):

Exercises:

14–4 Probability Distributions

WHAT YOU'LL LEARN

- Use random variables to compute probability.
- Use probability distributions to solve real-world predictions.

EXAMPLE Random Variable

1 The owner of a pet store asked customers how many pets they owned. The results of this survey are shown in the table.

Number of Pets	Number of Customers
0	3
1	37
2	33
3	18
4	9

a. Find the probability that a randomly chosen customer has at most 2 pets.

There are 3 + 37 + 33 or ____ outcomes in which a customer owns at most 2 pets. There are ____ survey results.

$P(X \leq 2) =$ ____ or ____

b. Find the probability that a randomly chosen customer has 2 or 3 pets.

There are ____ + ____ or ____ outcomes in which a customer owns 2 or 3 pets.

$P(X = 2 \text{ or } 3) =$ ____ or ____

Your Turn **A survey was conducted concerning the number of movies people watch at the theater per month. The results of this survey are shown in the table.**

Movies (per month)	Number of People
0	7
1	23
2	30
3	29
4	11

a. Find the probability that a randomly chosen person watched at most 1 movie per month.

b. Find the probability that a randomly chosen person watches 0 or 4 movies per month.

BUILD YOUR VOCABULARY (page 325)

The probability of every possible ______ of the random variable X is called a **probability distribution**.

The ______ distribution for a random variable can be given in a table or in a **probability histogram**.

EXAMPLE Probability Distribution

KEY CONCEPT

Properties of Probability Distributions

1. The probability of each value of X is greater than or equal to 0 and less than or equal to 1.
2. The probabilities of all the values of X add up to 1.

2 The table shows the probability distribution of the number of students in each grade at Sunnybrook High School. If a student is chosen at random, what is the probability that he or she is in grade 11 or above?

X = grade	$P(X)$
9	0.29
10	0.26
11	0.25
12	0.2

The probability of a student being in grade 11 or above is the sum of the probability of grade 11 and the probability of grade 12.

$P(X \geq 11) = P(X = 11) + P(X = 12)$ Sum of individual probabilities

$= \square + \square$ or $\square$ $P(X = 11) = \square$

$P(X = 12) = \square$

The probability is ______.

Your Turn The table shows the probability distribution of the number of children per family in the city of Maplewood. If a family was chosen at random, what is the probability that they have at least 2 children?

X = Number of Children	$P(X)$
0	0.11
1	0.23
2	0.32
3	0.26
4	0.08

HOMEWORK ASSIGNMENT

Page(s):

Exercises:

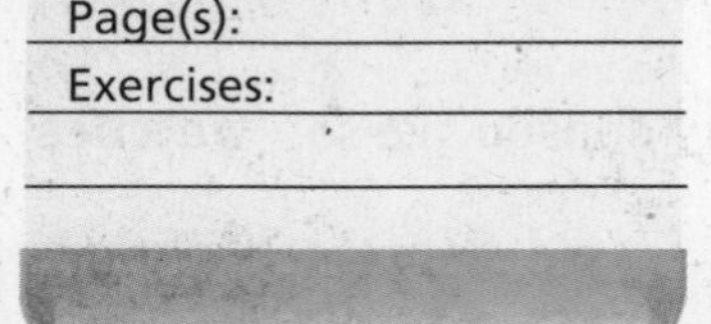

14–5 Probability Simulations

WHAT YOU'LL LEARN

- Use theoretical and experimental probability to represent and solve problems involving uncertainty.
- Perform probability simulations to model real-world situations involving uncertainty.

BUILD YOUR VOCABULARY (pages 324–325)

Theoretical probabilities are determined ________ and describe what should happen.

Experimental probability is determined using data from tests or ________.

Experimental probability is the ________ of the number of times an outcome occurred to the ________ number of events or trials. This ratio is also known as the **relative frequency.**

WRITE IT

Where is data obtained for experimental probability?

____

____

____

____

EXAMPLE Experimental Probability

1 Miguel shot 50 free throws in the gym and found that his experimental probability of making a free throw was 40%. How many free throws did Miguel make?

Miguel made ________ out of every 100 free throws.

experimental probability = 40% or ________ ← number of success ← total number of free throws

Miguel shot 50 free throws. Write and solve a proportion.

experimental sucesses → ________ = ________ ← Miguel's successes

experimental total free throws → ________ ← Miguel's total free throws

$50(40) = 100(x)$ Find the cross products.

________ = ________ Simplify.

$x =$ ________ Divide each side by 100.

Miguel made ________ free throws.

Your Turn Nancy was testing her serving accuracy in volleyball. She served 80 balls and found that the experimental probability of keeping it in bounds was 60%. How many serves did she keep in bounds?

BUILD YOUR VOCABULARY (pages 324–325)

When you ______ an experiment repeatedly, collect and combine the ______, and ______ the results, this is known as an **empirical study**.

A **simulation** allows you to use objects to act out an ______ that would be difficult or impractical to perform.

EXAMPLE Empirical Study

2 A pharmaceutical company performs three clinical studies to test the effectiveness of a new medication. Each study involved 100 volunteers. The results of the studies are shown in the table.

Result	Study 1	Study 2	Study 3
Expected Success Rate	70%	70%	70%
Condition Improved	61%	74%	67%
No Improvement	39%	25%	33%
Condition Worsened	0%	1%	0%

What is the experimental probability that the drug showed no improvement in patients for all three studies?

The number of outcomes with no improvement for the three studies was 39 + 25 + 33 or ______ out of the 300 total patients.

experimental probability = ______ or about ______

Your Turn A new study is being developed to analyze the relationship between heart rate and watching scary movies. A researcher performs three studies, each with 100 volunteers. Based on similar studies, the researcher expects that 80% of the subjects will experience a significant increase in heart rate. The table shows the results of the study. What is the experimental probability that the movie would cause a significant increase in heart rate for all three studies?

Result	Study 1	Study 2	Study 3
Expected Success Rate	80%	80%	80%
Rate increased significantly	83%	75%	78%
Littler or no increase	16%	24%	19%
Rate decreased	1%	0%	0%

EXAMPLE Simulation

3 In the last 30 school days, Bobbie's older brother has given her a ride to school 5 times.

a. What could be used to simulate whether Bobbie's brother will give her a ride to school?

Bobbie got a ride to school $\frac{5}{30}$ or $\frac{1}{6}$ days. Since a die has ____ sides, you could use one side of a die to represent a ride to school.

b. Describe a way to simulate whether Bobbie's brother will give her a ride to school in the next 20 school days.

Choose the side of the die that will be used to represent a ride to school. Let the 1-side of the die equal a ride to school. Toss the die ____ times and record each result.

REMEMBER IT

A spinner should simulate the possible outcomes of the event.

HOMEWORK ASSIGNMENT

Page(s):

Exercises:

Your Turn In the last 52 days, it has rained 4 times. What could be used to simulate whether it will rain on a given day? Describe a way to simulate whether it will rain in the next 15 days.

CHAPTER 14

BRINGING IT ALL TOGETHER

STUDY GUIDE

FOLDABLES™	VOCABULARY PUZZLEMAKER	BUILD YOUR VOCABULARY
Use your **Chapter 14 Foldable** to help you study for your chapter test.	To make a crossword puzzle, word search, or jumble puzzle of the vocabulary words in Chapter 14, go to www.glencoe.com/sec/math/t_resources/free/index.php	You can use your completed **Vocabulary Builder** (pages 324–325) to help you solve the puzzle.

14–1 Counting Outcomes

Use the tree diagram above for Exercises 1–3.

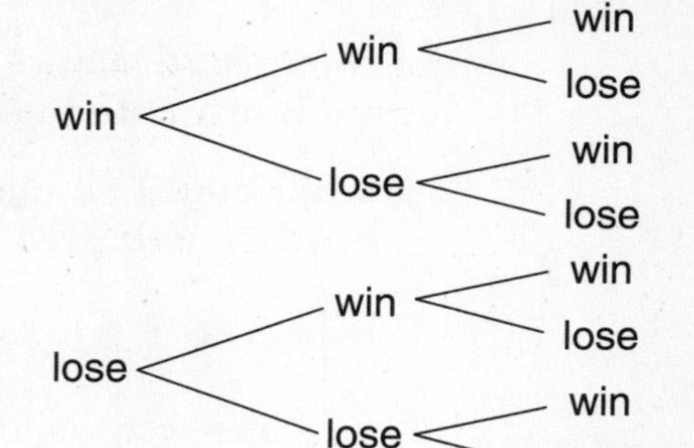

1. What is the sample space?

2. Name two different outcomes.

3. Use the Fundamental Counting Principle to find the possible outcomes shown above.

	Game 1		Game 2		Game 3		Number of Outcomes
Number of Choices		·		·		=	

14–2 Permutations and Combinations

4. Three of seven students are chosen to go to a job fair. How many different groups of students could be selected?

5. In how many ways can you arrange 3 pieces of art from a total of 9 pieces?

14–3

Probability of Compound Events

A die is rolled and a card is drawn from a standard deck of 52 cards. Find each probability.

6. P(6 and queen)

7. P(even and red)

8. P(less than 3 and a spade)

9. P(greater than 1 and red ace)

14–4

Probability Distributions

The table shows the probability of various family sizes in the United States.

10. For each value of X, is the probability greater than or equal to 0 and less than or equal to 1?

11. What is the sum of the probabilities?

12. Is the probability distribution valid?

Family Size (United States)	
X = Size of Family	Probability
2	0.42
3	0.23
4	0.21
5	0.10
6	0.03
7	0.01

14–5

Probability Simulations

13. Choose the manipulative you would use to simulate the problem. Explain your choice.

Situation	Simulation method
58% of drivers (commercial and private vehicles) have a cell phone in their car. Simulate whether or not the next 10 drivers you meet on the road will have a cell phone.	• die • coins • marbles • spinner

ARE YOU READY FOR THE CHAPTER TEST?

Visit **www.algebra1.com** to access your textbook, more examples, self-check quizzes, and practice tests to help you study the concepts in Chapter 14.

Check the one that applies. Suggestions to help you study are given with each item.

☐ **I completed the review of all or most lessons without using my notes or asking for help.**

- You are probably ready for the Chapter Test.
- You may want to take the Chapter 14 Practice Test on page 793 of your textbook as a final check.

☐ **I used my Foldable or Study Notebook to complete the review of all or most lessons.**

- You should complete the Chapter 14 Study Guide and Review on pages 789–792 of your textbook.
- If you are unsure of any concepts or skills, refer back to the specific lesson(s).
- You may also want to take the Chapter 14 Practice Test on page 793.

☐ **I asked for help from someone else to complete the review of all or most lessons.**

- You should review the examples and concepts in your Study Notebook and Chapter 14 Foldable.
- Then complete the Chapter 14 Study Guide and Review on pages 789–792 of your textbook.
- If you are unsure of any concepts or skills, refer back to the specific lesson(s).
- You may also want to take the Chapter 14 Practice Test on page 793.

Student Signature

Parent/Guardian Signature

Teacher Signature